HAER

Biomathematics

Volume 14

SEP/ae
MATH

SD 12/11/85 jml

C. J. Mode

Stochastic Processes in Demography and Their Computer Implementation

With 49 Figures and 80 Tables

Springer-Verlag
Berlin Heidelberg NewYork Tokyo

Charles J. Mode

Department of Mathematics and
Computer Science
College of Science, Drexel University
Philadelphia, PA 19104
U.S.A.

52880692

Mathematics Subject Classification (1980):
60J85, 60K05, 60K15, 62M01, 62M09, 62P05, 62P25, 65C20, 92–02, 95A15

ISBN 3-540-13622-3 Springer-Verlag Berlin Heidelberg New York Tokyo
ISBN 0-387-13622-3 Springer-Verlag New York Heidelberg Berlin Tokyo

Library of Congress Cataloging in Publication Data.
Mode, Charles J., 1927- . Stochastic processes in demography and their computer implementation. (Biomathematics; v. 14) Bibliography: p.
1. Population – Mathematical models. 2. Population – Data processing. I. Title. II. Series.
HB849.51.M635 1985   304.6′028′54   84-26707

Typesetting: H. Stürtz AG, Würzburg. Offsetprinting: Saladruck, Berlin.
Bookbinding: Lüderitz & Bauer, Berlin
2141/3020-543210

To my wife Eleanore

Preface

According to a recent report of the United States Census Bureau, world population as of June 30, 1983, was estimated at about 4.7 billion people; of this total, an estimated 82 million had been added in the previous year. World population in 1950 was estimated at about 2.5 billion; consequently, if 82 million poeple are added to the world population in each of the coming four years, population size will be double that of 1950. Another way of viewing the yearly increase in world population is to compare it to 234 million, the estimated current population of the United States. If the excess of births over deaths continues, a group of young people equivalent to the population of the United States will be added to the world population about every 2.85 years. Although the rate of increase in world population has slowed since the midsixties, it seems likely that large numbers of infants will be added to the population each year for the foreseeable future.

A large current world population together with a high likelihood of substantial increments in size every year has prompted public and scholarly recognition of population as a practical problem. Tangible evidence in the public domain that population is being increasingly viewed as a problem is provided by the fact that many governments around the world either have or plan to implement policies regarding population. Evidence of scholarly concern is provided by an increasing flow of publications dealing with population. Even among scholars and scientists, however, the view that population is a problem is not universally held. In conversations, some of my eminent colleagues, who are of world renown for their works, rejoice in the phenomenon of human life and hold the view that the world would be better off with more people. Such views are understandable in terms of value for human life, but other colleagues express concern about overpopulation.

Concern about overpopulation usually centers around two propositions. In the long run, a rapidly growing human population ultimately threatens the very existence of the human species, since, even if a means to colonize space on a large scale were to develop within the foreseeable future, the vast majority of humans would still be confined to the planet earth with a finite, but unknown, capacity to produce food, clothing, and shelter. In the short term, overpopulation is viewed as a multiplier of social and economic ills and a barrier to realizing the fruits of development aimed at human betterment.

There is also a nonhuman aspect of overpopulation, most forcefully pointed out by biologists. As the human population of the earth expands, more of the earth's resources and land area need to be devoted to providing food, clothing, and shelter. Increased land use is exemplified by the continued felling of tropical forests on a large scale and the destruction of the habitats of many living plant and animal species. Some biologists, particularly ecologists, point to growing evidence that the earth may, within the next 100 to 200 years, experience an extinction episode rivaling that of 65 million years ago when many species, including the fabled dinosaurs, large and small, became extinct. It appears to some that the chain of events leading to such massive extinctions, a chain that may be unique in geologic history, could be set in motion by the overpopulation of a single species – Homo sapiens.

Even if a society recognizes population as a problem, it may not be easy to design and implement family planning programs aimed at decreasing fertility. For having children is an intensely personal experience closely bound to human emotions, which may not be quickly influenced by pressures imposed by society at large. In a developed society, for example, with a large financial investment in each child, a social security program for its members in old age, and low infant mortality, the argument for a small number of children per family is persuasive. But, to people in a society with high infant mortality and with no visible means of support in their old age, other than that they receive from children – usually sons – the argument carries much less weight. Indeed, the argument may not be comprehensible to a person with such a view of life or time horizon. In the minds of many thinkers, it is such time horizons, coupled with the lack of social and economic development, that underlie the population problem.

On the other hand, there is historical evidence, some of which will be sketched in this book, that the time horizons of people can be changed through social and economic development, leading to the widespread belief that further development seems to be the best hope for coming to grips with the population problem. Furthermore, it is widely held that, in addition to technological innovation contributing to human betterment, family planning should be an integral part of that development. Part of this technological development centers around research activity directed towards the development and computer implementation of mathematical models designed to aid in the formulation and evaluating of population policies. This book is part of the continuing saga of such research activity.

From the biological point of view, the events constituting a human life are set in motion by conception. It thus seemed natural to begin the book with a chapter on fecundability, the monthly probability that a sexually active fertile human female conceives. Following Chap. 1 on fecundability is a chapter on survivorship, dealing with the concepts of life table methodology widely used in the study of human mortality after birth. In addition to giving a review of methodology commonly used in constructing period or current life tables, attention is also focused on cohort or generational models of survivorship needed in the studying evolutionary changes in human mortality. Chapter 2 is followed by a chapter on theories of competing risks and multiple decrement life tables used in describing causes of human mortality. Theories of competing risks are not only useful in the study of human mortality but also arise in semi-Markovian processes used in the development of models of maternity histories. Some evolutionary aspects of human mortality are also illustrated in Chaps. 2 and 3 through the medium of computer generated graphs based on historical U.S. and Swedish data.

Chapters 4, 5, and 6 are concerned with fertility, the factors and circumstances underlying the flow of births into a population. Briefly, Chap. 4 deals with models of maternity histories starting with marriage and continuing through successive live births at the level of waiting times among live births. Special attention in this chapter is also given to the problem of developing a formalism with a capacity for distinguishing between cohort and period effects, a formalism that is also helpful in studying the evolution of fertility in a population. Chapter 5, which contains extensive numerical results, is devoted primarily to the description and utilization of a computer software design used to implement a class of models of maternity histories. This class is sufficiently general to include models in which waiting times among live births may be decomposed into events representing the use of contraceptives and induced abortions to limit births. Chapter 6 is devoted to the analysis of a large data set collected in connection with a family planning experiment in Taichung, Taiwan. Such analyses are essential to the development of methodologies useful in evaluating population policies but, at the same time, forcefully demonstrate the challenges facing investigators charged with the responsibility of analyzing large data sets.

Chapter 7, which is concerned with the development of population projection methodologies based on stochasic population processes, stands apart from the rest of the book in that it provides a synthesis of material presented in the foregoing chapters. Much of the material contained in this chapter is new, particularly those parts dealing with population processes with time inhomogeneous laws of evolution. Among the new developments is a methodology, based on random sums of random functions, for deriving covariance functions describing stochastic variability in the evolution of a population. The framework developed in this chapter is sufficiently general to give rise to a Leslie matrix as a computing algorithm and to permit the derivation a of many formulas from stable population theory as asymptotic results appearing in population processes with time homogeneous laws of evolution. Also included in this chapter is an overview of a computer software design, implementing population processes with either time homogeneous or time inhomogeneous laws of evolution. Just as in other chapters, information coming out of the computer in the form of large digital arrays is frequently presented in the form of computer generated graphs.

Although the book is intended primarily for the lone reader interested in population as a field of research, or perhaps a reader with eclectic interests, portions of the material have been used in courses at both the undergraduate and graduate levels. Parts of Chaps. 1, 2, and 6, for example, were used in a semester course given to graduate students in demography in the Population Studies Center of the University of Pennsylvania. Chapters 4 and 5 also provided core material for a one-quarter graduate course in the computer implementation of applied probability models given at Drexel University. Selected parts of Chaps. 2 and 3 were also used in a one-quarter undergraduate course given at Drexel University to computer science and mathematics majors interested in the applications of mathematics and computers.

The diversity of the above audience naturally raises a question regarding the technical background material needed to read the book. Of all the mathematical themes used in the book, that of renewal theory, as practiced by applied probabilists, is the most persistent. In a class room situation, renewal ideas can be readily communicated to students with a limited background, because the central notions are simple and intuitive. For a lone reader, a good place to learn something about the central ideas of renewal theory, as well as others used in this book, is W. Feller's volume I on probability theory and its applications, a well-known work that has been cited several times in the text. Work on the computer implementation of the type of applied probability models that occur throughout the book is of more recent origin; there are thus no sources, other than the references cited, that can be recommended for study. Fortunately, the algebraic nature of the algorithms used in the computer implementation of the models makes them relatively easy to understand.

Most authors owe an intellectual debt to those who have traveled related pathways. As a lone reader, it is a pleasure to acknowledge that the books by N. Keyfitz, *An Introduction to the Mathematics of Population*, by A.J. Coale, *The Growth and Structure of Human Populations*, and by M.C. Sheps and J.A. Menken, *Mathematical Models of Conception and Birth,* significantly influenced the thinking behind this book. Had these pioneering works not been written, my work, as well as that of others, would have been made much more difficult. A word of appreciation for the book, *World Population – An Analysis of Vital Data*, by Keyfitz and Flieger, is also appropriate. Without books of this nature, information on the historical development of populations would be inaccessible to many workers. An intellectual debt is also owed to a man-made device, the computer, which motivated and helped discover many of the results presented in this book.

Even though the use of stochastic processes is becoming widespread in science and technology, there are some who question their need, particularly as models concerned with large populations. Questions regarding how one should proceed in order to perceive and understand the real world rightfully belong to the philosophical domain of epistemology, a domain that will not be entered into here. Many casual observers of a human population would tend to agree, however, that there is much variation among the life cycles of individuals in a population and that at any given time the future course of even large populations is far from certain. Stochastic processes offer no panacea for coming to grips with this variation and uncertainty. But, as part of man's quest for methodologies needed to interpret and understand the real world in the face of variation and uncertainty, stochastic processes seem to be a promising step forward. It was this promise that guided the author as the pages of the book were composed.

Philadelphia, PA, April 1985   Charles J. Mode

Acknowledgements

The development of a book involving computers would not be possible without the cooperation and help of many people. A particularly labor intensive aspect of the research underlying the book was the development of computer software by my colleague, Dr. Robert C. Busby. Nearly all of the interactive and userfriendly software designs, represented by thousands of lines of FORTRAN and some APL code, are either his inventions or have been developed with his advice and assistance. A highly visible manifestation of one of these designs are computer generated graphs from which the figures presented in Chaps. 2 through 7 were drawn. All the userfriendly FORTRAN code was written for the PRIME operating system. Some of this code has been tested for portability and with additional labor can be made to run on other operating systems, including those for recent IBM machines.

Three Ph.D. students, Gary S. Littman, Michael G. Soyka, and Gary T. Pickens, who completed dissertations under the author's guidance, have made fundamental contributions both to the theory and the computer implementation of the models. Littman and Soyka were students in mathematics at Drexel University; while Pickens was a student in demography and statistics at the University of Pennsylvania. As a fellow colleague at Drexel, Pickens has been a strong motivating force helping to shape the contents of Chap. 7 on population projections. Two other students in mathematical sciences at Drexel, Andrzej Zawodniak and Marc E. Jacobson, have made significant contributions in the development of evolutionary models of human mortality. Mark Cain, manager of the computer graphs laboratory at Drexel, has provided valuable advice in writing the software and operating the hardware.

Many people in the demographic community have been most helpful. Throughout the years, Dr. Robert G. Potter of Brown University has been a faithful correspondent who offered comments on Chaps. 1, 2, and 3. Dr. Samuel H. Preston of the University of Pennsylvania has also offered valuable comments on Chaps. 2 and 3 along with welcomed words of encouragement. Without the cooperation of Drs. Albert Hermalin and Ronald Freedman of the University of Michigan, the analysis of the Taichung Medical IUD data presented in Chap. 6 would not have been possible. A special word of appreciation is also due Dr. Roger Avery of Ithaca, New York, for his superb job in constructing the Taichung data tape. Dorothy Nortman of the Population Council in New York, a recognized authority on family planning programs throughout the world, has also given valuable counsel. Exchanges of letters with Professor Nathan Keyfitz, formerly of Harvard University, have been very stimulating. Dr. Joel Cohen of Rockefeller University was most encouraging by suggesting that preliminary notes be expanded into a book.

No book of this sort, representing several years of reasearch effort, could have been written without financial support. A grant from the Population Council gave initial impetus to start work in demographic research. Anonymous peers in the demographic community, through a review process, recommended that grant proposals be funded by the National Institute for Child Health and Human Development, NICHD Grant RO1 HD 09571. Other anonymous colleagues at Drexel University recommended that two quarters faculty leave in 1982–83, free of teaching, be granted in order to complete work on the book. The author is most grateful to his peers and colleagues for their support.

Two excellent technical typists, Ms. Carol Kieth and Mrs. Elizabeth Curtis, are also due a warm word of thanks for transforming notes into a readable manuscript. Finally, my wife, Eleanore, gave much needed help in reading the proofs.

Contents

Chapter 1. Fecundability

1.1 Introduction

From the biological point of view, the life of every human being begins with a single cell, called a zygote, which arises from the union of an egg cell from his mother and a sperm cell from his father. When a zygote forms, a conception is said to have occurred and a pregnancy begins. Even among couples who engage in sexual intercourse regularly, a conception in any particular month is far from certain. To take these uncertainties into account, it seems necessary to introduce probabilistic concepts. The term, fecundability, has come into wide usage in discussions of stochastic models of conception and birth as an expression of the uncertainties surrounding conception. A tentative definition of fecundability is the probability a married woman conceives in a given month.

   This definition, although clear in its biological meaning, is not a workable one from the point of view of data analysis, because all conceptions are not recognizable. For instance, if a zygote is not viable, then it may abort a few days following its formation. The only indication of such a spontaneous abortion might be a delayed menstrual period, which could not be distinguished from the natural variation in the length of the menstrual cycle. In this book, fecundability will be the probability a married woman experiences a recognizable pregnancy in a given month, an event that could be openly declared on the occasion of an interview.

1.2 A Model of Constant Fecundability – The Geometric Distribution

The simplest case to consider is that in which fecundability is not only constant among women (couples) but also from month to month. Let θ, $0 < \theta < 1$, be the constant probability of conceiving in any month and assume months represent independent trials. The month in which conception occurs will be denoted by the random variable T. Measuring the waiting times to conception in this way implies T is a discrete-valued random variable taking values in the set $[t: t = 1, 2, ...]$ of positive integers.

Our first task in constructing and analyzing a model of the waiting times to conception in the case of constant fecundability is to deduce a formula for

$$h(t) = P[T = t], \tag{1.2.1}$$

the probability density function, p.d.f., of the random variable T. It is clear that the probability $T = 1$, a conception occurs in the first month, is $h(1) = \theta$. If $T = t > 1$, then, in the preceding $t - 1$ months a conception has not occurred with probability $(1 - \theta)^{t-1}$ but a conception occurs in month t with probability θ. Therefore,

$$h(t) = \theta(1 - \theta)^{t-1}, \tag{1.2.2}$$

$t = 1, 2, \ldots$, is the desired formula for the p.d.f. of T. The p.d.f. in (1.2.2) is, of course, that of the well-known geometric distribution.

Having derived a p.d.f. of a random variable, it is useful to form the habit of checking whether the random variable is proper. A random variable T, taking values in the set of positive integers, is said to be proper if

$$\sum_{t=1}^{\infty} P[T = t] = 1. \tag{1.2.3}$$

By putting $\eta = 1 - \theta$ and substituting (1.2.2) into (1.2.3), we are led to ask if the equation

$$\theta(1 + \eta + \eta^2 + \ldots) = 1 \tag{1.2.4}$$

is valid? The infinite series within the parentheses on the left in (1.2.4) is the well-known geometric series which converges when $|\eta| < 1$, a condition satisfied in our model.

Because the idea of summing a geometric series will arise elsewhere in this book, it will be useful to go through an often used argument for deriving a formula for the desired sum. Put

$$S_n(\eta) = 1 + \eta + \ldots + \eta^n.$$

Then, it is well-known that $S_n \to S$, the sum of a geometric series, as $n \uparrow \infty$. The arrow, $\to$, means "converges to". Some simple algebra leads to the conclusion that the equation

$$S_n(\eta)(1 - \eta) = (1 - \eta)S_n(\eta) = 1 - \eta^{n+1} \tag{1.2.5}$$

is valid for all $n \geq 1$. But, if $|\eta| < 1$, then $\eta^n \to 0$ as $n \uparrow \infty$. By letting $n \uparrow \infty$ in (1.2.5), it follows that S satisfies the equation

$$(1 - \eta)S = S(1 - \eta) - 1. \tag{1.2.6}$$

Hence,

$$S = (1 - \eta)^{-1} = 1 + \eta + \eta^2 + \ldots. \tag{1.2.7}$$

From (1.2.7) and the definition $\eta = 1 - \theta$, it is easy to see that Eq. (1.2.4) is indeed valid. Later on we shall see that the formula in (1.2.7) holds for more general kinds of algebra than that for the real numbers considered in this section.

When applying a distribution to data, it is useful to have a good under-standing of some of its associated properties and functions. Among the proper-ties of interest is the shape of the p.d.f. From (1.2.2), it is clear that $h(t)$ is monotone decreasing on $t=1, 2, ...$, with a mode, the most probable value, at $t=1$. Another function of interest, especially for waiting time distributions such as the geometric, is the survival function $S(t)$ defined by $S(t)=P[T>t]$ for $t\geq0$. In the model under consideration, it is clear that $T>t$ if, and only if, a conception does not occur in the first t months. Hence,

$$S(t)=(1-\theta)^t.\tag{1.2.8}$$

The distribution function of the random variable T is given by

$$H(t)=P[T\leq t]=1-(1-\theta)^t.\tag{1.2.9}$$

Two additional properties, the mean and variance of the geometric distribu-tion, are also of interest. By definition, the mean is

$$E[T]=\sum_{t=1}^{\infty} tP[T=t]=\theta(1+2\eta+3\eta^2+...)$$

$$=\theta\frac{d}{d\eta}\left(\frac{1}{1-\eta}\right)=\frac{\theta}{(1-\eta)^2}=\frac{1}{\theta}.\tag{1.2.10}$$

It can also be shown that the variance of the geometric distribution is given by

$$\mathrm{Var}\,[T]=E[(T-E[T])^2]=E[T^2]-(E[T])^2$$

$$=\frac{1-\theta}{\theta^2},\tag{1.2.11}$$

see Problem 1 for details of the derivation.

1.3 Applying the Geometric Distribution to Data

Having developed a simple probability model of the waiting times till con-ception, it is natural to ask whether the proposed model is in agreement with data. Suppose we have a random sample of $n\geq1$ waiting times till conception. Let the waiting times in such samples be denoted by the random variables $T_1, T_2, ..., T_n$. In mathematical terms, the idea of a random sample corresponds to the assumption that these random variables are independent and identically distributed, their common distribution being that of the geometric random variable T introduced in Sect. 1.2.

It should be emphasized at the outset that a highly idealized situation is being considered. In practice, it is difficult to observe a complete random sample of n waiting times till conception. For example, suppose we begin observing a sample of n newly married and non-pregnant women at $t=0$ and continue observation until some fixed time $t_1>0$, a cut-off time determined by practical limitations. During the time interval $(0, t_1] = [t : 0 < t \leq t_1]$, all of the women in the sample may not have conceived. Such data are said to be truncated. There may also be difficulties in precisely identifying the month in which conception occurs. For the time being, however, such problems with the data, as well as others, will be ignored.

The geometric distribution depends on one unknown parameter θ. Two widely used parametric methods of estimation are frequently considered in estimating unknown parameters, the method of moments and the method of maximum likelihood. In the one-parameter case, one obtains a moment estimator of a parameter by equating the theoretical mean to the observed sample mean and solving the resulting equation. Thus, if we put

$$\bar{T} = \frac{1}{n} \sum_{i=1}^{n} T_i,$$

then $\tilde{\theta}_n$, the moment estimator of θ, is obtained by solving the equation

$$\frac{1}{\tilde{\theta}_n} = \bar{T},$$

yielding the result

$$\tilde{\theta}_n = \frac{1}{\bar{T}}. \tag{1.3.1}$$

To find $\hat{\theta}_n$, the maximum likelihood estimator of θ, one maximizes the likelihood function in θ for every set of sample values. It is shown in Problem 2 that $\hat{\theta}_n = \tilde{\theta}_n$. Hence, for the geometric distribution, the two estimators coincide.

A simple alternative to the estimator in (1.3.1) is based on the observation that $\theta = P[T = 1]$, suggesting that an estimator of θ may be constructed by considering only those conceptions occurring during the first month. Let the random variable Z_n be the number of women who conceive in the first month in random samples of size n. Then θ_n^*, an estimator of θ based on Z_n, is given by the relative frequency

$$\theta_n^* = \frac{Z_n}{n}. \tag{1.3.2}$$

An investigator might rightfully suspect that the estimator in (1.3.2) may not be as good as the one in (1.3.1) because all the data are not used. The data analysis which follows will supply evidence that these suspicions are well founded.

Jain (1968) used some data collected in a fertility survey conducted in Taichung, Taiwan towards the end of 1962 to study the distribution of the waiting times till conception. The survey was based on a sample of 2,443 married women between the ages 20–29 and living with their husbands and was conducted just before the start of a year-long family planning action program. Jain's analysis used only a small fraction of the information gathered in the survey. Out of a total of 2,443 women interviewed, 158 were premaritally pregnant, 29 had used contraception prior to the first pregnancy, and 66 had never conceived by the time of interview. These latter women were excluded, leaving a sample size of 2,190. Since only a small fraction of the women had not conceived by the time of the interview, it seems reasonable to assume that the data used in the analysis were not seriously biased by truncation effects.

In each month following marriage, the frequencies of women conceiving were computed. These frequencies, which may be viewed as an estimate of the underlying p.d.f., were monotone decreasing, suggesting that the data could have come from a geometric distribution. The sample mean and variance were 8.2 and 138.82, respectively. Using the value $\bar{T}=8.2$, formula (1.3.1) yields $\hat{\theta}_n = \tilde{\theta}_n = 0.12195$ as the moment and maximum likelihood estimate of θ. The observed frequency of women conceiving in the first month was 0.171, leading to $\theta_n^* = 0.171$ as another estimate of θ.

One quick check on whether the sample may be viewed as coming from a geometric distribution is to compare estimates of the theoretical mean and variance based on formulas (1.2.11) with the observed values. Presented in columns 2 and 3 of Table 1.3.1 are estimates of the mean and variance corresponding to $\hat{\theta}_n$ and θ_n^*. The estimates in column 2 based on $\theta_n^* = 0.171$ differ markedly from the observed mean and variance. The estimate of the mean based on $\hat{\theta}_n = 0.12195$ necessarily coincides with the observed mean, but the estimate of the variance based on $\hat{\theta}_n$ again differs significantly from the observed variance. These negative results suggest that the geometric distribution may not be a good representation of the data.

Comparing point estimates of means and variances based on competing estimators is not always informative, because estimates of other functions of the distribution are ignored. Presented in Table 1.3.2 are estimated survival functions of the distribution for the first 12 months. Column 1 is the empirical survival function calculated directly from Jain's data; while columns 2 and 3 were calculated from the estimates θ_n^* and $\hat{\theta}_n$, using formula (1.2.8). The estimates of the survival function based on $\theta_n^* = 0.171$ in column 2 are consistently lower than the empirical survival function in column 1. This result suggests that one could go seriously wrong by using θ_n^* as an estimate of θ. For at 12 months, the estimate of the fraction of women not pregnant is only about half that of the observed value. The estimate of the survival function in column 3 based on the estimate $\hat{\theta}_n = 0.12195$ is much closer to the empirical survival function but consistently over estimates it. The fact that the survival

Table 1.3.1. Estimates of mean and variance

	1	2	3	4
Mean	8.20	5.84	8.20	8.20
Variance	138.82	28.35	59.04	138.82

1 Sample values
2 Estimates based on θ_n^*
3 Estimates based on $\hat{\theta}_n$
4 Beta-Geometric distribution

Underlying distribution assumed to be geometric.

Table 1.3.2. Estimates of survival functions

Month	1	2	3	4
1	0.829	0.829	0.878	0.837
2	0.716	0.687	0.771	0.707
3	0.613	0.570	0.677	0.602
4	0.521	0.472	0.594	0.516
5	0.455	0.392	0.522	0.445
6	0.405	0.325	0.458	0.386
7	0.363	0.269	0.402	0.337
8	0.323	0.223	0.353	0.296
9	0.288	0.185	0.310	0.261
10	0.256	0.153	0.272	0.231
11	0.229	0.127	0.239	0.205
12	0.207	0.105	0.210	0.183

1 Empirical survival function
2 Estimates based on θ_n^*
3 Estimates based on $\hat{\theta}_n$
4 Beta-Geometric distribution

Underlying distribution assumed to be geometric.
Parameters estimated by method of moments.

function in column 3 is closer to the empirical one than that in column 2 suggests that the properties of the estimator $\hat{\theta}_n$ are superior to those of θ_n^*. There are also good theoretical reasons for preferring the maximum likelihood estimator $\hat{\theta}_n$ to θ_n^*, see Problem 2.

Column 4 in Tables 1.3.1 and 1.3.2 refer to the Beta-Geometric distribution, a model of the waiting times till conception that will be developed in the next section. The fit of this model to Jain's data is such that the estimated mean and variance coincide with the sample values, see column 4 in Table 1.3.1. From column 4 of Table 1.3.2, it can be seen that the survival function based on this distribution is close to the empirical survival function in column 1 for the first five months but thereafter the values in column 4 are consistently lower than those in column 1.

1.4 A Model of Heterogeneous Fecundability

An assumption underlying the developments in Sects. 1.2 and 1.3 was that fecundability was constant among women. There are reasons for believing, however, that this assumption cannot be justified. In subsequent sections of this chapter, the biological and behavioral mechanisms underlying heterogeneity in fecundability will be investigated more thoroughly. But, in this section, we will content with an interesting generalization of the geometric distribution, accommodating variable fecundability among women.

Let Θ be a random variable representing variation in fecundability among women in a population. The range of Θ is the open interval $(0, 1)$, written $(0, 1)$

$=[\theta:0<\theta<1]$ in set notation. The fecundability θ of a particular woman, which is assumed constant from month to month, may be thought of as a realization of the random variable Θ. As before, let T be a random variable representing the waiting times till conception. Given that $\Theta=\theta$, the conditional p.d.f. of T will be denoted by

$$f(t|\theta)=P[T=t|\Theta=\theta], \tag{1.4.1}$$

$t=1,2,....$ A possible choice of the p.d.f. in (1.4.1) is that for the geometric distribution

$$f(t|\theta)=\theta(1-\theta)^{t-1}, \quad t=1,2,.... \tag{1.4.2}$$

To completely determine the distribution of T, it is necessary to specify $f(\theta)$, the p.d.f. of the random variable Θ. Then, by a conditioning argument, it follows that

$$h(t)=P[T=t]=\int_0^1 f(\theta)\,f(t|\theta)\,d\theta, \tag{1.4.3}$$

$t=1,2,...$, is the p.d.f. of the random variable T. It is easy to see from (1.4.3), that if $f(t|\theta)$ is a proper p.d.f. for each θ, then so is $h(t)$. Probability density functions, which may be represented in form (1.4.3), are frequently referred to as mixtures of distributions.

A judicious choice of $f(\theta)$, when $f(t|\theta)$ is the p.d.f. of the geometric distribution in (1.4.2), is the well-known two-parameter Beta distribution whose p.d.f. is

$$f(\theta)=\frac{1}{B(\alpha,\beta)}\,\theta^{\alpha-1}(1-\theta)^{\beta-1}, \tag{1.4.4}$$

where $0<\theta<1$ and α and β are positive parameters, $\alpha>0$ and $\beta>0$. The normalizing constant, $B(\alpha,\beta)$, in (1.4.4) is the famous beta function defined by the integral

$$B(\alpha,\beta)=\int_0^1 \theta^{\alpha-1}(1-\theta)^{\beta-1}d\theta \tag{1.4.5}$$

for $\alpha,\beta>0$. It is well-known that the beta function may be expressed in terms of another famous function, the gamma function, defined by

$$\Gamma(\alpha)=\int_0^\infty x^{\alpha-1}e^{-x}dx, \tag{1.4.6}$$

for $\alpha>0$. An integration by parts shows that the gamma function satisfies the recursion relation

$$\Gamma(\alpha+1)=\alpha\Gamma(\alpha), \quad \alpha>0. \tag{1.4.7}$$

The fundamental equation connecting the gamma and beta functions is

$$B(\alpha,\beta)=\frac{\Gamma(\alpha)\Gamma(\beta)}{\Gamma(\alpha+\beta)}. \tag{1.4.8}$$

Another useful property of the gamma function is that $\Gamma(\alpha+1)$ is a generalization of the factorial function defined on the positive integers in the sense that if $\alpha=1, 2, \ldots$, then

$$\Gamma(\alpha+1)=\alpha(\alpha-1)\ldots(2)(1)=\alpha!. \tag{1.4.9}$$

In fact, (1.4.9) follows immediately from (1.4.7).

When $f(t|\theta)$ and $f(\theta)$ are given by (1.4.2) and (1.4.4), respectively, the resulting mixture of distributions defined by the integral in (1.4.3) will be referred to as the Beta-Geometric distribution. Although this distribution is well known in probability and statistics circles, the French demographer, Louis Hénry, is credited with introducing it into models used in population research. The distribution is also known as the Type I Geometric, so named because of a classification system introduced by the British statistician, Karl Pearson. English translations of some of the works of Louis Hénry are presented in the book, Sheps and Lapierre-Adamcyk (1972). In particular, an account of Hénry's early use of the Beta-Geometric distribution is provided in Appendix 3, page 199, of this book. A very readable account of the mathematics underlying the beta and gamma functions has been provided by Artin (1964).

1.5 Some Properties of the Beta-Geometric Distribution

A good place to start our investigation of the Beta-Geometric distribution is with the distribution of fecundability in the population. The mean of a random variable Θ with a Beta distribution is

$$E[\Theta]=\int_0^1 \theta f(\theta) d\theta=\frac{\Gamma(\alpha+\beta)\Gamma(\alpha+1)\Gamma(\beta)}{\Gamma(\alpha)\Gamma(\beta)\Gamma(\alpha+\beta+1)}. \tag{1.5.1}$$

Using (1.4.7), we see the mean fecundability in the population is given by

$$E[\Theta]=\frac{\alpha}{\alpha+\beta}. \tag{1.5.2}$$

Similar calculations yield the formula

$$\mathrm{Var}[\Theta]=\frac{\alpha\beta}{(\alpha+\beta)^2(\alpha+\beta+1)} \tag{1.5.3}$$

for the variance of fecundability in the population. For $\alpha>1$ and $\beta>1$, the mode of the Beta distribution is given by

$$m[\Theta]=\frac{\alpha-1}{\alpha+\beta-2}, \tag{1.5.4}$$

see Problem 3 for further details.

It is desirable to express $h(t)$, the p.d.f. of T, in terms of the parameters α and β. For ease of computation, it is natural to look for a recursion formula for calculating the values $h(t)$ at $t=1, 2, \ldots$. From (1.4.2), it follows that $f(1|\theta)$

$= \theta$. Substituting this function into (1.4.3) and using (1.4.4) yields the result

$$h(1) = \int_0^1 \theta f(\theta) \, d\theta = E[\Theta] = \frac{\alpha}{\alpha + \beta}. \tag{1.5.5}$$

By noting that $f(2|\theta) = \theta(1-\theta)$, it follows that

$$h(2) = \int_0^1 \theta(1-\theta) f(\theta) \, d\theta = \frac{h(1)\beta}{\alpha + \beta + 1}. \tag{1.5.6}$$

In general, it can be shown that $h(t)$ satisfies the recursion relationship

$$h(t) = \frac{h(t-1)(\beta + t - 2)}{\alpha + \beta + t - 1} \tag{1.5.7}$$

for $t \geq 2$. Taken together, (1.5.5) and (1.5.7) completely determine $h(t)$ for each combination of parameter values α and β.

Although formula (1.5.7) is convenient for calculating the p.d.f. $h(t)$, it is not useful for calculating the mean and variance of T. To derive formulas for the mean and variance of the Beta-Geometric distribution one may proceed directly. By definition,

$$E[T] = \sum_{t=1}^{\infty} t h(t) = \int_0^1 f(\theta) \left[\sum_{t=1}^{\infty} t f(t|\theta) \right] d\theta = \int_0^1 f(\theta) E[T|\theta] \, d\theta. \tag{1.5.8}$$

Formula (1.5.3) has again been used in deriving (1.5.8). A reader with some acquaintance with integration theory will recognize that the monotone convergence theorem may be used to justify the interchange of summation and integration in (1.5.8).

If $f(t|\theta)$ is given by (1.4.2), then

$$E[T|\theta] = \frac{1}{\theta}. \tag{1.5.9}$$

Substituting (1.5.9) into (1.5.8) leads to

$$E[T] = E\left[\frac{1}{\Theta}\right] = \frac{\Gamma(\alpha + \beta)}{\Gamma(\alpha)\Gamma(\beta)} \int_0^1 \theta^{\alpha - 2}(1 - \theta)^{\beta - 1} \, d\theta$$

$$= \frac{\alpha + \beta - 1}{\alpha - 1}, \tag{1.5.10}$$

provided $\alpha > 1$. Observe that the condition, $\alpha > 1$, guaranteeing the nonnegativity and finiteness of $E[T]$, is not the same as the original condition, $\alpha > 0$, imposed on the parameter α. A formula for the variance of T may be derived by slightly extending the argument outlined above. The desired formula is

$$\text{Var}[T] = \frac{\alpha\beta(\alpha + \beta - 1)}{(\alpha - 1)^2(\alpha - 2)}, \tag{1.5.11}$$

provided $\alpha > 2$, see Problem 4 for details.

If one wishes to estimate the parameters α and β by the method of moments, it is useful to solve Eqs. (1.5.10) and (1.5.11) in terms of α and β. To simplify the notation put $\mu = E[T]$ and $\sigma^2 = \text{Var}[T]$. Then,

$$\alpha = \frac{2\sigma^2}{\sigma^2 - \mu^2 + \mu}$$

and   (1.5.12)

$$\beta = (\alpha - 1)(\mu - 1).$$

Given $\bar{T}$ and s^2, the sample mean and variance, moment estimates of α and β may be calculated from (1.5.12) by substituting $\bar{T}$ and s^2 for μ and σ^2. Moment estimates of the parameters in the Beta-Geometric distribution are of particular interest because they can be easily calculated.

A recursion formula for calculating the survival function of the Beta-Geometric distribution is useful in applying it to data. Applying (1.2.8), we see that

$$P[T > t \mid \Theta = \theta] = S(t \mid \theta) = (1 - \theta)^t.$$ (1.5.13)

A direct calculation then yields the formula

$$S(t) = P[T > t] = \int_0^1 f(\theta) S(t \mid \theta) d\theta$$

$$= \frac{\beta(\beta + 1) \dots (\beta + t - 1)}{(\alpha + \beta)(\alpha + \beta + 1) \dots (\alpha + \beta + t - 1)}$$ (1.5.14)

which is valid for $t \geq 1$. Finally, from (1.5.14) it is easy to derive the recursion formula

$$S(t) = \frac{S(t - 1)(\beta + t - 1)}{\alpha + \beta + t - 1},$$ (1.5.15)

where $S(0) = 1$ and $t = 1, 2, \dots$.

1.6 Applying the Beta-Geometric Distribution to Data

Jain (1968) applied the Beta-Geometric distribution extensively in analyzing the Taiwanese data described in Sect. 1.3. Throughout this analysis the method of moments was used to estimate the parameters α and β. An application of formulas (1.5.12), using the values 8.20 and 138.82 in column 1 of Table 1.3.1 for μ and σ^2, yields $\tilde{\alpha} = 3.48$ and $\tilde{\beta} = 17.86$ as moment estimates of the parameters α and β. These estimates, along with formula (1.5.15) could be used to calculate the values of the survival function appearing in column 4 of Table 1.3.2 to within rounding errors.

Also reported by Jain were the sample means and variances of the waiting times till conception classified according to twelve ages at marriage. These

Table 1.6.1. Estimates of the parameters in the Beta-Geometric distribution by age at marriages

Age at marriage (years)	Sample size	Sample mean	Sample variance	$\tilde{\alpha}$	$\tilde{\beta}$	Mean fecundability
12–15	69	13.4	265	5.36	54.09	0.090
16	110	11.7	151	11.70	114.50	0.093
17	211	10.4	215	3.67	25.08	0.128
18	292	9.2	95	9.00	65.58	0.121
19	385	8.7	146	3.70	20.76	0.151
20	374	7.2	100	3.61	16.20	0.182
21	234	6.4	84	3.40	12.95	0.208
22	186	6.4	127	2.75	9.44	0.226
23	125	6.0	51	4.86	19.29	0.201
24	72	5.3	103	2.57	6.74	0.276
25	62	6.4	98	3.09	11.28	0.215
≥ 26	70	8.9	444	2.38	10.87	0.180
Total*	2190	8.2	139	3.48	17.86	0.163

* Estimates for combined sample.

estimates are presented in the third and fourth columns of Table 1.6.1, the sample variances being rounded off to the nearest integer. Since Jain reported the sample mean and variance for each age classification, it is of interest to compute and compare twelve moment estimates of the parameters α and β together with estimates of mean fecundability, using formula (1.5.2). The results of these calculations are presented in the last three columns of Table 1.6.1.

From an inspection of the table, it can be seen that there was considerable variation in the moment estimates for both α and β. The estimates of α ranged form a low of 2.38 for the ages, ≥ 26, to a high of 11.70 at age 16. There was even greater variation in the estimates of β, ranging from a low of 6.74 at age 24 to a high of 114.50 at age 16. No doubt some of the variation in the estimates of α and β is attributable to the vagaries of sample size but there also appears to be a systematic effect of age. For, from an inspection of the last column in Table 1.6.1, it can be seen that there is a steady rise in the estimates of mean fecundability, despite the variation in the estimates of the parameters α and β. These estimates range from a low of 0.09 for the early teens to a high of 0.276 for age 24. Viewing the data from the perspective of Table 1.6.1 suggests that fecundability has a significant age component.

When working with retrospective survey data, such as that analyzed by Jain, one might rightfully suspect that the longer the duration of marriage at the date of interview, the greater would be the tendency to misreport events occurring early in marriage. At least two types of biases may result from deficient memories. Firstly, pregnancies not ending in a live birth could go unreported; secondly, there could be misreporting of times of marriage and first conceptions. Not reporting pregnancies would tend to lengthen the record- ed times of first conceptions. Without knowing the extent of misreporting of times of marriages and first conception, it is difficult to assess their impact.

Estimates of mean fecundability by duration of marriage are of interest, however, independent of any interpretations one wishes to associate with them.

Jain partitioned the data into 20 classifications of duration of marriage, ranging from 0 net years married to 20 or more. Within each classification mean fecundability was estimated by the procedure used in Table 1.6.1. The results of these computations were presented in Jain's Table 2. An interesting feature of this table was that estimates of mean fecundability declined with increasing age at marriage and exhibited an even wider range than those in Table 1.6.1. In Jain's Table 2, estimates of mean fecundability ranged from a high of 0.552 for marriage durations of less than a year to a low of 0.081 for about 19 years of marriage. A cross-classification of estimates of mean fecundability by wife's age at marriage and duration of marriage tended to preserve the picture of increasing fecundability with the age of the wife but decreasing fecundability with increasing duration of marriage, see Jain's Table 5.

Investigators may not completely agree on the interpretations of Jain's analysis of fecundability in Taiwanese women. Some tentative conclusions do seem possible, however, Firstly, within the age range considered in the data, mean fecundability tended to increase with age. Secondly, there seems to be some evidence that biases associated with memory deficiencies could lead to an under-estimation of mean fecundability, especially in marriages of long duration at the time of interview. Thirdly, it seems likely that decreasing mean fecundability with increasing duration of marriage cannot be entirely explained in terms of memory biases. For example, in a population of variable fecundability, pregnancies in marriages of short duration could select for highly fecund couples. Such selective effects could at least partially explain a mean fecundability of 0.552 for marriages of duration less than one year. Selective effects could also partially explain a decline in fecundability with increasing duration of marriage, because longer durations of marriage would tend to include less fecund couples in the samples.

Leridon (1977) has compiled and interpreted an extensive literature on fecundability. The interested reader should consult his Table 3.2, page 33, for a listing of estimates of the parameters α and β in the Beta-Geometric distribution based on various data sources. Even though they have less desirable properties than maximum likelihood estimates, moment estimates have been used in this section, primarily because they were easy to compute given the information available. Sheps and Menken (1973) have worked out the complicated details for estimating the parameters α and β by the method of maximum likelihood, see pages 96–107 of their book, for a discussion of procedures for the cases of truncated and untruncated data.

Since no attempt was made to estimate individual fecundabilities in Jain's data analysis, it was not possible to test whether fecundabilities among women followed a Beta distribution. Consequently, tests of the empirical validity of the Beta-Geometric distribution as a model of waiting times till first conceptions have been confined to Chi-square tests of goodness of fit such as those displayed by Jain in his Tables 1 and 3. Tests of goodness of fit, other than the visual one displayed in Table 1.3.2, have been deliberately avoided in this chapter, because they do not seem to be sufficiently informative in an exploratory analysis aimed at constructing explanatory models. In view of the phenomena observed in Jain's data analysis, a model based on variable fecundability among women (couples) seems to have greater explanatory power than one based on constant fecundability.

1.7 An Investigation of Selectivity

An indisputable conclusion that can be drawn from the data analysis discussed in Sect. 1.6 is that estimates of mean fecundability can vary greatly, depending on how the sample was selected. The purpose of this section is to study the phenomenon of selectivity within the framework of the Beta-Geometric distribution. Because our development of models is still rather primitive at this stage, it will be possible to study selectivity in only simple ways. For example, an analytic study of selectivity for long durations of marriage involving pregnancies among live births, is beyond the scope of the models considered in this chapter.

The idea of estimating fecundability as a function of duration of marriage goes back to the beginning of its study. Gini, who seems to be the first to study fecundability, presented estimates of mean fecundability according to a classification determined by months following marriage, see table on page 370 of Smith and Keyfitz (1977). The type of selectivity implied in Gini's work may be easily studied within the framework of the Beta-Geometric distribution.

From the probabilistic point of view, selectivity may be approached through conditioning. Suppose women marry at $t=0$ and the waiting times T till the first conception follow a Beta-Geometric distribution with parameters α and β. Let $f(\theta, t)$ be the joint p.d.f. of the random variables Θ and T. Then, according to (1.4.2) and (1.4.4), $f(\theta, t) = f(\theta) f(t|\theta)$, By definition, the conditional p.d.f. of the random variable Θ, given that $T = t \geq 1$, is

$$f(\theta|t) = \frac{f(\theta, t)}{h(t)}, \tag{1.7.1}$$

see (1.4.3). But,

$$f(\theta, t) = \frac{\Gamma(\alpha+\beta)}{\Gamma(\alpha)\Gamma(\beta)} \theta^\alpha (1-\theta)^{\beta+t-2}, \tag{1.7.2}$$

and

$$P[T=t] = \frac{\Gamma(\alpha+\beta)\Gamma(\alpha+1)\Gamma(\beta+t-1)}{\Gamma(\alpha)\Gamma(\beta)\Gamma(\alpha+\beta+t)}. \tag{1.7.3}$$

Therefore,

$$f(\theta|t) = \frac{\Gamma(\alpha+\beta+t)}{\Gamma(\alpha+1)\Gamma(\beta+t-1)} \theta^\alpha (1-\theta)^{\beta+t-2}, \tag{1.7.4}$$

for $0<\theta<1$. We thus reach the interesting conclusion that the conditional p.d.f. of Θ, fecundability, given that $T=t$, conception occurs in the t-th month following marriage, is a Beta distribution with parameters $\alpha+1$ and $\beta+t-1$. The conditional distribution in (1.7.4) has the mean

$$E[\Theta|T=t]=\frac{\alpha+1}{\alpha+\beta+t}, \tag{1.7.5}$$

for $t\geq 1$.

Like the effect observed in Jain's data analysis, mean fecundability in (1.7.5) decreases as the month t increases. The mean function in (1.7.5), although suggestive, can hardly be regarded as an explanation of the relationship between mean fecundability and duration of marriage observed by Jain, because, clearly, after the first conception the random variable T and duration of marriage no longer coincide. The selective effects of month of conception on mean fecundability can be quite pronounced. Suppose, for the sake of illustration, that α and β are the estimates $\hat{\alpha}=2.57$ and $\hat{\beta}=6.74$ corresponding to age 24 in Table 1.6.1. Then, the values of the conditional mean in (1.7.5) for t in the range $1\leq t\leq 6$ are: 0.346, 0.316, 0.290, 0.268, 0.249, 0.233. Recall the mean value of the unconditional distribution was 0.276.

It is also of interest to derive $f(\theta|T>t)$, the conditional p.d.f. of Θ given that $T>t$. This is the conditional distribution of fecundability among those women who have not conceived by month t. By definition,

$$f(\theta|T>t)=\frac{f(\theta)P[T>t|\Theta=\theta]}{P[T>t]}. \tag{1.7.6}$$

A computation similar to that used in the derivation of (1.7.4) yields the formula

$$f(\theta|T>t)=\frac{\Gamma(\alpha+\beta+t)}{\Gamma(\alpha)\Gamma(\beta+t)}\,\theta^{\alpha-1}(1-\theta)^{\beta+t-1}, \tag{1.7.7}$$

for $0<\theta<1$. Therefore, the conditional distribution of Θ, given that $T>t$, is that of Beta random variable with parameters α and $\beta+t$. Observe that large values of t have the effect of skewing the distribution to the left in the direction of small fecundabilities. Since not having conceived by month t tends to select for women of low fecundability, this type of result was expected.

Rather than conditioning on $T=t$, the conditional distribution of Θ, given that $T\leq t$, could also be derived. This conditional distribution is not of the Beta type, however, see Problem 5 for details. Potter et al. (1970) have also investigated selection for fecundability along the lines developed in this section.

1.8 Fecundability as a Function of Coital Pattern

Even though heterogeneity in fecundability among woman was taken into account in the Beta-Geometric distribution introduced in Sect. 1.4, it was assumed constant from month to month for each woman. There are also good reasons for believing that this latter assumption is not congruent with the actual processes affecting conception. If, for example, there were no coital acts in a given month, then fecundability would be zero. On *a priori* grounds it would be expected that fecundability could depend on coital patterns. Barrett and Marshall (1969) and Barrett (1971), in an interesting analysis of an unusual data set, have proposed a model of fecundability taking coital patterns into account explicitly.

The sample studied by Barrett and Marshall consisted of 241 British married couples who had demonstrated their fertility by the birth of at least one child. Roman Catholics, who had sought advice about the basal body temperature (B.B.T.) method of regulating births by the timing of coitus in relation to the time of ovulation, comprised the majority of the sample. The ages of the wives when entering the study ranged from 21 to 50 years, with the age groups 20–29 and 30–39 each containing about 45% of the sample. All couples in the sample were experienced in the use of the B.B.T. method of regulating births, and each wife had completed at least one ovulatory cycle since the birth of her last child. Those who regularly produced temperature charts that were difficult to interpret were excluded from the sample, resulting in further selection. On each day the couples marked on a chart the morning B.B.T. and whether coitus occurred on that day, returning their charts every three months. This procedure was followed even when they wished to have another child.

Each monthly cycle was divided into pre- and post-ovulatory days. The time of ovulation was determined by a rise in B.B.T. as judged by two specialists, see Barrett and Marshall (1969) for details. It is interesting to note from the charts displayed in Figs. 1 and 2 in the paper of Barrett and Marshall that ovulation occurred at about the 14-th day of the cycle. Altogether 1,898 cycles were considered in the data analysis. From inspection of Table 1 in Barrett and Marshall, one gets the impression that conceptions were associated with coital acts on only 9 pre- and 10 post-ovulatory days.

A coital pattern for a given cycle may be represented by a 19 dimensional vector $\mathbf{x}$ with elements x_i indexed by the set $A = [-9, -8, \ldots, -1, 1, 2, \ldots, 10]$, where the minus and plus signs represent pre- and post-ovulatory days. Let $x_i = 1$ if at least one coital act occurs on day $i \in A$ and let $x_i = 0$ otherwise. Let α_i be the probability of a conception as a result of a coital act on day $i \in A$ and assume days represent independent trials. Next let $\theta(\mathbf{x})$ be the conditional probability of a conception in a monthly cycle, given coital pattern $\mathbf{x}$. Then,

$$1 - \theta(\mathbf{x}) = \prod_i \beta_i^{x_i}, \tag{1.8.1}$$

where $\beta_i = 1 - \alpha_i$ and $i \in A$. Passing to logarithms in (1.8.1) leads to the linear regression model

$$\log_e(1 - \theta(\mathbf{x})) = \sum_i \gamma_i x_i, \tag{1.8.2}$$

where $\gamma_i = \log_e \beta_i$.

By using the method of maximum likelihood on the whole sample of cycles, it was possible to estimate the γ-parameters on the right in (1.8.2). Nelder and

Wedderburn (1972) have given an exposition of a general theory, including the model in (1.8.2) as a special case. Problem 6 contains an outline of the theory for the model under consideration.

All of the γ-parameters in (1.8.2) should have the same sign, which is usually not guaranteed by the estimation procedure. One way of attaining a commonality of signs in the γ's is to estimate parameters δ_i defined by $\gamma_i = -\exp[\delta_i]$. A question that naturally arises is whether the assumption of independence on different days of the menstrual cycle has empirical validity. Barrett and Marshall present evidence in their Table 4 that this assumption leads to predictions of fecundability that are in agreement with the data.

In addition to estimating the daily probabilities of conception in the combined sample of 1,898 cycles, Barrett (1971) also estimated these probabilities for women in the age groups under and over 30. The numbers of ovulatory cycles in the respective age groups were 701 and 1,012; 185 cycles were excluded from the analysis because the ages of the women were not known. Barrett's estimates of daily probabilities with indices in the set $[-7, -6, \ldots, -1, 1]$ are shown in Table 1.8.1. All daily probabilities with indices outside this set were judged to be zero. The equation connecting γ_i and α_i is $\alpha_i = 1 - \exp[-\gamma_i]$.

The estimates of the daily probabilities of conception in Table 1.8.1 have many interesting implications. Firstly, they suggest that if the rhythm method of contraception is to be effective, then there should be no coital acts in the preovulatory phase of the menstrual cycle. Furthermore, to be on the safe side, coitus should not be resumed until two to four days following ovulation. A second set of implications of the estimates in Table 1.8.1 concerns the variability in fecundability associated with coital patterns. If a zero or one is associated with each of the eight days, then the set coital patterns relative to the probabilities in Table 1.8.1 consists of $2^8 = 256$ vectors of dimension 8. Four of these vectors are presented in Table 1.8.2. The four associated fecundabilities, calculated by applying formula 1.8.1 in each group, are presented in Table 1.8.3.

Table 1.8.1. Estimates of daily probabilities of conception

		Age of wife			
		30 and under		Over 30	
Day		γ_i	α_i	γ_i	α_i
Pre-ovulatory	−7	0.058	0.056	0.039	0.038
	−6	0.000	0.000	0.050	0.049
	−5	0.091	0.086	0.158	0.146
	−4	0.364	0.305	0.022	0.022
	−3	0.134	0.125	0.120	0.113
	−2	0.383	0.317	0.224	0.200
	−1	0.118	0.111	0.088	0.084
Post ovulatory	1	0.086	0.082	0.100	0.095

Table 1.8.2. Four sample coital patterns

Pattern	Day							
	-7	-6	-5	-4	-3	-2	-1	1
1	1	1	1	1	1	1	1	1
2	1	0	1	0	1	0	1	0
3	0	1	0	1	0	1	0	1
4	1	0	0	1	0	0	1	0

Table 1.8.3. Estimates of fecundability by age and coital pattern

Coital pattern	Age of wife	
	30 and under	over 30
1	0.706	0.551
2	0.321	0.332
3	0.567	0.327
4	0.410	0.138

By way of illustration, consider the first coital patterns in Table 1.8.2, the case at least one coital act occurs on each of the eight days. Then, the associated estimate of fecundability for the age group over 30 is

$$\theta = 1 - (0.962)(0.951) \ldots (0.905) \simeq 0.551. \tag{1.8.3}$$

The other fecundabilities in Table 1.8.3 were calculated in a similar way. Even in the small sample of fecundabilities in Table 1.8.3, the range of values is rather large. The high estimates of fecundability, 0.706 and 0.551, associated with the first coital pattern seem plausible. It might be expected that coital patterns 2 and 3 would generate about the same fecundabilities. This expectation seems to be realized in the age group over 30; but, the difference in the two values 0.321 and 0.561 for the age group 30 and under is probably attributable to sampling errors. Since the estimates of fecundability for the age group over 30 are based on a larger sample of ovulatory cycles, they would seem more reliable than those in the younger age group.

1.9 A Distribution on the Set of Coital Patterns

Let χ be the set of coital patterns associated with the probabilities in Table 1.8.1. As mentioned before, the set contains 256 8-dimensional vectors $\mathbf{x} = \langle x_1, x_2, \ldots, x_8 \rangle$ of zeros and ones. Since with each $\mathbf{x} \in \chi$ there is associated a fecundability $\theta(\mathbf{x})$, a distribution on the set χ will determine a distribution of fecundabilities based on coital patterns. Such a distribution based on human behavioral factors should have greater explanatory power than the Beta-Geometric distribution introduced in Sect. 1.4 in a purely descriptive way.

One approach to putting a distribution on χ is to assume that every vector $\mathbf{x} \in \chi$ is a realization of a Markov chain constructed as follows. Suppose the menstrual cycle, starting with the first day of the menstrual period lasting four days, is thirty days in length. Thus, the intermenstrual period starts on the 5-th day of the cycle and continues through the 30-th. The Markov chain model of coital patterns will be assumed to operate during the intermenstrual period. It will also be supposed that ovulation occurs between the 14-th and 15-th days of the menstrual cycle. The assumptions regarding the lengths of the menstrual cycle as well as the timing of ovulation are only special cases of the variation observed in populations of women. It will also be assumed that couples stay together throughout a menstrual cycle. Once the model is understood in the special case under consideration, it should not be too difficult to extend it.

Consider a Markov chain with two states symbolized by 0 and 1. The chain will be said to be in state 0 if no coital act occurs on a given day and in state 1 if at least one coital act occurs on that day. To completely determine the Markov chain it suffices to specify $p_i(0)$, $i=0, 1$, the initial distribution, and the 2×2 matrix $\mathbf{P}=(p_{ij})$ of one-step transition probabilities. In terms of the model of coital patterns under consideration, $p_0(0) \geq 0$ is the probability no coital act occurs on the 5-th day of the menstrual cycle, $p_1(0)$ is the probability at least one coital act occurs on that day, and $p_0(0)+p_1(0)=1$. Similarly, $p_{01} \geq 0$ is the conditional probability that at least one coital act occurs on a given day, given there were no coital acts on the previous day, the other elements of the matrix $\mathbf{P}=(p_{ij})$ being defined similarly. Each row of the matrix $\mathbf{P}$ sums to one, i.e., $p_{i0} + p_{i1}=1$ for $i=0, 1$.

If it is assumed that ovulation occurs between the 14-th and 15-th days of the cycle, then the eight day susceptible period, suggested by the probabilities in Table 1.8.1, begins on the 8-th day of the menstrual cycle and lasts through the 15-th. Consequently, if the inter-menstrual period starts on the 5-th day of the cycle, then the Markov chain runs for two days prior to the start of the 8-day susceptible period. Let $p_{ij}^{(2)}$ be the conditional probability of the Markov chain in state $j=0, 1$ on its second day, given that it was in state $i=0, 1$ on the initial day. Then, under the assumption coital patterns follow a Markov chain,

$$p_{ij}^{(2)} = \sum_{k=0}^{1} p_{ik} p_{kj}. \tag{1.9.1}$$

Observe that $p_{ij}^{(2)}$ is the ij-th element in $\mathbf{P}^2$, the second power of the transition matrix $\mathbf{P}$.

For every $\mathbf{x} \in \chi$, let $p(\mathbf{x})$ be the probability of the coital pattern $\mathbf{x} = \langle x_1, x_2, \ldots, x_8 \rangle$ during an eight day susceptible period. Then, under Markovian assumptions,

$$p(\mathbf{x}) = \sum_{i=0}^{1} \sum_{j=0}^{1} p_i(0) p_{ij}^{(2)} p_{jx_1} p_{x_1 x_2} \cdots p_{x_7 x_8}. \tag{1.9.2}$$

An immediate consequence of the assumption that coital patterns follow a Markov chain is that the equation

$$\sum_{\mathbf{x}} p(\mathbf{x}) = 1 \tag{1.9.3}$$

is satisfied, where the sum extends over all vectors $\mathbf{x} \in \chi$. Hence, the proposed distribution of coital patterns is proper.

When fecundability is viewed over a population of women, it may be regarded as a random variable Θ taking values in the set $[\theta(\mathbf{x}): \mathbf{x} \in \chi]$. The mean of Θ is

$$E[\Theta] = \sum_{\mathbf{x}} p(\mathbf{x}) \theta(\mathbf{x}), \tag{1.9.4}$$

and its second moment is

$$E[\Theta^2] = \sum_{\mathbf{x}} p(\mathbf{x}) \theta^2(\mathbf{x}). \tag{1.9.5}$$

The variance of Θ may be computed using the formula

$$\text{Var}[\Theta] = E[\Theta^2] - (E[\Theta])^2. \tag{1.9.6}$$

James (1971) has studied coital rates during the menstrual period as a percentage of available days. A day is available for coitus if neither spouse is absent or ill. The theoretical analogue of the coital rate studied by James seems to be the function $p_1(t)$, the probability that at least one coital act occurs on day $t = 0, 1, 2, \ldots$. For $t = 0$, $p_1(0)$ is the initial probability; but, for $t \geq 1$,

$$p_1(t) = \sum_{i=0}^{1} p_i(0) p_{i1}^{(t)}, \tag{1.9.7}$$

where $p_{i1}^{(t)}$ is an element of $\mathbf{P}^t$, the t-th power of the matrix $\mathbf{P}$.

In order to compute the conditional probability $p_{i1}^{(t)}$ on the right in (1.9.7), it will be necessary to compute all elements in the matrix $\mathbf{P}^{t-1}$. For any $t \geq 2$, the elements of $\mathbf{P}^t$ may be calculated recursively according to the formula

$$p_{ij}^{(t)} = \sum_{k=0}^{1} p_{ik}^{(t-1)} p_{kj}, \tag{1.9.8}$$

with $p_{ij}^{(1)} = p_{ij}$. Unlike the probability $p(\mathbf{x})$ in (1.9.2), depending on the length of the susceptible period, $p_1(t)$ in (1.9.7) covers the entire 26 day intermenstrual period.

An excellent source of background material on finite Markov chains is the book by Kemeny and Snell (1976). Initially published nearly 20 years ago, this book was among the first to treat Markov chains from a computational point of view.

1.10 Some Implications of Markov Chain Model of Coital Patterns

In some surveys, married women are asked how many times they had sexual intercourse during the four week period preceding the day of interview, see for example Westoff (1974). The reliability of answers to such questions depends, among other things, on the veracity of the respondents. Even if they are inclined to tell the truth, the accuracy of the answer to such questions depends on their abilities to recall events. When a respondent has a memory problem, there may be a tendency to report a perceived mean value rather than the actual number. As part of a program for validating and interpreting the Markov chain model of coital patterns introduced in the previous section, it is of interest to derive expression for the mean number of coital acts during a menstrual cycle.

Let the 26 days of the intermenstrual period be indexed by the numbers t $=0, 1, 2, ..., 25$; and, for each t, let $X(t)$ be a random variable defined as follows. Put $X(t)=1$ if at least one coital act occurs on the t-th day and let $X(t)=0$ otherwise. Then,

$$Z = \sum_{t=0}^{25} X(t) \tag{1.10.1}$$

is a random variable representing the number of coital acts during a menstrual cycle. From the definition of $X(t)$, it follows that $E[X(t)]=p_1(t)$, the function defined in (1.9.7). Therefore, the mean number of acts during a menstrual cycle takes the form

$$E[Z] = \sum_{t=0}^{25} p_1(t) = 26\bar{p}_1, \tag{1.10.2}$$

where

$$\bar{p}_1 = \tfrac{1}{26} \sum_{t=0}^{25} p_1(t). \tag{1.10.3}$$

In the class of regular Markov chains, see Chap. IV of Kemeny and Snell (1976), an approximation to the mean in (1.10.3) may be easily calculated.

A Markov chain with transition matrix $\mathbf{P}$ is said to be regular if there is a positive integer t such that all the elements in the matrix $\mathbf{P}^t$ are positive. The two-state Markov model of coital patterns under consideration will always be regular. Therefore, according to a theorem, see page 70 of Kemeny and Snell (1976), there are positive numbers π_0 and π_1 such that $\pi_0+\pi_1=1$ and

$$\lim_{t \uparrow \infty} \mathbf{P}^t = \begin{bmatrix} \pi_0 & \pi_1 \\ \pi_0 & \pi_1 \end{bmatrix}. \tag{1.10.4}$$

Furthermore, the vector $\boldsymbol{\pi}=\langle \pi_0, \pi_1 \rangle$ of probabilities π_0 and π_1 satisfies the equation

$$\boldsymbol{\pi} = \boldsymbol{\pi}\mathbf{P}, \tag{1.10.5}$$

see Problem 7 for details.

From Eqs. (1.9.7) and (1.10.4), it also follows that $\lim_{t \uparrow \infty} p_1(t) = \pi_1$. That is, π_1 is the long-run probability of being in state 1, π_0 being interpreted similarly. Because of these properties and Eq. (1.10.5), the vector $\pi = \langle \pi_0, \pi_1 \rangle$ is frequently called the stationary distribution of the Markov chain.

It can be shown that $p_1(t) \to \pi_1$ as $t \uparrow \infty$ implies

$$\lim_{n \uparrow \infty} \frac{1}{n} \sum_{t=0}^{n-1} p_1(t) = \pi_1. \tag{1.10.6}$$

Hence, the mean in (1.10.3) is approximately π_1; and, by (1.10.2), $26\pi_1$ is approximately the mean number of coital acts during a menstrual cycle. For the two-state Markov chain under consideration, Eq. (1.10.5) may be easily solved. In fact, the solution is

$$\pi_0 = \frac{p_{10}}{p_{01} + p_{10}} \quad \text{and} \quad \pi_1 = \frac{p_{01}}{p_{01} + p_{10}}. \tag{1.10.7}$$

Barrett (1971) has analyzed coital patterns in the data set described in Sect. 1.8 for the combined age groups. Two types of coital patterns were distinguished; namely, persisters and alternators. Briefly, coital patterns in persisters are characterized by runs of coital acts on consecutive days followed by runs of no coital acts; while those of alternators tend to have coital acts on alternate days. In his Table 1, Barrett presents the data for two couples illustrating persister and alternator patterns for a six-day susceptible period within the intermenstrual period. The reason for using a six-day susceptible period was that in the regression analysis for the combined age groups only six regression coefficients were judged significantly different from zero.

Under the assumption that coital patterns among persisters and alternators follow a Markov chain, it is possible to get estimates of the transition matrices from the data presented by Barrett in his Table 1. Thus, to estimate the probabilities p_{00} and p_{01}, it suffices to count the number of times the transitions $0 \to 0$ and $0 \to 1$ occurred, following entrance into state 0, and then divide these counts by the number of entrances into state 0. The probabilities p_{10} and p_{11} were estimated in a similar way. The estimate of the transition matrix for persisters obtained in this way was

$$\hat{P} = \begin{bmatrix} 0.586 & 0.414 \\ 0.196 & 0.804 \end{bmatrix}; \tag{1.10.8}$$

while that for alternators was

$$\hat{P} = \begin{bmatrix} 0.366 & 0.634 \\ 0.962 & 0.038 \end{bmatrix}. \tag{1.10.9}$$

It should be re-emphasized that the estimates in (1.10.8) and (1.10.9) are based on incomplete data for only two couples. They should, therefore, be regarded as indicative of coital patterns among persisters and alternators in only a tentative way.

Despite the limitations of the estimates in (1.10.8) and (1.10.9), they appear useful in succinctly summarizing coital patterns for the two couples. The concentration of probability mass in (1.10.8) on the principal diagonal, running from the upper left to the lower right, summarizes the tendency of persisters to remain in a state. On the other hand, the concentration of probability mass in (1.10.9) on the diagonal, running from the lower left to the upper right, illustrates the tendency of alternators to change states on alternate days. The stationary probabilities for persisters are

$$\pi_0 = 0.321 \quad \text{and} \quad \pi_1 = 0.679, \tag{1.10.10}$$

and those for alternators are

$$\pi_0 = 0.604 \quad \text{and} \quad \pi_1 = 0.396. \tag{1.10.11}$$

Consequently, based on the estimates in (1.10.8) and (1.10.9), the mean number of coital acts during a menstrual cycle are $26 \times (0.679) = 17.65$ and $26 \times (0.396) = 10.30$ for persisters and alternators, respectively. Both of these values fall within the range of coital frequencies in the four weeks prior to interview reported by Westoff (1974).

1.11 Computer Implementation and Numerical Examples

A computer programming language called APL greatly facilitates the computer implementation of the Markov chain model of coital patterns set forth in Sect. 1.9. A number of APL functions were written with a view towards further exploring the implications of the estimated transition matrices in (1.10.8) and (1.10.9) for persisters and alternators. The purpose of this section is to present some of the computer output of these APL functions. Two initial probability vectors $p(0) = \langle p_0(0), p_1(0) \rangle$ were considered; namely $p(0) = \langle 0.5, 0.5 \rangle$ and $p(0) = \langle 0, 1 \rangle$. The first vector corresponds to the case at least one coital act occurs on the first day of the intermenstrual period with probability 0.5; the second to the case there is always a coital act on the first day.

Considered first was the function $p_1(t)$ defined in (1.9.7), representing the probability that at least one coital act occurs on the t-th day of the 26-day intermenstrual period. Presented in Table 1.11.1 are the values $p_1(t)$, $t = 0, 1, 2, \ldots, 25$, for persisters and alternators under the two chosen initial vectors. From the values of $p_1(t)$ for persisters, it can be seen that the approach of $p_1(t)$ to $\pi_1 = 0.679$, the limiting value, is quite rapid. Thus, for $p_1(0) = 0.5$ and $p_1(0) = 1$, the limiting value is attained on days 8 and 7 respectively. For $p_1(0) = 0.5$, the convergence of $p_1(t)$ to π_1 is nearly monotone increasing; but for $p_1(0) = 1$, the convergence to π_1 is monotone decreasing.

The convergence of $p_1(t)$ to $\pi_1 = 0.396$ for alternators presents quite a different picture. For both initial values of $p_1(0)$, the values of $p_1(t)$ fluctuate before they settle down to π_1, the fluctuations being greatest for the initial value $p_1(0) = 1$. When $p_1(0) = 0.5$, $p_1(t)$ hits the limiting value 0.396 for the first time on day 11; for $p_1(0)=1$, the corresponding time is day 15. James (1971) in his Text-Figure 1 presents graphs showing fluctuations of daily coital rates for various lengths of the menstrual cycle. Apart from sampling fluctuations, some of these graphs appear to be those of a function decreasing monotonely to a limit; others exhibit fluctuations similar to the values of $p_1(t)$ for alternators. The values in Table 1.11.1 are not directly comparable to the graphs displayed by James but there are qualitative similarities. Consequently, the data presented by James do not falsify a working hypothesis that they could have been generated by a mixture of Markov chain models of coital patterns of the type under consideration, with the components of the mixture representing various gradations of persisters and alternators.

The sum of the columns at the bottom of Table 1.11.1 are estimates of the expected value, $E[Z]$, in 1.10.2, the mean number of coital acts during a menstrual cycle. Within both persisters and alternators the two estimates of $E[Z]$ are very close, indicating that the initial distribution has little impact on $E[Z]$. The approximations, 17.65 and 10.30, of $E[Z]$ for persisters and alternators derived in Sect. 1.10 are very close to the actual values presented in Table 1.11.1. It would appear, therefore, that for the model of coital patterns under consideration, the suggested approximation of $E[Z]$ is a useful one.

Considered next was the computation of the mean and variance in fecundability, $E[\Theta]$ and $\text{Var}[\Theta]$, defined in formulas (1.9.4), (1.9.5), and (1.9.6). This computation proceeded in steps. Firstly, the computer generated $2^8=256$ coital patterns $\mathbf{x}$ in the set χ relative to the α-probabilities in Table 1.8.1. For each

Table 1.11.1. Probabilities of at least one coital act on days of intermenstrual period

Day	Persisters		Alternators	
0	0.5	1.0	0.5	1.0
1	0.609	0.804	0.334	0.034
2	0.652	0.728	0.434	0.614
3	0.668	0.698	0.374	0.266
4	0.675	0.686	0.410	0.474
5	0.677	0.682	0.388	0.349
6	0.678	0.680	0.401	0.424
7	0.678	0.679	0.393	0.379
8	0.679	0.679	0.398	0.407
9	0.679	0.679	0.395	0.390
10	0.679	0.679	0.397	0.400
11	0.679	0.679	0.396	0.394
12	0.679	0.679	0.396	0.398
13	0.679	0.679	0.396	0.395
14	0.679	0.679	0.396	0.397
15	0.679	0.679	0.396	0.396
16	0.679	0.679	0.396	0.396
$\vdots$	$\vdots$	$\vdots$	$\vdots$	$\vdots$
25	0.679	0.679	0.396	0.396
Total	17.359	18.179	10.364	10.677

coital pattern $\mathbf{x} \in \chi$, a fecundability $\theta(\mathbf{x})$ was computed using equation (1.8.1), within each age group. In the age group under 30, the α-probability associated with the preovulatory day, -6, is zero. This zero estimate, under the assumption that it could be a statistical artifact, was replaced by the intermediate value 0.071 but the remaining α-probabilities were left unchanged throughout the calculations. The α-probabilities for the age group, over 30, were also left unchanged. After computing the fecundabilities and the probabilities, $p(\mathbf{x})$, $\mathbf{x} \in \chi$, given by (1.9.2), $E[\Theta]$ and $\mathrm{Var}[\Theta]$ were calculated for persisters and alternators within each age group, under 30 and over 30, for the initial probabilities used in Table 1.11.1. The results of these calculations are presented in Table 1.11.2.

A comparison of the upper and lower panels in Table 1.11.2, corresponding to the initial probabilities, $p_1(0)=0.5$ and $p_1(0)=1$, shows that the initial distribution had little impact on the mean and variance in fecundabilities. Since the model is based on a Markov chain, where the evolution of the process depends on the immediate past, this result was partially anticipated. Within each age group, the mean fecundabilities for persisters and alternators differ markedly, suggesting that fecundability has an important behavioral component. There are also marked differences between age groups, reaffirming the notion that there is a significant age component in fecundability. Thus, for the age group under 30, the mean fecundabilities for persisters and alternators are about 0.56 and 0.39. Correspondingly, the two mean fecundabilities for age group, over 30, are about 0.41 and 0.26.

Jain (1969) in his Table 2 reports estimates of mean fecundabilities by duration of marriage at the date of interview for a sample of Taiwanese women. For those married less than a year, a mean fecundability of 0.552 was reported. Because the vast majority of the Taiwanese women in the survey married before age 30, the ages of these women were predominately under 30. It is of interest to note that, despite the differing cultural backgrounds of the Taiwanese and British couples, Jain's estimate, 0.552, is close to the mean fecundability, 0.560, for persisters under 30 in Table 1.11.2. Among the competing explanations of Jain's estimate of fecundability for women married less than one year, one may add the possibility that there was selectivity for coital patterns similar to those of the persisters considered by Barrett (1970).

Table 1.11.2. Means and variances of fecundability

Initial probability $p_1(0)$	Under 30		Over 30	
	Persisters	Alternators	Persisters	Alternators
0.5	0.558*	0.388	0.405	0.266
	0.030**	0.018	0.017	0.007
1.0	0.560	0.390	0.470	0.261
	0.030	0.019	0.017	0.007

  * Mean.
** Variance.

The mean and variance of a distribution are not by themselves vary informative, since they do not provide any insights into its shape. Although determining shapes of distributions can lead to computational complexities, especially for distributions arising in some stochastic processes, the implementation of the necessary calculations can lead to useful insights. For the Markov chain model under consideration, the necessary calculations were implemented in APL, using the following ideas.

Let $y_0, y_1, \ldots, y_n$ be numbers such that $0 = y_0 < y_1 < \ldots < y_{n-1} < y_n = 1$ and let $B_i = [y_{i-1}, y_i)$, $1 \le i \le n$, be the partition of the interval $[0, 1)$ determined by these numbers. This partition, in turn, induces a partition, $A_1, A_2, \ldots, A_n$, of the set χ determined by putting

$$A_i = [\mathbf{x} \in \chi : \theta(\mathbf{x}) \in B_i] \qquad (1.11.1)$$

for $1 \le i \le n$. With each set in this partition, there is associated a probability

$$PA_i = \sum_{\mathbf{x} \in A_i} p(\mathbf{x}), \qquad (1.11.2)$$

where $p(\mathbf{x})$ is defined in Eq. (1.9.2). By matching PA_i with B_i, $1 \le i \le n$, a histogram, representing a distribution of fecundabilities, is determined. Four of these histograms are presented in Fig. 1.11.1. In these histograms, the interval $[0, 1)$ was partitioned into ten subintervals of length 0.1, which are represented on the horizontal axes labeled fecundability. The vertical axes, labelled frequency, represent the probabilities PA_i, $1 \le i \le n$, in (1.11.2). The distributions represented in Fig. 1.11.2 are those whose means and variances are displayed in the upper panel of Table 1.11.2.

The same set of fecundabilities, those determined by α-probabilities for the age group under 30 in Table 1.8.1, were used in calculating the histograms for persisters and alternators in that age group in Fig. 1.11.1. Similarly, the α-probabilities for the age group over 30 in Table 1.8.1 were used to calculate the histograms for persisters and alternators in that age group in Fig. 1.11.1. The

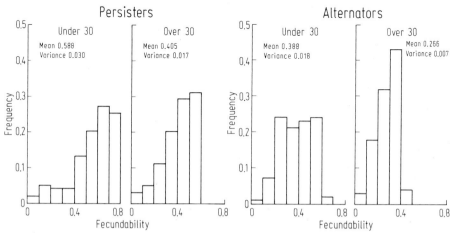

Fig. 1.11.1. Histograms of fecundability distributions

estimated transition matrix for persisters in (1.10.8) was used to calculate the histograms for persisters in both age groups in the left half of Fig. 1.11.1; while that for alternators was used to calculate the histograms in the right half of the figure. Consequently, the differences in the histograms in the first and third and those in the second and fourth quarters of Fig. 1.11.1 are attributable to the effects of age; those in the left and right halves of the figure are attributable to differing coital patterns among persisters and alternators.

Differences between the histograms of persisters and alternators are quite marked, especially in the age group under 30. The distribution of fecundabilities for persisters in this age group is skewed to the right, the essential support of the distribution is the interval [0, 0.8), and the most frequent or modal class is the interval [0.6, 0.7). By contrast, the distribution of fecundabilities for alternators under 30 is not skewed to the right; its essential support is the interval [0, 0.7), and there are two modal intervals [0.2, 0.3) and [0.5, 0.6). More verbal comparisons among histograms in Fig. 1.11.1 could be set forth; but, further verbal discussion seems unnecessary, because, in some ways, visual comparisons of histograms are more informative than words. In making comparisons among these histograms, it should be kept in mind that they represent only a fleeting glimpse of a range of plausible possibilities, since the transition matrices for persisters and alternators are based on data from only two couples.

Some mention of the validity of the Beta distribution as a model for variation in fecundability seems appropriate in light of the histograms in Fig. 1.11.1. As before, let α and β be the parameters of the Beta distribution. If $\alpha > 1$ and $\beta > 1$, the p.d.f. is unimodal. Should further data analyses turn up bimodal histograms such as that in the lower left in Fig. 1.11.1, the validity of the Beta as an approximation would be brought into question. The remaining histograms in Fig. 1.11.1 do not, however, rule out the possibility that they could be approximated by a Beta p.d.f. For if a Beta random variable is transformed from (0, 1), the canonical interval of support, to a subinterval (0, γ), representing the essential range of fecundabilities, then it seems plausible that some Beta p.d.f.'s could at least crudely approximate these histograms.

1.12 Conclusions and Further Research Directions

In addition to references cited in previous sections, other papers have been influential in developing the ideas presented in this chapter. Variation in the timing of ovulation within the menstrual cycle was ignored in the Markov chain model of coital patterns developed in Sect. 1.9. Also ignored was variation in the length of the menstrual cycle. Lachenburgh (1967), in his Table 1, presents some empirical distributions on the timing of ovulation within the menstrual cycle. The previously cited paper, James (1971), also contains some quantitative information on the variability in the lengths of the menstrual cycle.

Other authors have approached the problem of modeling coital patterns in a way differing from that in Sect. 1.9. Glaser and Lachenburgh (1969) modeled such patterns as a renewal process, with coital acts representing "points" of renewal. Among the limitations of this approach was that the computations presented were based on a long-run or equilibrium distribution of a renewal process. The use of this equilibrium distribution, although convenient for calculations, ignored the transitory nature of the coital process. For example, among many couples, coitus is discontinued during the menstrual period. A desire to accommodate both coital patterns and the transitory nature of the coital process in one formulation lead to the construction of the Markov chain model introduced in Sect. 1.9.

A number of factors, ignored in the models developed in this chapter, could be expected to influence coital patterns with their implied mean coital frequencies during a monthly menstrual cycle. Included in these factors would be the age of women and whether they are pregnant, wanting to become pregnant, or are in a postpartum period.

Westoff (1974) has studied the impact of a number of factors on coital frequencies for U.S. women interviewed in 1965 and 1970. There was a marked association of mean coital frequency with the age of the wife. The highest frequencies, 11 to 12 times per month, were observed in the youngest age group ≤ 19; while the lowest, 5 to 6 times per month, were observed in the oldest age group, 43–44.

The state of a woman at the time of interview can also influence coital frequency. Westoff reports that, among women trying to get pregnant, mean coital frequencies in the four-week period prior to the interview ranged between 8 and 11. For pregnant women and those in a postpartum period the corresponding ranges were 6 to 6.4 and 0.1 to 0.9, respectively. The method of contraception being used can also influence coital frequency. In fact, among users of contraception, the highest coital frequencies, 9 to 10 times per month, were associated with the most medically effective methods, the pill and IUD (intrauterine device).

Superimposed over any states women may occupy at any time was a secular increase in coital frequencies across all age groups between 1965 and 1970. Westoff suggests that this increase, 21%, could, in part, be due to women's greater willingness to discuss sexual matters in household interviews. For the most part, however, the reported increase was believed to be real.

Many methods of contraception have been developed in attempts to lower fecundability to zero, thus controlling births. In practice, however, fecundability is rarely zero even when a method of contraception is used. The capability of a method of contraception to lower fecundability is referred to as its effectiveness. Problem 8 contains a discussion centering around measuring contraceptive effectiveness. Tietze (1970) has given a general review of the subject.

Not considered in the models developed in this chapter was a distribution of the waiting times till conception when fecundability not only varies from month to month within couples but also among couples. Contained in Problem 9 is a discussion of a general point of view for deriving such distributions. Computer simulation is often a useful tool for generating realizations of a stochastic process. Displays of samples of such realizations are often helpful in arriving at a deeper understanding of the implications of a process. Problem 10 contains displays of realizations of coital patterns, during the intermenstrual period, generated by the transition matrices for persisters and alternators in (1.10.8) and (1.10.9).

Problems and Miscellaneous Complements

1. If a random variable T, taking values in the positive integers, has the p.d.f. $h(t)$, $t=1, 2, \ldots$, then its generating function is defined by

$$g(s) = E[s^T] = \sum_{t=1}^{\infty} h(t)s^t, \quad |s| \leq 1.$$

(a) Show that if T follows a geometric distribution, then

$$g(s) = \frac{\theta s}{1 - \eta s}.$$

(b) Show that if $E[T]$ and $E[T^2]$ are finite, then

$$E[T] = g^{(1)}(1)$$

and

$$E[T(T-1)] = g^{(2)}(1),$$

where $g^{(1)}$ and $g^{(2)}$ stand for the first and second derivatives of g.

(c) Use the results in (b) to derive formulas (1.2.10) and (1.2.11). Hint: $E[T^2] = E[T(T-1)] + E[T]$.

2. Let $T_1, T_2, \ldots, T_n$ be random variables based on random samples of size n from a geometric distribution. For a particular sample $\langle T_1, T_2, \ldots, T_n \rangle = \langle t_1, t_2, \ldots, t_n \rangle$, the likelihood function is

$$L = \prod_{i=1}^{n} h(t_i) = \theta^n (1 - \theta)^{\sum_{i=1}^{n} t_i - n}.$$

(a) Compute $d \log L / d\theta$, where the logarithm is base e. Then, solve the equation

$$\frac{d \log L}{d\theta} = 0$$

to show that $\hat{\theta}_n = 1/\bar{T}$, see (1.3.1), is the maximum likelihood estimator of θ based on random samples of size n.

(b) Show that Fisher's information function is

$$\vartheta(\theta) = -E\left[\frac{d^2 \log L}{d\theta^2}\right] = \frac{n}{\theta^2 (1 - \theta)}.$$

The information function is useful because in large samples $\hat{\theta}_n$ is approximately normally distributed with a mean of θ and variance $\theta^2(1-\theta)/n$. In symbols,

$$\hat{\theta}_n \sim N\left(\theta, \frac{\theta^2(1-\theta)}{n}\right).$$

(c) An estimator θ_n^* of a parameter θ is said to be unbiased if $E[\theta_n^*]=\theta$ for all values of θ. That is, the mean of θ_n^* is θ. Show that

$$E[\theta_n^*]=\theta$$

and

$$\text{Var}[\theta_n^*]=\frac{\theta(1-\theta)}{n}.$$

Hint: Let $X_1, X_2, \ldots, X_n$ be independent random variables defined as follows: Put $X_i=1$ if $T_i=1$ and $X_i=0$ if $T_i \neq 1$, $1 \le i \le n$. Then,

$$Z_n = X_1 + X_2 + \ldots + X_n.$$

In large samples, it is known that

$$\theta_n^* \sim N\left(\theta, \frac{\theta(1-\theta)}{n}\right).$$

Observe that the large sample variance of $\hat{\theta}_n$ is smaller than that of θ_n^*. This is an example of why maximum likelihood estimators are generally preferable to competing estimators.

3. (a) Derive the formula for the variance of the Beta distribution given in (1.5.3). Hint: Show that

$$E[\Theta^2]=\frac{\alpha(\alpha+1)}{(\alpha+\beta)(\alpha+\beta+1)}$$

and then use the formula

$$\text{Var}[\Theta]=E[\Theta^2]-(E[\Theta])^2.$$

(b) Let $f(\theta)$ be the p.d.f. of the Beta distribution. Compute $f^{(1)}(\theta)$, the derivative of $f(\theta)$, and solve the equation $f^{(1)}(\theta)=0$ to derive formula (1.5.4). Explain why it is necessary to impose the restrictions $\alpha>1$ and $\beta>1$.

4. Derive the formula given in (1.5.11) for the variance of the Beta-Geometric distribution. Hint: Show that

$$E[T^2]=\int_0^1 f(\theta)E[T^2|\theta]d\theta$$

and use part (c) of Problem 1 to find $E[T^2|\theta]$.

5. (a) With reference to the Beta-Geometric distribution show that the conditional p.d.f. of Θ, given that $T \le t$, is

$$f(\theta|T \le t)=c^{-1}\theta^{\alpha-1}(1-\theta)^{\beta-1}(1-(1-\theta)^t),$$

where

$$c = \frac{\Gamma(\alpha)\Gamma(\beta)}{\Gamma(\alpha+\beta)} - \frac{\Gamma(\alpha)\Gamma(\beta+t)}{\Gamma(\alpha+\beta+t)}$$

and $0 < \theta < 1$.

(b) Derive a useful computing formula for $E[\Theta|T \leq t]$. Show that

$$E[\Theta|T \leq t] = c^{-1}\left[\frac{\Gamma(\alpha+1)\Gamma(\beta)}{\Gamma(\alpha+\beta+1)} - \frac{\Gamma(\alpha+1)\Gamma(\beta+t)}{\Gamma(\alpha+\beta+t+1)}\right],$$

and then investigate forms that are useful computationally.

6. The following is a sketch of the theory underlying the maximum likelihood estimation of the γ-parameters in (1.8.2).

(a) Let Z be a random variable indicating whether a pregnancy occurs during a menstrual cycle. Put $Z = 1$ if a pregnancy occurs and let $Z = 0$ otherwise. The conditional p.d.f. of Z, given a coital pattern $\mathbf{x} \in \chi$ during a menstrual cycle, is

$$P[Z = z] = \theta^z(\mathbf{x})(1 - \theta(\mathbf{x}))^{1-z}, \quad z = 0, 1;$$

where, according to (1.8.2),

$$\theta(\mathbf{x}) = 1 - \exp\left[\sum_i \gamma_i x_i\right].$$

(b) Suppose there are m women in the sample and the j-th woman is observed for n_j menstrual cycles. Let Z_{jk} be an indicator random variable associated with the j-th woman and the k-th menstrual cycle of that woman; where $1 \leq j \leq m$ and, for each j, $1 \leq k \leq n_j$. For every pair of indices $\langle j, k \rangle$ a corresponding coital pattern $\mathbf{x}_{jk} \in \chi$ is observed. Assuming independence among women and cycles within women, the conditional likelihood of the sample values $Z_{jk} = z_{jk}$, given the coital patterns $\mathbf{x}_{jk}$, is

$$L = \prod_{j=1}^{m} \prod_{k=1}^{n_j} \theta^{z_{jk}}(\mathbf{x}_{jk})(1 - \theta(\mathbf{x}_{jk}))^{1-z_{jk}}.$$

The maximum likelihood estimates $\hat{\gamma}$ of the γ-parameters are found by maximizing the likelihood function in the γ's for a fixed set of sample values and coital patterns. As mentioned before, it may be useful to maximize the likelihood function in δ-parameters connected to the γ-parameters by the equation $\gamma = -\exp[\delta]$ in order to assure a commonality of signs in the γ's. Maximizing $\log_e L$ instead of L should also lead to simplifications.

7. In reference to the Markov chain model of coital patterns introduced in Sect. 1.9, let $p_0(t)$ be the probability of being in state 0 at time $t \geq 1$. The function $p_0(t)$ is determined by a formula analogous to (1.9.7). Let $\mathbf{p}(t) = \langle p_0(t), p_1(t) \rangle$. Show that $\mathbf{p}(t)$ may be calculated recursively, given the initial distribution $\mathbf{p}(0)$, according to the formula

$$\mathbf{p}(t) = \mathbf{p}(t-1)\mathbf{P}, \quad t \geq 1.$$

Since $\mathbf{p}(t) \to \pi$ as $t \uparrow \infty$, Eq. (1.10.5) may be derived by passing to the limit.

8. Many methods of contraception, designed to reduce natural fecundability, have been developed. The capability of a method of contraception to reduce fecundability is a measure of its effectiveness. For the case of constant fecundability studied in Sect. 1.2, let θ be natural fecundability and let $\theta_1 < \theta$ be fecundability under a method of contraception. Then, the effectiveness ε of a method of contraception is defined by

$$\varepsilon = \frac{\theta - \theta_1}{\theta}.$$

Equivalently, $\theta_1 = \theta(1 - \varepsilon)$. For the case of variable fecundability among women, it seems reasonable to replace the θ's in the definition of ε by mean values. Thus,

$$\varepsilon = \frac{E[\Theta] - E[\Theta_1]}{E[\Theta]}.$$

When fecundability follows a Beta distribution, it is easy to see from (1.5.2) that if the parameter α is held constant, then $E[\Theta] \to 0$ as $\beta \to \infty$. Suppose, for the sake of simplicity, that a method of contraception affects a Beta distribution of fecundabilities only by changing the parameter β to $\beta_1 > \beta$, the parameter α remaining constant.

(a) Show that the effectiveness of a method of contraception is given by

$$\varepsilon = \frac{\beta_1 - \beta}{\alpha + \beta_1}.$$

Alternatively,

$$\beta_1 = \frac{\varepsilon \alpha + \beta}{1 - \varepsilon}.$$

(b) If $\alpha = 3.48$ and $\beta = 17.88$, find values of β_1 such that $E[\Theta_1] = 0.05, 0.01$. Compute the effectiveness of the method of contraception for the two cases.

(c) Generalize the formula in part (a) for ε to the case contraception acts in such a way that the parametic point $\langle \alpha, \beta \rangle$ of the Beta distribution is transformed to the point $\langle \alpha_1, \beta_1 \rangle$ with $\alpha_1 < \alpha$ and $\beta_1 > \beta$.

9. The following is a sketch for a general approach to studying the distribution of the waiting times till conception when fecundability varies not only from month to month within couples but also among couples.

(a) As before, let T be a random variable representing waiting times till conception. We seek $h(t) = P[T = t]$, $t = 1, 2, \ldots$, for the case fecundability may vary from month to month within couples and also among couples. It is helpful to get a general overview of the problem. To this end, let S_1 and S_2 be two abstract-valued random variables, representing, respectively, variation within and among couples. The distribution of S_1 will be symbolized by $P(ds_1)$ and the conditional distribution of S_2 given that $S_1 = s_1$ by $P(s_1, ds_2)$.

The conditional p.d.f. of T given that $S_1 = s_1$ and $S_2 = s_2$ is

$$f(s_1, s_2; t) = P[T = t | S_1 = s_1, S_2 = s_2];$$

while the conditional p.d.f. of T given $S_1 = s_1$ is

$$f(s_1; t) = P[T = t | S_1 = s_1].$$

This last conditional p.d.f. is obtained by averaging $f(s_1, s_2; t)$ with respect to the conditional distribution $p(s_1, ds_2)$. Thus, symbolically

$$f(s_1; t) = \int P(s_1, ds_2) f(s_1, s_2; t).$$

Similarly, $h(t)$ is obtained by averaging $f(s_1; t)$ with respect to the distribution $P(ds_1)$. That is,

$$h(t) = \int P(ds_1) f(s_1; t), \qquad t = 1, 2, \dots .$$

In summary, to determine $h(t)$, it suffices to specify $f(s_1, s_2; t)$, $P(s_1, ds_2)$, and $P(ds_1)$. In the above, measure-theoretic formalism has been used freely without explanation.

(b) Let Θ be a random variable representing fecundability. Given $S_1 = s_1$ and $S_2 = s_2$, assume the fecundabilities $\Theta_1, \Theta_2, \dots$ in successive months are conditionally independent realizations of the random variable Θ. Under this assumption, the conditional probability $f(s_1, s_2; t)$ may be chosen as

$$f(s_1, s_2; t) = \prod_{i=1}^{t-1} (1 - \Theta_i) \Theta_t \tag{b.1}$$

for $t \geq 1$. In (b.1), the product is one if $t = 1$.

For the sake of definiteness, think of the conditional distribution $P(s_1, ds_2)$ as being determined by the Markov chain model of coital patterns introduced in Sect. 1.9 and the conditional independence of the random variables $\Theta_1, \Theta_2, \dots$, given $S_1 = s_1$ and $S_2 = s_2$. Let $\theta = E[\Theta_i | S_1 = s_1]$, $i \geq 1$, be their common conditional expectation. Then, calculating $f(s_1; t)$ from $f(s_1, s_2; t)$ is equivalent to taking the conditional expectation on the right in (b.1), using conditional independence. In symbols,

$$f(s_1; t) = \int P(s_1, ds_2) f(s_1, s_2; t) = (1 - \theta)^{t-1} \theta. \tag{b.2}$$

Finally,

$$h(t) = \int P(ds_1) f(s_1; t) = \int_0^1 f(\theta)(1 - \theta)^{t-1} \theta \, d\theta,$$

where $f(\theta)$ is some p.d.f. on $(0, 1)$ induced by variability in fecundability among couples. In particular, if $f(\theta)$ is a Beta density, the above assumptions recover the Beta-Geometric distribution. Among other things, the p.d.f. $f(\theta)$ on $(0, 1)$ could reflect variation in coital patterns among couples.

10. Displays of realizations of coital patterns for persisters and alternators, generated by the transition matrices, in (1.10.8) and (1.10.9), are of interest. An APL function was written to simulate realizations of these Markov chains. Realizations of coital patterns for five 26-day intermenstrual periods for persisters and alternators are presented in Tables 10.A and 10.B respectively. The initial distribution $\langle 0.5, 0.5 \rangle$ was used throughout these simulations. It is of interest to observe that in intermenstrual period 5 for alternators in Table 10.B,

Table 10.A. Realizations of coital patterns for persisters

Period	Realizations
1	0 1 1 1 1 1 1 1 1 1 1 0 0 1 1 1 1 1 1 1 1 0 0 1 1 1
2	1 1 1 1 1 0 1 1 1 0 0 1 1 0 0 1 1 1 1 0 0 0 0 0 0 1
3	0 0 1 0 1 1 0 0 1 1 1 1 1 1 1 1 1 1 1 1 1 1 1 1 0 0 1
4	1 1 1 1 1 1 0 0 0 1 1 1 0 0 0 0 0 1 1 1 0 1 1 1 1 1
5	0 0 0 0 0 0 1 1 1 1 1 0 1 1 0 1 0 0 1 0 0 1 1 1 1 1

Table 10.B. Realizations of coital patterns for alternators

Period	Realizations
1	1 0 0 1 0 1 0 1 0 1 0 1 0 1 1 0 1 0 1 0 1 0 1 0 1 0 1 0
2	0 1 0 0 1 0 1 0 0 0 1 0 1 0 0 1 0 0 1 0 1 0 0 0 1 0
3	1 0 1 0 0 0 1 0 1 0 1 0 0 0 1 0 1 0 1 0 1 0 0 1 0 0
4	1 0 0 0 0 1 0 1 0 0 0 0 1 1 0 0 1 0 1 0 1 0 0 0 0 0
5	1 0 0 1 0 0 1 0 1 0 1 1 0 0 0 0 0 1 0 1 1 0 0 1 0

there was a run of six zeros, indicating there were no coital acts for six consecutive days. One is naturally led to wonder whether this realization of a Markov chain is a plausible coital pattern for alternators. By comparing simulated realization of Markov chains with real data, when it is available, a researcher should be able to get insights into whether a Markov chain is a valid model of coital patterns.

References

Artin, E. (1964): *The Gamma Function.* Holt, Rinehart, and Winston, New York, Chicago, London

Barrett, J.C. (1970): An Analysis of Coital Patterns. *Jour. of Biosocial Science* 2: 351–357

Barrett, J.C. (1971): Fecundability and Coital Frequency. *Population Studies* 25: 309–313

Barrett, J.C. and Marshall, J. (1969): The Risk of Conception on Different Days of the Menstrual Cycle. *Population Studies* 23: 455–461

Glasser, J.H. and Lachenbrugh, P.A. (1968): Observations on the Relationship Between Frequency and Timing of Intercourse and the Probability of Conception. *Population Studies* 22: 399–407

Jain, A.K. (1969): Fecundability and Its Relation to Age in a sample of Taiwanese Women. *Population Studies* 23: 69–85

James, W.H. (1971): The Distribution of Coitus within the Human Intermenstrum. *Journal of Biosocial Science* 3: 159–171

Kemeny, J.G. and Snell, J.L. (1976): *Finite Markov Chains.* Springer-Verlag, New York, Heidelberg, Berlin

Lachenbrugh, P.A. (1967): Frequency and Timing of Intercourse – Its Relation to the Probability of Conception. *Population Studies* 21: 23–31

Leridon, H. (1977): *Human Fertility – The Basic Components.* The University of Chicago Press, Chicago and London

Nelder, J.A. and Wedderburn, R.W.M. (1972): Generalized Linear Models. *Jour. Royal Stat. Soc.* A 135: 370–384

Potter, R.G., McCann, B. and Sakoda, J.M. (1970): Selective Fecundability and Contraceptive Effectiveness. *Milbank Memorial Fund Quarterly* 48: 91–102

Sheps, M.C. and Lapierre-Adamcyk, E. (1972): *On the Measurement of Human Fertility – Selected Writings of Louis Henry.* Elsevier, Amsterdam, London, New York

Sheps, M.C. and Menken, J.A. (1973): *Mathematical Models of Conception and Birth.* The University of Chicago Press, Chicago and London

Smith, D. and Keyfitz, N. (1977): *Mathematical Demography.* Springer-Verlag, Berlin, Heidelberg, New York

Tietze, C. (1970): Ranking of Contraceptive Methods by Levels of Effectiveness. *Advances in Planned Parenthood-VI ; Excepta Medica International Congress Series* No. 224: 117–126

Westoff, C.F. (1974): Coital Frequency and Contraception. *Family Planning Perspectives* 6: 136–141

Chapter 2. Human Survivorship

2.1 Introduction

Among the possible outcomes of a conception is a live birth. In a subsequent chapter, other types of pregnancy outcomes will be considered, but here attention will be focused on the study of survival following a live birth. Although mortality is a commonly used heading for such a study, the term human survivorship will be used extensively, since survival of and by itself is a basic aspect of population dynamics.

A standard statistical tool used in the study of human survivorship is the life table. There exists an extensive literature on life table methodology. Consequently, only a brief overview of this methodology will be given here. Unlike most expositions on mortality, which center around the life table and its ramifications, the evolutionary or developmental aspects of human survivorship, as revealed by selected historical examples, will be emphasized in this chapter. Some attention will also be given the heterogeneity in survivorship, i.e., the variability among individuals in the capacity to survive.

When confronted with historical changes in human survivorship, it becomes clear that the model underlying life table methodology is deficient, primarily because risks of death that change with time are not accommodated. What seems to be needed in the development of evolutionary models of human survivorship is a two-variable force of mortality, depending not only on the epoch at which an individual is born but also on his age. According to the language of demography, cohort and period effects should be distinguished.

Evolutionary models in which cohort and period effects are distinguished would be of value in forecasting mortality and in providing systematic descriptions of changes in the mortality component underlying population projections. Furthermore, they are of basic interest in the biological and social sciences. They are of basic interest in the biological sciences, because they provide analytic descriptions of a fundamental evolutionary human trait, survivorship. From the point of view of the social sciences, they provide a framework for viewing social and economic development as mirrored through changes in human survivorship.

2.2 Mortality in a Cohort

Imagine a cohort of individuals born at some epoch $t=0$ and suppose each individual may be followed for his entire life span. For the moment, individuals will not be distinguished by sex. For the sake of simplicity, it will also be supposed that all variation in age at death may be described by a random variable X, taking values in the interval $R^+=[0, \infty)$.

Let $G(x)$ be the distribution function of X. Then,

$$P[X \leq x] = \begin{cases} G(x), & x \in R^+ \\ 0, & x \notin R^+. \end{cases} \tag{2.2.1}$$

From now on it will also be supposed that the condition, $\lim_{x \downarrow 0} G(x) = G(0) = 0$, is satisfied. The survival function for the cohort is defined by

$$S(x) = P[X > x] = 1 - G(x), \qquad x \in R^+. \tag{2.2.2}$$

Associated with $G(x)$ is the probability density function (p.d.f.) defined by $g(x) = G'(x)$, the derivative of G, at those points $x \in R^+$ for which it exists.

A useful way of characterizing a life span distribution is through $\mu(x)$, the force of mortality at age x. To define this function consider

$$P[x < X \leq x+h \mid X > x] = \frac{G(x+h) - G(x)}{S(x)}, \tag{2.2.3}$$

the conditional probability that death occurs during the interval $(x, x+h]$, $h>0$, given survival to age $x \in R^+$. The force of mortality is defined by

$$\mu(x) = \lim_{h \downarrow 0} \frac{G(x+h) - G(x)}{hS(x)}$$
$$= \frac{g(x)}{S(x)} \tag{2.2.4}$$

for those points $x \in R^+$ for which the p.d.f. $g(x)$ exists.

But, (2.2.4) is equivalent to the differential equation

$$\frac{d}{dx} \log S(x) = -\mu(x) \tag{2.2.5}$$

with the initial condition $S(0)=1$. Solving (2.2.5) yields the formula

$$S(x) = \exp\left[-\int_0^x \mu(s)\,ds \right], \qquad x \in R^+. \tag{2.2.6}$$

Equivalently,

$$G(x) = 1 - \exp\left[-\int_0^x \mu(s)\,ds \right]. \tag{2.2.7}$$

Differentiating (2.2.7) with respect to x and using (2.2.6) leads to

$$g(x) = \mu(x)S(x), \qquad x \in R^+, \tag{2.2.8}$$

a formula for the p.d.f. of X. In order that the distribution function $G(x)$ be proper, i.e., $G(x)\uparrow 1$ as $x\uparrow\infty$, it is necessary and sufficient that the integral in the exponent in (2.2.7) diverge as $x\uparrow\infty$.

Of basic interest in the construction of life tables is the expectation of life at birth. Mathematically, this quantity is merely

$$E[X]=\int_0^\infty xg(x)dx, \qquad (2.2.9)$$

the expected value of the random variable X. By integrating by parts, see Problem 1, it can be shown that (2.2.9) is equivalent to

$$E[X]=\int_0^\infty S(x)dx. \qquad (2.2.10)$$

For every $x\in R^+$, let $Y(x)$ be a random variable representing the life span remaining having survived till age x. The survival function of the random variable $Y(x)$ is clearly

$$P[Y(x)>y\,|\,X>x]=S(x,y)=\frac{S(x+y)}{S(x)}, \qquad (2.2.11)$$

for $x\in R^+$ and $y\in R^+$. In analogy with (2.2.10), the expectation of life remaining having survived till age x is the conditional expectation

$$E[Y(x)\,|\,X>x]=\int_0^\infty S(x,y)dy. \qquad (2.2.12)$$

Formula (2.2.12), which specializes to (2.2.10) when $x=0$, is also applied widely in the construction of life tables. Sometimes the conditional expectation in (2.2.12) is referred to as the expectation of future life having survived till age x.

2.3 Simple Parametric Examples of the Force of Mortality

A special subfield of investigation is the design of parametric forces of mortality and the development of procedures for estimating the parameters from data. Because further attention will be devoted to this subfield in subsequent sections of this chapter, only three simple parametric examples will be considered in this section.

Example 2.3.1. *The Exponential Distribution*

Let α be a positive constant and suppose $\mu(x)=\alpha$ for all $x\in R^+$. Then,

$$\int_0^x \mu(s)ds=\alpha x, \qquad x\in R^+. \qquad (2.3.1)$$

Hence, formula (2.2.6) for the survival function takes the form

$$S(x)=\exp[-\alpha x], \qquad x\in R^+. \qquad (2.3.2)$$

An application of formula (2.2.1) shows that

$$S(x, y) = \exp[-\alpha y] \tag{2.3.3}$$

for all $x \in R^+$ and $y \in R^+$. The functions in (2.3.2) and (2.3.3) are identical. Therefore, if life spans in a cohort follow an exponential distribution, there is no aging in the sense that an individual aged $x > 0$ and a newly born individual have the same survival function. This nonaging property of the exponential distribution clearly makes it an unrealistic model for survivorship in biological populations.

Example 2.3.2. *The Weibull Distribution*

Let α and β be two positive parameters. The Weibull distribution is characterized by a force of mortality, risk function, of the form

$$\mu(x) = \alpha \beta x^{\beta-1}, \qquad x \in (0, \infty). \tag{2.3.4}$$

If $\beta > 1$, then $\mu(x)$ is increasing on $(0, \infty)$; if $0 < \beta < 1$, then $\mu(x)$ is decreasing on $(0, \infty)$; if $\beta = 1$, then (2.3.4) reduces to the risk function of the exponential distribution.

From (2.3.4) it follows that

$$\int_0^x \mu(s)ds = \alpha x^\beta, \qquad x \in R^+. \tag{2.3.5}$$

Thus, the survival function of the Weibull distribution has the form

$$S(x) = \exp[-\alpha x^\beta], \qquad x \in R^+. \tag{2.3.6}$$

Example 2.3.3. *The Gompertz Distribution*

This distribution is characterized by a force of mortality that increases exponentially with age. Let α and β be two positive parameters. Then, the force of mortality for the Gompertz distribution has the form

$$\mu(x) = \alpha \beta \exp(\beta x), \qquad x \in R^+. \tag{2.3.7}$$

Therefore,

$$\int_0^x \mu(s)ds = \alpha(\exp(\beta x) - 1), \qquad x \in R^+. \tag{2.3.8}$$

Consequently, the survival function of the Gompertz distribution is

$$S(x) = \exp[-\alpha(\exp(\beta x) - 1)], \qquad x \in R^+. \tag{2.3.9}$$

For the case of the exponential distribution, the expectation in (2.2.12) can be evaluated by elementary means. The evaluation of (2.2.12) for both the Weibull and Gompertz distributions, however, requires special techniques; see Problem 2 for further details. Smith and Keyfitz (1977), in their chapter on parameterization and curve fitting, have given excerpts from historical papers on attempts to find parameteric forms for a law of human mortality.

2.4 Period Mortality – A Simple Algorithm

People often move about during their lifetimes, making it impossible, or at best very difficult, for collectors of mortality statistics to obtain survival data by following birth cohorts throughout their life spans. It has become necessary, therefore, to develop methods for estimating a survival function, as well as expectations of life remaining at given ages, which utilize readily available period data. In many developed countries, two types of period data are usually available from official sources.

A first type of data is the number of persons dying during a given year, or during a period consisting of several years, classified according to age at last birthday. A second type consists of estimates by age of the average number of persons who lived in a population during a given period. The purpose of this section is to outline a simple algorithm which forms the basis for using period data to obtain estimates of the survival function at a given set of ages.

Let r be the maximum age in a population; let $0 = x_0 < x_1 < \dots < x_m = r$ be $m \geq 1$ numbers determining a partition $I_v = (x_{v-1}, x_v]$, $1 \leq v \leq m$ of the interval $(0, r]$. A person belongs to age group I_v if his age at last birthday fell in the interval I_v. Let q_v be the conditional probability of dying during the age interval I_v, given survival till age x_{v-1}. Then,

$$q_v = \frac{S(x_{v-1}) - S(x_v)}{S(x_{v-1})},$$

(2.4.1)

for $1 \leq v \leq m$. The conditional probability of surviving till age x_v, given that a person was alive at age x_{v-1}, is

$$p_v = 1 - q_v = \frac{S(x_v)}{S(x_{v-1})}, \quad 1 \leq v \leq m.$$

(2.4.2)

Because $x_0 = 0$ and $S(0) = 1$, it follows from (2.4.2) that

$$S(x_v) = p_1 p_2 \dots p_v$$

(2.4.3)

for $1 \leq v \leq m$.

Equation (2.4.3) is the basic life table algorithm for estimating the values of the survival function at the ages x_v, $1 \leq v \leq m$. For if estimates $\hat{p}_v$ of the conditional probabilities p_v are available, then they may be substituted for the p_v on the right in (2.4.3) to yield estimates $\hat{S}(x_v)$ of the values of the survival function $S(\cdot)$ at ages x_v, $1 \leq v \leq m$.

Given sufficient information, these estimates are, in principle, easy to calculate. Consider a population during a given period of observation and let D_v be the number of deaths to persons belonging to the age interval I_v. Let N_v be the number of persons of age x_{v-1} who were alive at beginning of the period. Then $\hat{q}_v = D_v / N_v$ is a natural estimate of q_v and $\hat{p}_v = 1 - \hat{q}_v$ is the desired estimate of p_v, $1 \leq v \leq m$.

Before moving on to a deeper discussion of methods for estimating a conditional probability q_v, it should be mentioned that r may be interpreted as the greatest age in a life table. With this interpretation, the interval $(0, r]$ could be partitioned into $m \geq 2$ subintervals as before; but, an interval (r, ∞) would also be added to cover ages beyond r.

2.5 Transforming Central Death Rates into Probabilities and Expectations

Although reliable estimates of the number D_v of deaths occurring at ages in the interval I_v can frequently be obtained from official death registration data, it is more difficult to get estimates of the integer N_v. What is usually estimated is the central death rate associated with an age interval I_v. From now on, it will be convenient to deal with the length of the interval I_v defined by $h_v = x_v - x_{v-1}$, $1 \leq v \leq m$. To lighten the notation, the subscript v will be dropped and the discussion will center on an age interval $(x, x+h]$, $h > 0$.

If age is measured on a yearly scale, then the average number of years lived in the interval $(x, x+h]$ is

$$L(x, h) = \int_x^{x+h} S(y) \, dy = \int_0^h S(x+y) \, dy. \tag{2.5.1}$$

The probability of dying during the age interval $(x, x+h]$ is

$$d(x, h) = G(x+h) - G(x). \tag{2.5.2}$$

Let $m(x, h)$ be the central death rate associated with the interval $(x, x+h]$. Then, $m(x, h)$ is defined by

$$m(x, h) = \frac{d(x, h)}{L(x, h)}. \tag{2.5.3}$$

A basic problem in transforming central death rates into conditional probabilities is that of finding a formula connecting $m(x, h)$ with

$$q(x, h) = \frac{d(x, h)}{S(x)}, \tag{2.5.4}$$

the conditional probability of dying during the age interval $(x, x+h]$, given survival till age x. Observe that $q(x, h)$ is merely another way of writing q_v in (2.4.1) with $x_{v-1} = x$ and $x_v = x+h$ for some v.

To deduce the desired formula first observe that

$$L(x, h) = hS(x+h) + \int_0^h (S(x+y) - S(x+h)) \, dy. \tag{2.5.5}$$

But,

$$\frac{S(x+y) - S(x+h)}{d(x, h)} \tag{2.5.6}$$

is the conditional probability of surviving till age $x+y$, given that death occurs in the age interval $(x, x+h]$, see Problem 3 for further details. Hence, in analogy with (2.2.10),

$$a(x, h) = \frac{1}{d(x, h)} \int_0^h (S(x+y) - S(x+h)) dy \qquad (2.5.7)$$

is the conditional mean number of years lived in the interval $(x, x+h]$, given that death occurs at an age in this interval. Therefore, Eq. (2.5.5) may be written in the form

$$
\begin{aligned}
L(x, h) &= hS(x+h) + a(x, h)d(x, h) \\
&= h(S(x) - d(x, h)) + hf(x, h)d(x, h) \\
&= hS(x) - h(1 - f(x, h))d(x, h),
\end{aligned} \qquad (2.5.8)
$$

where $f(x, h) = a(x, h)/h$. If Eq. (2.5.8) is solved for $S(x)$ and the result is substituted into (2.5.4), then, after some algebraic reductions, it can be seen that

$$q(x, h) = \frac{hm(x, h)}{1 + h(1 - f(x, h))m(x, h)} \qquad (2.5.9)$$

is the desired formula connecting the central rate in (2.5.3) with the conditional probability in (2.5.4).

Formula (2.5.9) is exact. To apply it, estimates of $m(x, h)$ and $f(x, h)$ are needed. These estimates are usually calculated for two types of life tables. Reverting back to a general age interval I_v, a life table is said to be complete if $x_v - x_{v-1} = h_v = 1$ for all $1 \leq v < m$, i.e., the length of each age interval is one year. A table is said to be abridged if $x_1 - x_0 = h_1 = 1$, $x_2 - x_1 = h_2 = 4$, and $x_v - x_{v-1} = h_v = 5$ for $3 \leq v \leq m$. Included in most complete and abridged life tables is an interval (r, ∞), representing ages greater than some highest age r in the table. In most life tables either $r = 80$ or $r = 85$. If, for example, $r = 80$, then $m = 80$ or $m = 17$, depending on whether the table is complete or abridged. For the case of an abridged life table and $r = 85$, there is no danger of confusion if the subscripts on the x's are again dropped and the age intervals are symbolized by $(x, x+h]$ for $x = 0, 1, 5, 10, \ldots, 80$ and by $(85, \infty)$ for ages over 85.

Chiang (1972) estimated the fractions $f(x, h)$ for abridged life tables with $r = 85$, using death certificates from California in 1960 and a 10% sample of U.S. deaths in 1963. Elandt-Johnson and Johnson (1980) have presented Chiang's results, along with other estimates, in their Table 4.2 on page 106. Yang (1977) has given a theoretical discussion of the function $f(x, h)$.

Central death rates are usually estimated as follows. Let $K(x, h)$ be the number of persons in the age interval $(x, x+h]$ exposed to risk of death in a given calender year or period; if the period is a year, $K(x, h)$ is often taken as the midyear census population for the age group $(x, x+h]$. Let $D(x, h)$ be the observed number of deaths in the age interval $(x, x+h]$ during a given calendar year or period. Then, $m(x, h)$ is estimated by

$$\hat{m}(x, h) = \frac{D(x, h)}{K(x, h)}. \qquad (2.5.10)$$

An estimate $\hat{q}(x, h)$ of the conditional probability in (2.5.9) is then obtained by using $\hat{m}(x, h)$ in (2.5.9) along with $\hat{f}(x, h)$, an estimate of $f(x, h)$.

Upon substituting $\hat{p}(x, h) = 1 - \hat{q}(x, h)$ into (2.4.3) and properly identifying subscripts with the age intervals $(x, x+h]$, $x = 0, 1, 5, \ldots, 80$, estimates $\hat{S}(x)$ of $S(x)$ for $x = 0, 1, 5, 10, \ldots, 85$ may be calculated. Given these estimates of the survival function, the probability of death and the mean number of years lived in the interval $(x, x+h]$ are estimated by

$$\hat{d}(x, h) = \hat{S}(x) - \hat{S}(x+h)$$

and

$$\hat{L}(x, h) = h[\hat{S}(x) - (1 - \hat{f}(x, h))\hat{d}(x, h)] \tag{2.5.11}$$

for $x = 0, 1, 5, 10, \ldots, 80$.

The average number of years lived beyond age 85 is $L(85, \infty) = \lim L(85, h)$ as $h \to \infty$; the central death rate for the age interval $(85, \infty)$ is $m(85, \infty) = S(85)/L(85, \infty)$. Hence,

$$\hat{L}(85, \infty) = \frac{\hat{S}(85)}{\hat{m}(85, \infty)}, \tag{2.5.12}$$

where $\hat{m}(85, \infty) = D(85, \infty)/K(85, \infty)$, is an estimate of $m(85, \infty)$.

Define $T(x)$ by

$$T(x) = L(x, h) + L(x+h, h) + \ldots + L(85, \infty) \tag{2.5.13}$$

for $x = 0, 1, 5, 10, \ldots, 80$. Then, from (2.2.11) and (2.2.12), it follows that

$$E[Y(x)|X > x] = \frac{T(x)}{S(x)}. \tag{2.5.14}$$

Let $\hat{T}(x)$ be an estimate of $T(x)$ obtained by substituting $\hat{L}$s on the right in (2.5.13). In life table notation,

$$\mathring{e}(x) = \frac{\hat{T}(x)}{\hat{S}(x)} \tag{2.5.15}$$

is an estimate of the conditional expectation on the left in (2.5.14) at the ages $x = 0, 1, 5, 10, \ldots, 85$.

The formulas derived in this section have been widely applied in computing period life tables. Chiang (1968), Chap. 9, has given a detailed example. An extensive discussion can also be found in Shryock et al. (1973), Chap. 15. Further discussion may be found in Keyfitz (1977), Chap. 2. More recently, Elandt-Johnson and Johnson (1980) have also given an extensive discussion along with examples. Commonly used simplifications of formulas (2.5.8) and (2.5.9) for abridged and complete life tables are discussed in Problem 4.

Life table formulas are usually applied under the assumption that the registration of deaths is complete. In many developing countries, this is not the case. Preston et al. (1980), Preston and Hill (1980), and Bennett and Horiuchi (1981) have reviewed the literature on coping with incomplete death registration statistics and have proposed methods for estimating the completeness of registration.

2.6 Evolutionary Changes in Expectation of Life

The formulas, forming the basis for calculating a period life table, were derived by considering a stationary probability distribution of age at death in a hypothetical cohort. In reality, however, such distributions are not stationary. Indeed, significant changes in mean human life span have occurred in recorded history. A knowledge of these historical changes in expectation of life is not only of intrinsic interest in its own right but also useful in designing more realistic computer simulation scenarios with changing mortality rates. Presented in this section is a brief outline of the evolution of human life span. An extensive account of the subject has been provided by Ascádi and Nemeskéri (1970).

Among the lowest estimates of expectation of life at birth for both sexes is $\mathring{e}(0) = 15.7$ years for the Taforalt population, see Table 42, page 155, in Ascádi and Nemeskéri. This value of life expectancy in prehistoric man was based on estimating age at death from the remains of nearly 200 individuals found in a cave on the Maghreb peninsula in Morocco.

A considerable amount of information on human life span during the Roman empire has been obtained from the study of inscriptions on grave stones and other epitaphs. Despite the controversial nature of deriving estimates from such materials, interesting differences in average life spans by social status, occupation, and location of residence do emerge. Those living in the city of Rome tended to have shorter life spans on the average than those living outside Rome. Slaves living in Rome had an average life of 17.5 years, based on data for both sexes. Outside Rome the corresponding estimate for slaves was 25.5 years. Estimated average life span for clergy living outside Rome was 58.6 years for both sexes. Table 78 on page 224 of Ascádi and Nemenskéri, as well as their Chap. VI, may be consulted for further details and a critical discussion. Unlike many contemporary populations in which the expectation of life at birth for females exceeds that of males, the reverse was true for populations living at the time of the Roman empire. According to Table 122 and 123 in Ascádi and Nemenskéri $\mathring{e}(0)$ was 30.35 and 22.92, respectively, for males and females of the Intercisa and Brigetio population.

Expectations of life at birth for populations living in medieval Europe differed little from those living at the time of the Roman empire. Ascádi and Nemenskéri report a value of $\mathring{e}(0) = 28.73$ for Hungarian populations of the 10-th to 12-th centuries. Russel (1958) has made an extensive study of the British medieval population. Displayed in Fig. 2.6.1 is a computer generated graph representing the evolution of life expectancy, based on combined data for both sexes, in English cohorts born during the period 1275 to 1450. The markings on the curves represent the actual data points taken from table 95 in Acsádi and Nemeskéri and plotted in increments of a quarter century. To aid the eye in connecting data points belonging to expectations of life remaining at ages 0, 1, 10, and 30, points corresponding to each age were connected by straight lines.

At least three conclusions are apparent from an inspection of Fig. 2.6.1.

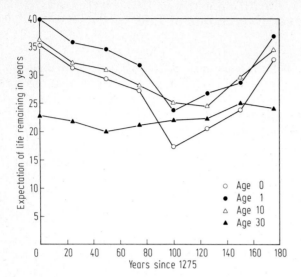

Fig. 2.6.1. Evolution of expectation of life in medieval England 1275 to 1450

First, the curve for $\mathring{e}(0)$, the expectation of life at birth, declined from a high of about 35 in 1275 to a low in the range 17–20 during the period 1370–1400. Historical research suggests that this decline was attributable to recurring epidemics of the plague accompanied by wars and famine, the lowest point being associated with the great plague. Recovery was slow; even by 1450 $\mathring{e}(0)$ $= 32.8$ was lower than in 1275. Second, the curves $\mathring{e}(1)$ and $\mathring{e}(10)$, representing expectations of life at one and ten years of age, are uniformly higher than that for $\mathring{e}(0)$, reflecting the high rates of infant mortality in medieval England. Third, throughout the period of 1275 to 1450, the expectation $\mathring{e}(30)$ remained within the interval $(22, 25]$, suggesting that those surviving till age 30 represented a selected group having a capacity, somewhat independent of environmental conditions, to survive.

Among the countries with longest record of mortality statistics is Sweden. Used to generate the computer graph in Fig. 2.6.2, representing the evolution of expectation of life for Swedish females by five year intervals during the period 1780 to 1965, were some historical life tables reported by Keyfitz and Flieger (1968). The markings on the curves represent the actual data points. At least three properties of the curves in Fig. 2.6.2 are of interest.

First, apart from temporary decreases in expectations of life which were probably associated with epidemics, starting with $\mathring{e}(0) = 38.52$ in 1780 there is a steady increase in expectation of life at birth to $\mathring{e}(0) = 76.14$, the value reported in 1965. Throughout the period, the expectation of life at one year, $\mathring{e}(1)$, exceeded $\mathring{e}(0)$ but the gap between the two curves narrowed progressively until they converged in 1965, indicating a steady decrease in infant mortality. Second, like the curves for medieval England in Fig. 2.6.1, the curve for the expectation of life at ten years, $\mathring{e}(10)$, remained above that $\mathring{e}(0)$ in the last two decades of the 18-th century and throughout the 19-th century. Only at about the turn of the 20-th century did the $\mathring{e}(10)$-curve cross the $\mathring{e}(1)$-curve and remain below it. Starting from a value of 48.32 in 1780, the $\mathring{e}(10)$-curve reached a value of 67.32

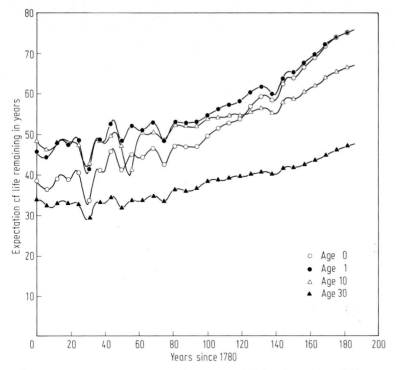

Fig. 2.6.2. Evolution of expectation of life for Swedish females 1780 to 1965

in 1965. Third, compared with the $\overset{\circ}{e}(0)$- and $\overset{\circ}{e}(10)$-curves, that for $\overset{\circ}{e}(30)$ increased more slowly. In 1780, the value of $\overset{\circ}{e}(30)$ was 34.0; by 1965 this value had risen to 47.78. It is of interest to note that the value for $\overset{\circ}{e}(30)$ in medieval England by 1450 was 24.1.

Sexual difference in the evolution of life expectancy have been recognized throughout the history of recorded mortality statistics. An example of this phenomenon is displayed in Fig. 2.6.3 for U.S. white males and females during the period 1900 to 1977. In 1900 the expectation of life at birth for white males was 48.23; that for white females was 51.08. As is apparent from the $\overset{\circ}{e}(0)$-curves for males and females in Fig. 2.6.3, this difference has diverged as the 20-th century progressed. By 1977, the expectation of life at birth for white males was 70.0; while that for white females was 77.7. It is also apparent from Fig. 2.6.3 that the $\overset{\circ}{e}(30)$-curves for white males and females diverged; but, like that for the Swedish females, the rate of increase in these curves was less than that for the $\overset{\circ}{e}(0)$-curves.

Racial differences in the evolution of life expectancy have long been apparent in official U.S. mortality statistics. Displayed in Fig. 2.6.4 are the $\overset{\circ}{e}(0)$- and $\overset{\circ}{e}(30)$-curves for U.S. nonwhite males and females during the period 1900 to 1977. According to official U.S. racial classifications, the vast majority of nonwhites are blacks. By comparing Fig. 2.6.3 and 2.6.4, it can be seen that both the $\overset{\circ}{e}(0)$- and $\overset{\circ}{e}(30)$-curves for nonwhite males and females followed a

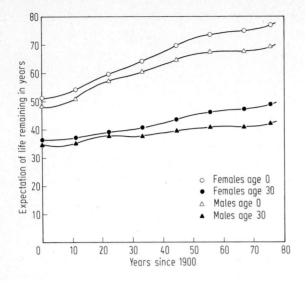

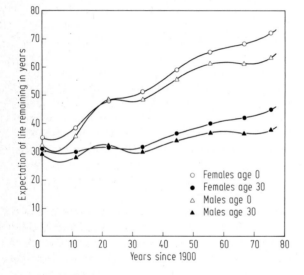

Fig. 2.6.4. Evolution of expectation of life for U.S. nonwhite males and females 1900 to 1977

divergent pattern similar to that for whites but the overall rate of increase for both these curves was greater than that for whites during the period 1900 to 1977.

Further insights into this rate of increase in life expectancy may be gained by comparing the $\overset{\circ}{e}(0)$-curve for Swedish females in Fig. 2.6.2 with that for nonwhites in Fig. 2.6.4. In 1900, the value of $\overset{\circ}{e}(0)$ for U.S. nonwhite females was 35.04; that for Swedish females in 1780 was 28.52. By 1977 the value of $\overset{\circ}{e}(0)$ for U.S. nonwhite females was 73.1. A comparable value for Swedish females, 72.76, was not reached until about 1950. Thus, an increase in life expectancy that had taken Swedish females 170 years to accomplish was accomplished by U.S. nonwhite females in 77 years. This more rapid rate of improvement in life

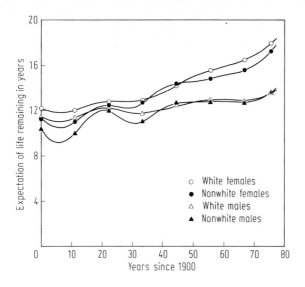

Fig. 2.6.5. Evolution of expectation of life at age 65 for U.S. males and females 1900 to 1977

expectancy for U.S. nonwhite females is probably attributable to a higher level of development contributing to good health that existed in the U.S. at the turn of the twentieth century as compared to the level of development in Sweden as of 1780.

Although racial differences in the evolution of life expectancy are apparent in Figs. 2.6.3 and 2.6.4, a question that naturally arises is whether these marked differences persist at the older ages. Presented in Fig. 2.6.5 is the evolution of the expected years of life remaining at age 65 for U.S. males and females, during the period 1900 to 1977. As can be seen from Fig. 2.6.5, sexual differences persist but the differences by race are less pronounced than at younger ages. One of the most striking properties of the curves in Fig. 2.6.5 is that the $\mathring{e}(65)$-curves for white and nonwhite males seem to have converged as the 20-th century progressed.

Throughout Figs. 2.6.3, 2.6.4, and 2.6.5 the markings represent the actual data points, which, aside from 1977, are spaced at equal ten-year intervals. The data were taken from the life tables published by the National Center for Health Statistics, DHEW Publication No. (PHS) 80-1104, Hyattville, Maryland, 1980. Data points in these figures as well as those in Fig. 2.6.2 were connected by the computer, using a cubic spline interpolation routine, see Forsythe et al. (1977), page 70, for further details.

2.7 An Evolutionary Process Governing Survivorship

From the foregoing brief historical overview of mortality, at least two features of an evolutionary process governing survivorship become apparent. One is longitudinal in the sense that cohorts born at different epochs may have markedly different average life spans. A second, but related feature, is by nature cross-sectional in that risks of death at any age may vary with time. Setting down mathematical formulations of an evolutionary process governing survivorship is not an easy task. In this section a simple descriptive formulation accommodating longitudinal and cross-sectional effects will be considered. Longitudinal effects will be described in terms of time, the real line $R = (-\infty, \infty)$; age will be a nonnegative number $x \in R^+$. A point $t \in R$ will be referred to as an epoch.

To accommodate possibly different experiences in survivorship among birth cohorts, they will be indexed by the epoch $t \in R$ at which they were born. Let $X(t)$ be a random variable representing the variation in life span for a cohort born at epoch $t \in R$. Then, the survival function for this cohort is

$$S_c(t, x) = P[X(t) > x], \qquad x \in R^+. \tag{2.7.1}$$

The distribution function of $X(t)$ is $G_c(x, t) = 1 - S_c(x, t)$.

Cross-sectional effects will be described in terms of a force of mortality $\mu(t, x)$ defined for all epochs $t \in R$ and ages $x \in R^+$ in terms of the cohort survival function in (2.7.1). If a cohort is born at epoch t, then survivors are age x at epoch $t + x$. Therefore, in analogy with (2.2.5), the nonnegative function $\mu(\cdot, \cdot)$ is defined by

$$\frac{\partial}{\partial x} \log S_c(t, x) = -\mu(t + x, x) \tag{2.7.2}$$

at the point $\langle t + x, x \rangle$, assuming the partial derivative on the left exists. Solving Eq. (2.7.2) for $S_c(t, x)$ under the condition, $S_c(t, 0) = 1$ for all $t \in R$, yields the relationship

$$S_c(t, x) = \exp\left[-\int_0^x \mu(t + s, s) \, ds \right], \qquad x \in R^+. \tag{2.7.3}$$

Equation (2.7.3) expresses the simple intuitive notion that survival till age x in a cohort depends not only on the epoch $t \in R$ at birth but also on the force of mortality at all epochs $t + s$, $0 \le s \le x$, following t.

As illustrated in Sect. 2.5, period life tables are estimated from data across age groups in a given period. Among other things a period survival function is estimated. In analogy with such life tables, a cross-sectional or period survival function at epoch $t \in R$ will be defined by

$$S_p(t, x) = \exp\left[-\int_0^x \mu(t, s) \, ds \right], \qquad x \in R^+. \tag{2.7.4}$$

In order that $S_c(t, x)$ and $S_p(t, x)$ be proper survival functions, the condition, $\lim_{x \uparrow \infty} S_c(t, x) = \lim_{x \uparrow \infty} S_p(t, x) = 0$, must be satisfied for all $t \in R$. For this condition to be satisfied, it is necessary and sufficient that

$$\lim_{x \uparrow \infty} \int_0^x \mu(t + s, s) \, ds = \infty \tag{2.7.5}$$

and

$$\lim_{x \uparrow \infty} \int_0^x \mu(t, s)\, ds = \infty$$

for all $t \in R$.

An observation that emerges from the foregoing discussion is that neither cohort nor period survival functions are the primary objects of interest in developing mathematical models of an evolutionary process governing survivorship. Rather, it is the force of mortality $\mu(t, x)$ defined for $t \in R$ and $x \in R^+$. In subsequent sections, a specific model of an evolutionary process governing survivorship will be formulated by constructing particular parametric forms of the function $\mu(t, x)$ satisfying conditions (2.7.5).

Relative to the period survival function in (2.7.4),

$$p(t, x, h) = \frac{S_p(t, x+h)}{S_p(t, x)} = \exp\left[-\int_x^{x+h} \mu(t, s)\, ds \right] \tag{2.7.6}$$

is the conditional probability of surviving to age $x+h$, given that an individual is alive at age x, based on cross-sectional information at epoch t. If $h=1$ and x is replaced by $x-1$ in (2.7.6), then

$$p(t, x-1, 1) = \exp\left[-\int_{x-1}^{x} \mu(t, s)\, ds \right], \qquad x \geq 1. \tag{2.7.7}$$

In terms of the discussion in Sect. 2.5, $p(t, x-1, 1)$ would be estimated by $\hat{p}(x, h) = 1 - \hat{q}(x, h)$, where $\hat{q}(x, h)$ is an estimate of the conditional probability $q(x, h)$ in (2.5.4) at epoch t for $h=1$.

Now consider a cohort born at epoch $t \in R$. In this cohort, $p(t, 0, 1)$ is the probability of surviving to age one. Given that an individual is alive at age one, $p(t+1, 1, 1)$ is the conditional probability of surviving to age two. By continuing in this way, it is of interest to consider

$$S_c^*(t, x) = \prod_{v=0}^{x-1} p(t+v, v, 1), \qquad x \geq 1 \tag{2.7.8}$$

as an approximation to the cohort survival function $S_c(t, x)$ in (2.7.3). That $S_c^*(t, x)$ is indeed an approximation to $S_c(t, x)$ may be seen from the relations

$$S_c^*(t, x) = \exp\left[-\sum_{v=1}^{x} \int_{v-1}^{v} \mu(t+v-1, s)\, ds \right]$$

$$\simeq \exp\left[-\int_0^x \mu(t+s, s)\, ds \right] = S_c(t, x). \tag{2.7.9}$$

In general, the more finely grouped the period information, the more closely $S_c^*(t, x)$ approximates $S_c(t, x)$.

Keyfitz and Flieger (1968) have complied on extensive collection of cohort life tables by rearranging period information as suggested by formula (2.7.8). Because Sweden had the longest record of vital statistics among the countries considered by these authors in their volume on world population, the list of Swedish cohort life tables is most extensive. Swedish period life tables were

estimated from data collected over four year intervals. A period life table
labeled 1780 in what follows, for example, was based on data collected during
1778–1782; one labeled 1785 was computed from data collected during 1783–
1787, and so on. Cohort life tables were computed by rearranging such period
tables, computed every five years and covering a time interval from 1780 to
1965, according to a modification of formula (2.7.8) to accommodate the age
groupings of abridged life tables. A cohort table labeled 1780 was based on
individuals age 0–4 in 1780; one labeled 1785 was based on individuals age 0–4
in 1785, and so on.

By comparing such cohort and period life tables, further insights into an
evolutionary process governing survivorship in Sweden may be gained. Presented in Table 2.7.1 are expectations of life remaining at ages 0 and 10 years
at epochs 1780, 1800, 1830, 1865, and 1900 taken from cohort and period life
tables as calculated by Keyfitz and Flieger (1968), using Swedish data. The
estimates in the table seem to fall into two sets. Comprising a first set are those
estimates for the epochs 1780 and 1800; a second set consists of those estimates for the epochs 1830, 1865, and 1900.

With respect to the first set, within each sex there are no marked differences
in estimates of life remaining either between periods and cohorts or between
epochs, indicating that the evolutionary process governing survivorship in
Sweden was relatively stationary for cohorts born around 1780 and 1800. For
estimates belonging to the second set, however, within each sex there are not
only marked differences between periods and cohorts but also differences
among epochs, indicating that the evolutionary process governing survivorship
in Sweden for cohorts born around 1830, 1865, and 1900 was nonstationary.
Differences between sexes in expectations of life remaining at ages 0 and 10 are
present at all epochs in Table 2.7.1, with those for females being uniformly
higher.

Table 2.7.1. Swedish cohort and period expectations of life remaining at ages zero and ten

Epoch	Males		Females	
	Period	Cohort	Period	Cohort
1780	35.999[a]	34.359	38.523	37.893
	46,399[b]	44.088	48.321	47.581
1800	35.990	35.515	38.885	39.431
	45.863	44.628	47.973	48.497
1830	36.767	42.110	41.273	46.097
	43.244	50.647	47.184	53.497
1865	43.347	47.092	47.177	49.978
	49.160	54.065	52.120	55.638
1900	51.528	55.669	54.257	59.254
	52.928	57.688	54.706	60.347

[a] Expectation of years remaining at age zero.
[b] Expectation of years remaining at age ten.

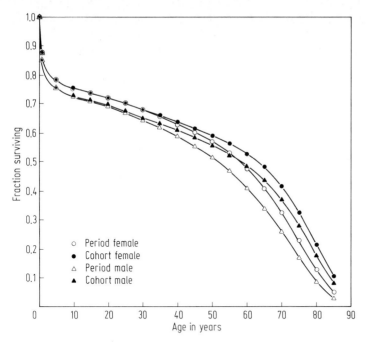

Fig. 2.7.1. Swedish period and cohort survival functions – epoch 1865

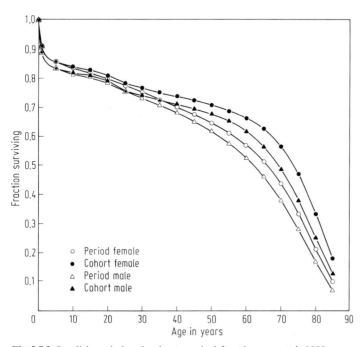

Fig. 2.7.2. Swedish period and cohort survival functions – epoch 1900

It is also of interest to compare the graphs of survival functions for periods and cohorts. Presented in Figs. 2.7.1 and 2.7.2 are graphs of the Swedish period and cohort survival functions for both sexes at the epochs 1865 and 1900. One criterion for improvement in survivorship is to compare the differences between period and cohort survival functions. The higher the cohort survival function lies above that for the period, the more rapid has been the pace of improvement in survivorship. In both Figs. 2.7.1 and 2.7.2, the cohort survival functions for each sex lie above the corresponding ones for the period, particularly at higher ages. Judging the curves in the two figures from the perspective of this criterion, it can be seen that the pace of improvement in survivorship was more rapid for Swedish cohorts born around 1900 than for those born around 1865.

Markings on the curves in Figs. 2.7.1 and 2.7.2 represent data points taken from Keyfitz and Flieger (1968). To aid the eye, these points were connected by a monotone cubic interpolation procedure due to Fritsch et al. (1980).

2.8 Historical Attempts at Modeling Survivorship

Motivated by the hope that the voluminous information available in life tables could be represented in succinct mathematical forms, many investigators have searched for parametric models of human survivorship. The 19-th century works of Gompertz (1825) and Makeham (1867) are frequently cited as the earliest attempts to find a simple and useful parametric form for the force of mortality. It has been found that the Gompertz-Makeham force of mortality works well, in many populations, for ages 30 and beyond. Thiele (1872) proposed a parametric formula for the force of mortality applicable throughout the whole of life in an attempt to overcome some of the shortcomings of the Gompertz-Makeham function. Keyfitz (1981) has given an informative review of the choice of function for mortality analysis, suggesting that effective forecasting depends on a minimum of parameter estimation.

When considering the problem of forescasting mortality, many investigators have recognized that the force of mortality should be function of not only the epoch t under consideration but also an age variable x, as suggested in Sect. 2.7. Cramér and Wold (1935) reviewed the early 20-th century literature on methods of forecasting mortality in terms of two variables, t and x, and also proposed a parametric form for the function $\mu(t, x)$ introduced in Sect. 2.7 for $x \geq 30$.

The particular form of $\mu(t, x)$ chosen by these authors was the function

$$\mu(t, x) = \alpha + \beta c^x = \alpha + \beta e^{\gamma x}, \tag{2.8.1}$$

of the Gompertz-Makeham type, where the parameters α, β, and γ are functions of the epoch $t \in R$. Specific parametric forms were proposed for these parameters based on an analysis of Swedish historical mortality data for age 30 and over during the period 1800–1930. The function chosen for the parameter

α was the straight line

$$10^3 \alpha(t) = \delta_1 - \delta_2(t - 1860), \qquad t \in R, \tag{2.8.2}$$

depending on the unknown parameters δ_1 and δ_2.

The parameters β and γ were expressed in terms of the logistic function

$$f(t) = \frac{\theta_1 + \theta_2 \exp[\theta_3(t - \tau)]}{1 + \exp[\theta_3(t - \tau)]}, \qquad t \in R, \tag{2.8.3}$$

depending on four unknown parameters $\theta_1, \theta_2, \theta_3$, and τ. When $\theta_3 > 0$, the function $f(t)$ is increasing on R and the parameters $\theta_1 < \theta_2$ are the asymptotes. The parameter τ is a point of inflection, having the property $f''(\tau) = 0$, see Problem 5 for further properties of the logistic function.

Let $f_1(t)$ and $f_2(t)$ be two logistic functions corresponding to different assignments of the four parameters. Then, the function chosen to describe the variation in the function $\beta(t)$, $t \in R$, was of the form

$$\beta(t) = 10^{-5} \exp[f_1(t)]; \tag{2.8.4}$$

while that for $\gamma(t)$, $t \in R$, was of the form

$$\gamma(t) = 10^{-3} f_2(t). \tag{2.8.5}$$

At this point, it should be observed that the force of mortality $\mu(t, x)$ in (2.8.1) depends on ten parameters, two for the function $\alpha(t)$ and four for each of the functions $\beta(t)$ and $\gamma(t)$.

Cramér and Wold used a two-stage procedure to estimate the ten parameters for each sex. The first stage consisted of estimating the parameters α, β, and γ for each period by a modified minimum Chi-square procedure. This procedure could also be described as one of weighted non-linear least squares. Having estimated the parameters α, β, and γ for twenty six periods centered at five year intervals ranging from 1800 to 1930, a search was made for expressions to fit period estimates as functions of time. This search resulted in the functions in (2.8.2), (2.8.4), and (2.8.5).

A least squares procedure used to estimate the parameters in each of the functions $\alpha(t)$, $\beta(t)$, and $\gamma(t)$. Because the functions in (2.8.4) and (2.8.5) are nonlinear in the parameters, a nonlinear least squares procedure was used to estimate the four parameters in $\beta(t)$ and $\gamma(t)$. A critical step in implementing this procedure was the choice of initial estimates used in the iteration scheme. Problem 6 contains further details on the procedure used by Cramér and Wold to obtain initial estimates of the four parameters in the logistic function.

Even by the standards of today, the results reported by Cramér and Wold required considerable computation; the fact that the work was done prior to the age of fast electronic computers makes it remarkable and of historical interest. Presented in Table 2.8.1 are the estimates of the ten parameters by sex as reported by Cramér and Wold. These estimates have several uses. For example, in a cohort born at epoch t, let $S_c(30; t, x)$ be the conditional probability of surviving to age $x > 30$, having survived till age 30. Then,

$$S_c(30; t, x) = \exp\left[-\int_{30}^{x} \mu(t + s, s)\, ds\right]. \tag{2.8.6}$$

Table 2.8.1. Estimates of ten parameters in the Cramér-Wold Model by sex-based on Swedish mortality 1800–1930

Function	$10^3 \alpha(t)$			
Parameter	Male		Female	
δ_1	7.0616		6.5954	
δ_2	0.042102		0.031257	

Function	$\beta(t)$		$\gamma(t)$	
Parameter	Male	Female	Male	Female
θ_1	1.39443	1.01023	35.63	39.92
θ_2	0.48422	0.18118	44.53	47.58
θ_3	0.116	0.108	0.136	0.124
τ	1871.0	1872.0	1871.0	1872.5

The corresponding period function at epoch t is

$$S_p(30; t, x) = \exp\left[-\int_{30}^{x} \mu(t, s)\,ds \right].$$ (2.8.7)

In principle, given the estimates in Table 2.8.1, the formula for $\mu(t, x)$ in (2.8.1) and the functions for the parameters $\alpha(t)$, $\beta(t)$, and $\gamma(t)$ in (2.8.2), (2.8.4), and (2.8.5), computer programs could be written to compute the functions in (2.8.6) and (2.8.7) for selected values of t and x.

Cramér and Wold were primarily concerned with forecasting Swedish mortality based on 130 years of recorded experience. Unlike Sweden, however, many countries, especially those in the developing world, do not have a long time series of mortality data. Indeed, an investigator, wishing to make a population projection to test the implications of a population policy, frequently has available only some scanty information such as some idea about the expectation of life at birth, perhaps for both sexes. Coale and Demeny (1966) approached the problem of dealing with scanty information by constructing model life tables based on a selected collection of 326 abridged historical period life tables from Africa, North America, South America, Asia, Europe, and Oceania.

To construct their model tables, Coale and Demeny made extensive use of simple linear regression. The first step was to partition the set of 326 tables into four regional subsets labeled East, West, North, and South. Within each region and for each sex, the simple linear equation

$$\mathring{e}(10) = \alpha + \beta \mathring{e}(10)$$ (2.8.8)

was used to connect the expectation of life at birth with that at ten years. Then, the regression equations

$$q(x, h) = \alpha(x) + \beta(x)\mathring{e}(10),$$ (2.8.9)

for $x = 0, 1, 5, 10, \ldots, 80$, were used to connect the conditional probabilities $q(x, h)$ with $\overset{\circ}{e}(10)$. The variable $\log_{10}(10^5 q(x, h))$ was also sometimes used on the left in (2.8.9). High product moment correlations between the variables $\overset{\circ}{e}(0)$ and $\overset{\circ}{e}(10)$ as well as the variables $q(x, h)$ *and* $\overset{\circ}{e}(10)$ for each x made the use of the linear regressions in (2.8.8) and (2.8.9) feasible. Coale and Demeny may be consulted for further technical and historical details.

To implement a system of model life tables based on Eqs. (2.8.8) and (2.8.9), within each region and sex, the two parameters $\alpha(x)$ and $\beta(x)$ in (2.8.9) must be estimated for 18 values of x, yielding a total of 36 estimates. In addition; equation (2.8.8) requires two estimates, making a total of 38. These 38 estimates are not particularly informative by themselves; but, when stored in a computer, they may be used to calculate a model period life table from scanty information.

Thus, if an investigator has some notion of the value of $\overset{\circ}{e}(0)$, then $\overset{\circ}{e}(10)$ may be calculated from (2.8.8). Given $\overset{\circ}{e}(10)$, the 18 conditional probabilities $q(x, h)$, $x = 0, 1, 5, 10, \ldots, 80$, may be calculated, using Eq. (2.8.9). Finally, given values of these 18 conditional probabilities, standard life table algorithms may be programmed to have the computer produce a model period life table. The U.S. Bureau of the Census has released a computer program on a magnetic tape whereby, if a user specifies the region, sex, and expectation of life at birth, the computer will return a Coale-Demeny period model life table. Altogether, about 304 regression coefficients are used in this program, representing four regions and two sexes within each region.

By using a system of many simple linear regressions, Coale and Demeny found a solution to the problem of computing model abridged life tables for 18 age groups, representing the whole of life. Moreover, their system seems to work well for expectations of life at birth ranging from 20 to 80 years. Although very useful, the Coale-Demeny system not only lacks the mathematical parsimony of that of Cramér and Wold but also makes no provision for forecasting cohort survivorship form period information. By assuming that the evolution of survivorship in a population could be represented by a time series of period expectations of life at birth lying in the interval $[20, 80]$, the latter shortcoming could be easily overcome by developing appropriate computer programs. In some populations, expectations of life at birth for females are approaching 80 years. A possible limitation of the Coale-Demeny system is that the linear regression equations may not be valid for $\overset{\circ}{e}(0)$-values beyond 80 years.

2.9 Modeling a Force of Mortality for the Whole of Life

A recurrent theme in designing forces of mortality for the whole of life, as exemplified by the work of Thiele (1872), is that of viewing $\mu(x)$ as composed of three functions, $\mu_0(x)$, $\mu_1(x)$, and $\mu_2(x)$. The component $\mu_0(x)$ characterizes risks of death in early life; while $\mu_1(x)$ and $\mu_2(x)$ characterize these risks in middle and later life. A discussion of the work of Thiele may be found in Problem 7. Recent additions to the literature on developing parametric formulas for the three components are Heligman and Pollard (1981) and Mode and Busby (1982). The work of Heligman and Pollard, which is essentially a discrete time formulation, is discussed in Problem 8. In this section, attention will be confined to continuous time formulations, incorporating desirable features of the two papers just cited.

The problem is to find a family of three nonnegative functions $\mu_0(x)$, $\mu_1(x)$, and $\mu_2(x)$, depending on a minimal number of parameters, such that if $\mu(x)$ is the total force of mortality, then

$$\mu(x) = \mu_0(x) + \mu_1(x) + \mu_2(x), \tag{2.9.1}$$

for $x \in (0, \infty)$. Among the necessary properties of these functions are that they should be sufficiently flexible to fit a variety of historical data sets, yet simple enough so that the parameters can be estimated repeatedly over a set of period life tables with relative ease. Other than tending to reduce the number of parameters to be estimated, there is nothing essential about decomposing the total force of mortality into three functions. As will become clear in the discussion that follows, a variety of schemes could be considered.

It has been widely observed in infant mortality statistics, that the risk of death decreases rapidly with age during early life. Accordingly, the function $\mu_0(x)$ should decrease as x increases. It would also be nice, although not necessary, that the function $\mu_0(x)$ remain finite as x approaches zero. A four-parameter family, having these properties, has the form

$$\mu_0(x) = e^{\alpha_0} \beta_0 \delta_0 (x + \gamma_0)^{\delta_0 - 1} \exp[-\beta_0 (x + \gamma_0)^{\delta_0}], \tag{2.9.2}$$

where $x \in (0, \infty)$. Observe that this function has a form similar to that of the density of the Weibull distribution, see Sect. 2.3.

It is not always easy to give compelling biological interpretations of the parameters in (2.9.2). Consequently, for the time being, the discussion will be limited to their mathematical properties. The parameter $\alpha_0 \in R = (-\infty, \infty)$ determines a weighting factor expressed in exponential form $\exp[\alpha_0]$ so that it is always nonnegative. To prevent the nonnegative function $\mu_0(x)$ from vanishing identically, the parameters β_0 and δ_0 must be positive, $\beta_0 > 0$ and $\delta_0 > 0$. Because $\mu_0(x)$ must be nonnegative for all $x > 0$, the parameter γ_0 must be nonnegative, $\gamma_0 \geq 0$.

Biological considerations lead to a further restriction on the parameter δ_0 and interpretations of the parameters β_0 and γ_0. If $0 < \delta_0 \leq 1$, then, as statistical observations would require, $\mu_0(x)$ is a decreasing function of $x > 0$. If $\delta_0 > 1$, then $\mu_0(x)$ increases to a maximum before it decreases. Therefore, the biologically relevant values of the parameter δ_0 lie in the interval $(0, 1]$. When $0 < \delta_0 < 1$, the value of the parameter γ_0 determines the magnitude of $\mu_0(x)$ as x approaches zero. In fact, $\mu_0(x)$ becomes infinite as $x \to 0$ when $\gamma_0 = 0$. On the other hand, the parameters β_0 and δ_0 affect the rate of decrease in the function $\mu_0(x)$ as x becomes large and thus reflect the rate of mortality decline in early life as age increases.

As can be seen from (2.2.6), the integral of each function on the right in (2.9.1) plays a basic role in determining the survival function. For the case of the μ_0-function, this integral takes the form

$$H_0(x) = \int_0^x \mu_0(s)\,ds = \exp\left[\alpha_0 - \beta_0\gamma_0^{\delta_0}\right] - \exp\left[\alpha_0 - \beta_0(x+\gamma_0)^{\delta_0}\right], \qquad (2.9.3)$$

for $x \in R^+$. From (2.9.3), it can be seen that

$$\lim_{x\uparrow\infty} H_0(x) = \exp\left[\alpha_0 - \beta_0\gamma_0^{\delta_0}\right], \qquad (2.9.4)$$

the ultimate weight attributed to infant mortality.

Some sets of mortality data contain evidence of a so-called accident hump somewhere between the ages 20 and 30 years. To accommodate such phenomena, Heligman and Pollard (1981) introduced a force of mortality of the form

$$\mu_1(x) = \exp\left[\alpha_1 - \beta_1(\log\gamma_1 x)^2\right] \qquad (2.9.5)$$

for middle life. Like the parameter α_0, $\alpha_1 \in R$ determines an exponential weight factor $\exp[\alpha_1]$. The parameters β_1 and γ_1 are positive and γ_1 is such that the maximum of the μ_1-function occurs at $1/\gamma_1$. At this point, the maximum value of the μ_1-function is $\mu_1(\gamma_1^{-1}) = \exp[\alpha_1]$.

Among the desirable mathematical properties of the function in (2.9.5) is that its integral may be expressed in terms of the standard normal distribution function. Thus, it can be shown that

$$H_1(x) = \int_0^x \mu_1(s)\,ds = c\,\Phi(\xi(x)), \qquad x \in R^+, \qquad (2.9.6)$$

where

$$c = \exp\left[\alpha_1 + \frac{1}{4\beta_1}\right]\gamma_1^{-1}\left(\frac{\pi}{\beta_1}\right)^{1/2} \qquad (2.9.7)$$

$$\xi(x) = \sqrt{2\beta_1}\left(\log\gamma_1 x - \frac{1}{2\beta_1}\right),$$

the logarithm being taken to base e, and

$$\Phi(x) = \frac{1}{\sqrt{2\pi}}\int_{-\infty}^x e^{-\frac{s^2}{2}}\,ds. \qquad (2.9.8)$$

From the above, it follows that $H_1(0) = \lim_{x\downarrow0} H_1(x) = 0$, $\lim_{x\uparrow\infty} H_1(x) = c$, and $H_1(x) \le c$ for all $x \in R$. Hence, the factor c is the ultimate weight attributed to middle life. Problem 9 may be consulted for further technical details. Many computer facilities have libraries containing programs for calculating the standard normal distribution function in (2.9.8). A discussion of algorithms underlying such programs has been provided by Kennedy and Gentle (1980) in their Sect. 5.3.

As age increases, particularly during late life, the risk of death increases rapidly. A risk function, having this property and belonging to the same family as the Weibull density, has the form

$$\mu_2(x) = e^{\alpha_2} \beta_2 \delta_2 (x+\gamma_2)^{\delta_2-1} \exp[\beta_2(x+\gamma_2)^{\delta_2}], \tag{2.9.9}$$

where $x \in (0, \infty)$. Like the parameters in (2.9.2), $\alpha_2 \in R$, β_2 and δ_2 are positive, and γ_2 is nonnegative. If $\delta_2 = 1$ and $\gamma_2 = 0$, then $\mu_2(x)$ reduces to a Gompertzian risk function; if $\delta_2 > 1$ then $\mu_2(x)$ increases with x at a rate faster than an exponential.

A function that is not monotone increasing arises when $0 < \delta_2 < 1$. In this case, the derivative $\mu_2'(x)$ vanishes at the point

$$x_0 = \left(\frac{1-\delta_2}{\beta_2 \delta_2}\right)^{1/\delta_2} - \gamma_2. \tag{2.9.10}$$

Furthermore, $x_0 > 0$ if, and only if, the right hand side of (2.9.10) is positive. Therefore, the parameters β_2, γ_2, and δ_2 cannot be chosen independently when $0 < \delta_2 < 1$. If $x < x_0$, then $\mu_2(x)$ decreases as x increases; while if $x > x_0$, then $\mu_2(x)$ increases as x increases. This observation suggests that a total force of mortality of the form $\mu(x) = \mu_0(x) + \mu_2(x)$, omitting the function $\mu_1(x)$. could be U-shaped, the shape characteristic of some period human mortality. Whether such a total force of mortality might prove to be useful could only be determined by actually fitting the function to data. It is of interest to observe that if $\gamma_2 = 0$ and $0 < \delta_2 < 1$, then $\mu_2(x)$ becomes infinite as x approaches zero.

An advantage of risk functions belonging to the same family as the Weibull density is that their integrals may be expressed in terms of elementary functions. For the function in (2.9.9) this integral is

$$H_2(x) = \int_0^x \mu_2(s)ds = \exp[\alpha_2 + \beta_2(x+\gamma_2)^{\delta_2}] - \exp[\alpha_2 + \beta_2\gamma_2^{\delta_2}], \tag{2.9.11}$$

for $x \in R^+$. From the expression on the right, it is clear that $\lim_{x \downarrow 0} H_2(x) = 0 = H_2(0)$ and $H_2(x) \uparrow \infty$ as $x \uparrow \infty$.

It is clear, therefore, from the foregoing discussion that if a function $H(x)$ is defined by

$$H(x) = H_0(x) + H_1(x) + H_2(x) \tag{2.9.12}$$

for $x \in R^+$, then the survival function $S(x)$ defined by

$$S(x) = \exp[-H(x)], \quad x \in R^+, \tag{2.9.13}$$

is proper. The function $S(x)$ so defined is proper in the sense that $S(0) = 1$ and $S(x) \downarrow 0$ as $x \uparrow \infty$.

The survival function in (2.9.13) belongs to an eleven-parameter family of functions. It is always of interest to search for a minimal set of parameters when fitting a survival function to data. Among the possibilities is to set the weight parameters α_0, α_1, and α_2 equal to zero so that the function in (2.9.13) depends on only eight parameters. Another possibility for reducing the number of parameters in the H_2-function is to put $\gamma_2 = 0$ when $\delta_2 > 1$. Other choices of the functions on the right in (2.9.12) could also be made. For example, if $\mu_2(x)$

is a force of mortality of the Gompertz-Makeham type,

$$\mu_2(x) = \alpha_2 + \beta_2 \gamma_2 e^{\gamma_2 x}, \tag{2.9.14}$$

then the H_2-function takes the form

$$H_2(x) = \alpha_2 x + \beta_2(e^{\gamma_2 x} - 1), \quad x \in R^+. \tag{2.9.15}$$

In this case, the parameters β_2 and γ_2 are positive but $\alpha_2 \geq -\beta_2 \gamma_2$, because the function in (2.9.14) must be nonnegative for all $x \geq 0$.

2.10 Computer Experiments in Fitting Survivorship Models to Swedish Historical Data

Due to the long time span of official Swedish mortality statistics, they are well suited as a source of data for computer experiments in fitting parametric evolutionary models of human survivorship to data. Accordingly, thirty eight sets of Swedish period life tables, as reported by Keyfitz and Flieger (1968), were used in computer experiments designed to explore the practical feasibility of the modeling schemes set forth in Sect. 2.9. The thirty eight sets of period life tables were equally spaced in five-year time units and centered at epochs ranging from 1780 to 1965. A nonlinear least squares procedure, with a capability for approximating all derivatives in the computer, was used throughout all experiments. The quantity minimized as a function of the unknown parameters was

$$\sum_x (S_p(t, x) - \hat{S}_p(t, x))^2, \tag{2.10.1}$$

the sum of the squares of derivations of an empirical period survival function, $\hat{S}_p(t, x)$, from the theoretical one, $S_p(t, x)$. In (2.10.1) the sum extends over $x = 0, 1, 5, 10, \ldots, 85$, the ages used in the empirical life tables.

A commonly encountered difficulty when attempts are made to fit multiparameter models by nonlinear least squares is that successive iterates do not converge to satisfactory fits, when as many as ten or more parameters are considered simultaneously. Such difficulties were encountered when attempts were made to explore the practical feasibility of fitting some of the multiparameter models discussed in Sect. 2.9. Previous work with the model outlined in Mode and Busby (1982) had, however, demonstrated that the Gompertz-Makeham force of mortality led to very good fits between conditional and theoretical survival functions for ages over thirty in historical Swedish data, for periods ranging from the early nineteenth to past middle of the twentieth century. Although the model outlined in Mode and Busby (1982) worked well throughout all ages for periods in the middle quarters of the twentieth century, unsatisfactory fits were obtained for ages thirty and under throughout most of the nineteenth century. To overcome these difficulties, the following fitting scheme, based on conditioning, was introduced.

Suppose the parameters in the forces of mortality in (2.9.2) and (2.9.5) depend on the epoch $t \in R$; and, let $H_0(t, x)$ and $H_1(t, x)$ be the integrals of these functions as defined in (2.9.3) and (2.9.6). Define the function $H_{01}(t, x)$ at epoch t by

$$H_{01}(t, x) = H_0(t, x) + H_1(t, x), \qquad x \geq 0. \tag{2.10.2}$$

Then, the period survival function at epoch t is defined by

$$S_p(t, x) = \exp[-H_{01}(t, x)] \tag{2.10.3}$$

for $0 \leq x \leq 30$. Observe that the survival function in (2.10.3) depends on seven parameters.

To accommodate ages thirty and over, suppose the parameters in the Gompertz-Makeham force of mortality in (2.9.14) depend on the epoch $t \in R$. Let $H_2(t, x)$ be the integral of this function at t as defined by (2.9.15). Then, the conditional period survival function at epoch t, having survived to age 30, is defined by

$$S_p(30; t, x) = \exp[-H_2(t, x)] \tag{2.10.4}$$

for $x \geq 0$. The function in (2.10.4) depends upon three parameters and is interpreted as the period probability of surviving to age $30 + x$, given survival to age 30. The period survival function for ages thirty and over at epoch t is defined by

$$S_p(t, x) = S_p(t, 30) S_p(30; t, x - 30) \tag{2.10.5}$$

for $x \geq 30$.

To avoid nonconvergence problems when attempts were made to simultaneously fit the ten parameters in the model specified by (2.10.3), (2.10.4), and (2.10.5), a two-state fitting procedure was adopted. The first stage consisted of fitting the seven parameter survival function in (2.10.3) at the eight ages $x = 0, 1, 5, 10, \ldots, 30$. Seven parameter models fitted to eight points may not always lead to acceptable fits. For the model specified in (2.10.3), however, the fits were very good at all epochs. The second stage consisted of fitting the three-parameter function in (2.10.4) to the empirical conditional survival function $\hat{S}_p(30; t, x) = \hat{S}_p(t, 30 + x) / \hat{S}_p(t, 30)$ at the values $x = 0, 5, 10, \ldots, 55$. After all ten parameters were estimated by this two-stage procedure, an over-all measure of goodness of fit was calculated for each period by computing $S_p(t, x)$ in (2.10.1), according to formulas (2.10.3) and (2.10.5), and then computing an error mean square based on the nineteen ages, $x = 0, 1, 5, 10, \ldots, 85$, going into the sum in (2.10.1).

Presented in the upper panel of Table 2.10.1 are the maxima, minima, and ranges in the error mean squares so calculated and rounded to four decimals for the thirty-eight period life tables considered for each sex. According to Table 2.10.1, the maximum and minimum error mean squares for males were 0.1031×10^{-4} and 0.4215×10^{-6}. Error mean squares of these orders of magnitude suggest that the theoretical and empirical survival functions agree to at least two decimal places at a majority of points. Computer printout, not presented here, did indeed confirm such close agreement in the vast majority of

Table 2.10.1. Error mean squares for fitting then parameter period survival function to Swedish mortality data – 1780 to 1965

Summary statistic	Summary measures	
	Males	Females
Maximum	0.1031×10^{-4}	0.8859×10^{-4}
Minimum	0.4215×10^{-6}	0.5739×10^{-6}
Range	0.9888×10^{-5}	0.8802×10^{-4}
Epoch	Selected examples	
	Males	Females
1800	0.4157×10^{-5}	0.1674×10^{-5}
1900	0.2396×10^{-5}	0.8779×10^{-6}
1965	0.2948×10^{-5}	0.1046×10^{-5}

Table 2.10.2. Nonlinear least squares estimates of ten parameters in period survival function at selected epochs for Swedish males

Parameter	Epoch		
	1800	1900	1965
α_0	4.5948	3.1134	4.2076
β_0	4.6971	1.3904	3.8476
γ_0	0.0962	0.0570	0.0067
δ_0	0.0228	0.0051	0.0005
α_1	-4.8551	-5.1776	-6.7942
β_1	5.2352	4.9695	2.6200
γ_1	0.0312	0.0376	0.0353
α_2	0.0078	0.0062	0.0004
β_2	0.0244	0.0062	0.0060
γ_2	0.0894	0.1047	0.1027

cases. In fact, in many cases, the empirical and theoretical survival functions agreed to three decimal places. For reasons unknown, it can be seen from the values of the error mean squares in Table 2.10.1 that fits between theoretical and empirical survival functions for females were satisfactory but not quite as good as those for males.

Altogether the thirty eight sets of Swedish life tables yielded 760 estimates, 380 for each sex, of the ten parameters in the period survival function under consideration. Only a sample of these estimates can be presented here. Contained in Table 2.10.2 are the estimates of the ten parameters for males, rounded to four decimal places, at the epochs 1800, 1900, and 1965. Estimates of the ten parameters for females at the same epochs are presented in Table 2.10.3. The error mean squares for the six sets of estimates in Tables 2.10.2 and 2.10.3 are presented in the lower panel of Table 2.10.1. As can be seen

Table 2.10.3. Nonlinear least squares estimates of ten parameters in period survival function at selected epochs for Swedish females

Parameter	Epoch		
	1800	1900	1965
α_0	3.7688	2.3929	5.2117
β_0	4.3292	0.1825	2.0927
γ_0	0.1617	0.0940	0.0004
δ_0	0.0414	0.0236	0.0003
α_1	−5.0915	−5.3076	−7.7587
β_1	7.0071	7.5263	2.1202
γ_1	0.0342	0.0416	0.0407
α_2	0.0090	0.0061	0.0008
β_2	0.0090	0.0030	0.0016
γ_2	0.1056	0.1149	0.1207

from the latter table, the closest fits between empirical and theoretical survival functions for the six sets of estimates were those for females, the smallest error mean square, 0.8779×10^{-6}, being that for females at epoch 1900.

The estimates presented in Tables 2.10.2 and 2.10.3 are illustrative of some general properties of the estimates as viewed across the thirty eight epochs considered. With respect to the early, middle, and late life components of the model, the estimates of the parameters α_2, β_2, and γ_2 in the Gompertz-Makeham force of mortality, representing late life, displayed the most definite time trends across epochs. Some time trends were also apparent across epochs for estimates of the parameters α_1, β_1, and γ_1 in the middle life component of the model. The most erratic behavior of the estimates across epochs occurred for the parameters α_0, β_0, γ_0, and δ_0 in the early life component of the model.

Among the sets of estimates for these four parameters, those for β_0 seemed to fluctuate most erratically. Some idea about the magnitude of these fluctuations may be obtained by comparing the estimates of the parameter β_0 for each sex at epoch 1900 with the estimates of the other epochs, 1800 and 1965. The estimate of β_0 at 1900 for males was 1.3904; while that for females was 0.1825. Both of these estimates are relatively much smaller than the corresponding estimates at epochs 1800 and 1965. It is also of interest to observe, from the values of the error mean squares in the lower panel of Table 2.10.1, that the closest fits between the empirical and theoretical survival functions in both sexes occurred at epoch 1900. Despite the disconcerting nature of fluctuations in estimates across epochs from the point of view of fitting time trends, these fluctuations, together with the close fits of the model to the empirical survival functions at distinct points in the parameter space, actually demonstrate a type of robustness with respect to movement among a set of points in a 10-dimensional space.

A useful property of parameterized period survival functions is that of making it possible to impute period survival densities from abridged life tables.

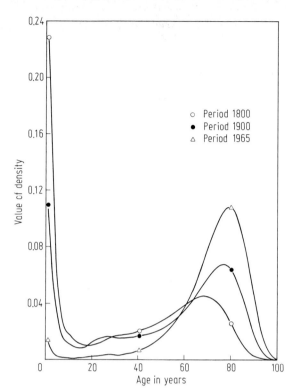

Fig. 2.10.1. Imputed period survival densities for Swedish males

An alternative perspective on the evolution of human mortality can thus be obtained with the help of computer generated graphs. Computer generated graphs of the imputed densities for males, determined by the parameter estimates in Table 2.10.2, are presented in Fig. 2.10.1; those for females, as determined by the parameter estimates in Table 2.10.3, are presented in Fig. 2.10.2. It should be pointed out that the actual densities of the period survival functions determined by differentiating Eqs. (2.10.3) and (2.10.5) with respect to x, are not graphed in Figs. 2.10.1 and 2.10.2. Rather, to ease computer implementation, the graphs in these figures represent crude approximations to these densities obtained by differencing the period survival functions on a grid of points equally spaced at three-year intervals and then connecting these differences by spline interpolation.

Highlighted in the graphs displayed in Figs. 2.10.1 and 2.10.2 are at least three aspects of the evolution of Swedish mortality in both sexes. Firstly, as would be reflected in the graphs of the associated period survival functions, marked decreases in infant mortality among the epochs 1800, 1900, and 1965 can also be seen from the left most portions of the graphs of the period densities. Secondly, as time progressed, the graphs of the densities became more skewed to the right, indicating that a substantial number of members of the population were living longer life spans. A third feature is the hump in the graphs between twenty and forty years of age, particularly those for females,

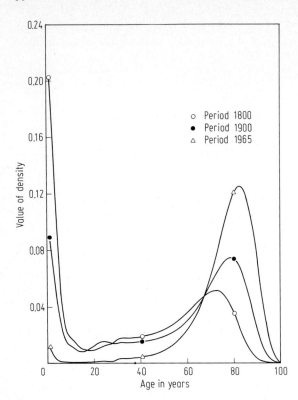

Fig. 2.10.2. Imputed period survival densities for Swedish females

○ Period 1800
● Period 1900
△ Period 1965

Value of density

Age in years

see graphs in Fig. 2.10.2, at the epochs 1800 and 1900. According to the theory developed in Sect. 2.9 for the middle life component of the model, the age at which this hump occurs is given by the value $\hat{\gamma}_1^{-1}$. From Table 2.10.3, it can be seen that these values are $\hat{\gamma}_1^{-1} = 29.24$ and $\hat{\gamma}_1^{-1} = 24.04$ years, respectively, for epochs 1800 and 1900. These values seem to agree quite well with those suggested by inspection of Fig. 2.10.2. Whether these humps reflect real features of female mortality or are merely artifacts induced by fitting the model to abridged life tables could only be ascertained by examining the original ungrouped data.

When the parameters in the forces of mortality in (2.9.2), (2.9.5), and (2.9.15) depend on the epoch $t \in R$, then expressions for the cohort survival functions may be written down. Define a force of mortality for $t \in R$ and $x \in (0, \infty)$ by

$$\mu_{01}(t, x) = \mu_0(t, x) + \mu_1(t, x), \tag{2.10.6}$$

where $\mu_0(t, x)$ and $\mu_1(t, x)$ are the forces of mortality in (2.9.2) and (2.9.5) with parameters depending on the epoch t. Then, for the model fitted to Swedish period life tables, the cohort survival function for individuals born at epoch $t \in R$ is given by

$$S_c(t, x) = \exp\left[-\int_0^x \mu_{01}(t + s, s)\, ds \right] \tag{2.10.7}$$

for $0 \le x \le 30$.

Let $\mu_2(t, x)$ be the force of mortality in (2.9.15), with the parameters depending on the epoch t. Then, for an individual born at epoch t who survives to age 30,

$$S_c(30; t, x) = \exp\left[-\int_0^x \mu_2(t+30+s, s)\,ds\right] \qquad (2.10.8)$$

is the conditional probability of surviving to age $30+x$. Therefore, for individuals born at epoch t,

$$S_c(t, x) = S_c(t, 30) S_c(30; t, x-30) \qquad (2.10.9)$$

is the cohort survival function for ages $x \geq 30$.

Before formulas (2.10.7), (2.10.8), and (2.10.9) could be implemented on a computer, it would be necessary to find expressions representing the ten parameters in the model as functions of the epoch $t \in R$. Finding appropriate expressions for these parameters as functions of t is a research exercise that is beyond the scope of this chapter. The period estimates of the ten parameters discussed in this section are, however, a necessary step in an empirical search for functions expressing the ten parameters in question as functions of the epoch $t \in R$.

The spline interpolation procedure used in the computer generation of Figs. 2.10.1 and 2.10.2 was the same as that described in previous sections. A mathematical description and analysis of the nonlinear least squares procedures utilized in obtaining period estimates of the ten parameters in the model has been provided by Gill and Murray (1978). Used in implementing these procedures was the collection of FORTRAN programs, E04FCF, in the National Algorithms Group (NAG) Library. Information concerning this library of programs may be obtained by writing NAG (USA) Inc., 1250 Grace Court, Downer's Grove, Illinois 60515.

2.11 Heterogeneity in Survivorship

Heterogeneity, variability among individuals in the capacity to survive, exists in human populations. Some of this variability is genetic as exemplified by bleeder's disease, hemophilia. A classic textbook on human genetics, such as Stern (1960), may be consulted for detailed discussions on genetic factors contributing to variability in survivorship. Other factors contributing to the capacity to survive among members of a population are less easy to understand. In this section, heterogeneity in survivorship will be treated from a simplistic point of view.

Consider a cohort of individuals born at some epoch with a stationary law governing survivorship and suppose individuals may be followed throughout their life spans. It will be supposed that survivorship in the cohort may be described in terms of a pair of random variables $\langle X, \Lambda \rangle$. As before X represents age at death; but, Λ is a random variable representing variation in the capacity to survive among members of the cohort. Given $\Lambda = \lambda > 0$, the conditional survival function is assumed to be

$$P[X \leq x | \Lambda = \lambda] = S(\lambda, x) = e^{-\lambda H(x)}, \tag{2.11.1}$$

where $x \in R^+$ and $H(x)$ is the integral of the force of mortality $\mu(x)$.

The H-function could be chosen as in (2.9.12); it could, moreover, be parameterized along the lines outlined in Sect. 2.9. From (2.11.1) it is clear that those individuals with smaller values of λ have a greater capacity to survive than those with large values. If $h_2(\lambda)$ is the p.d.f. of the random variable Λ, then the survival function of the cohort is

$$S(x) = \int_0^\infty h_2(\lambda) e^{-\lambda H(x)} d\lambda, \tag{2.11.2}$$

for $x \in R^+$.

It is apparent that the mixture of survival functions defined in (2.11.2) has a close connection with the theory of Laplace transforms. For, by definition, the Laplace transform associated with the random variable Λ is the function

$$h_2^*(x) = E[e^{-\Lambda x}] = \int_0^\infty e^{-\lambda x} h_2(\lambda) d\lambda, \tag{2.11.3}$$

defined for $x \in R^+$. An interesting observation is that every Laplace transform $h^*(x)$ of a p.d.f. $h(\lambda)$ on $(0, \infty)$ is a survival function in the sense that $h^*(0) = 1$, $h^*(x)$ decreases as x increases, and $h^*(x) \to 0$ as $x \to \infty$. From (2.11.2) and (2.11.3), it is easy to see that

$$S(x) = h_2^*(H(x)), \quad x \in R^+. \tag{2.11.4}$$

Tractable cases, amenable to mathematical manipulations, arise when $h_2^*(x)$ is an elementary function. A well known distribution, with positive parameters α and β and an elementary Laplace transform, is the gamma, whose p.d.f. is

$$h_2(\lambda) = \frac{\beta^\alpha}{\Gamma(\alpha)} \lambda^{\alpha-1} e^{-\beta\lambda}, \tag{2.11.5}$$

where $\lambda > 0$. The Laplace transform of the gamma distribution is

$$h_2^*(x) = \frac{\beta^\alpha}{(\beta+x)^\alpha}, \quad x \in R^+. \tag{2.11.6}$$

Therefore, for the case the random variable Λ follows a gamma distribution with parameters α and β, the survival function in (2.11.4) becomes

$$S(x) = \frac{\beta^\alpha}{(\beta+H(x))^\alpha}, \quad x \in R^+. \tag{2.11.7}$$

Either period or cohort estimates of the survival function on the left in (2.11.7) may be obtained, using life table methods. When such estimates of $S(x)$ are available and $H(x)$ has a parametric representation, then, in principle, the parameters α and β, as well as those in the H-function, could be estimated by nonlinear least squares. A goodness of fit criterion could also be used to judge the empirical validity of a proposed parameteric model. Evidently, as yet, such a procedure has not been attempted.

Another approach is to consider fixed values of the parameters α and β. Then, given estimates of the survival function $S(x)$ at selected values of x, the problem would be that of estimating the H-function. For fixed values of x and the parameters α and β, Eq. (2.11.7) may be solved for the H-function. Thus, the equation,

$$H(x) = \frac{\beta[1 - (S(x))^{1/\alpha}]}{(S(x))^{1/\alpha}}, \tag{2.11.8}$$

could be used to estimate the H-function at those values of x for which estimates of $S(x)$ are available. Given estimates of the H-function, the conditional survival function $S(\lambda, x)$ in (2.11.1) could also be estimated for assigned values of λ, the capacity to survive.

Conditional distributions of the random variable Λ, given conditions on the random variable X, are also of interest. For example, the conditional distribution of Λ, given that $X = x$, would describe the variability in the capacity to survive among individuals who are age x. Similarly, the conditional distribution of Λ, given that $X > x$, would describe this variability in a cohort among individuals who have survived to age x.

As a first step in deriving these conditional distributions, let $h(x, \lambda)$ be the joint p.d.f. of the random variables X and Λ. Let $h_1(\lambda, x)$ be the conditional p.d.f. of the random variable X, given that $\Lambda = \lambda$. Then, from (2.11.1), it follows that

$$h_1(\lambda, x) = -\frac{\partial S(\lambda, x)}{\partial x} = \lambda \mu(x) e^{-\lambda H(x)}, \tag{2.11.9}$$

where $x \in R^+$ and $\mu(x) = H'(x)$. Therefore, from (2.11.5) and (2.11.9), it follows that

$$h(x, \lambda) = h_2(\lambda) h_1(\lambda, x) = \frac{\beta^\alpha}{\Gamma(\alpha)} \mu(x) \lambda^\alpha \exp[-\beta(x)\lambda], \tag{2.11.10}$$

where $\beta(x) = \alpha + H(x)$, $x \geq 0$, and $\lambda > 0$. The marginal p.d.f. of the random variable X is, see Eq. (2.11.7),

$$h_1(x) = -\frac{dS(x)}{dx} = \alpha \beta^\alpha \mu(x) [\beta(x)]^{-(\alpha+1)}, \tag{2.11.11}$$

where $x \in R^+$.

Let $h_2(x, \lambda)$ be the conditional p.d.f. of the random Λ, given that $X = x$. Then, by definition,

$$h_2(x, \lambda) = \frac{h(x, \lambda)}{h_1(x)} = \frac{[\beta(x)]^{\alpha+1}}{\Gamma(\alpha+1)} \lambda^\alpha \exp[-\beta(x)\lambda], \tag{2.11.12}$$

where $x \geq 0$ and $\lambda > 0$. Observe that the conditional p.d.f. of the capacity to survive in a cohort of individuals age x is gamma p.d.f. with parameters $\alpha + 1$ and $\beta(x)$.

Let $f_2(x, \lambda)$ be the conditional p.d.f. of the random variable Λ, given that $X > x$. To derive this p.d.f. consider

$$P[\Lambda \le \lambda | X > x] = \frac{1}{S(x)} \int_0^\lambda \int_0^\infty h(t, s) \, dt \, ds. \qquad (2.11.13)$$

After substituting the p.d.f. in (2.11.10) in the double integral on the right in (2.11.13), it can be shown that

$$P[\Lambda \le \lambda | X > x] = \frac{[\beta(x)]^\alpha}{\Gamma(\alpha)} \int_0^\lambda s^{\alpha-1} \exp[-\beta(x)s] \, ds. \qquad (2.11.14)$$

Hence, the conditional p.d.f. $f_2(x, \lambda)$ takes the form

$$f_2(x, \lambda) = \frac{[\beta(x)]^\alpha}{\Gamma(\alpha)} \lambda^{\alpha-1} \exp[-\beta(x)\lambda], \qquad (2.11.15)$$

for $x \ge 0$ and $\lambda > 0$. From (2.11.5) it is clear that, given survival to age x in a cohort, the conditional distribution of the capacity to survive is again a gamma distribution with parameters α and $\beta(x)$.

Vaupel et al. (1979) made use of ideas of this section in their study of heterogeneity in human survivorship. Among other things, based on life table estimates of the survival function, these authors used ideas equivalent to Eq. (2.11.8) to study variation in the conditional survival function in (2.11.1) for assigned values of the parameters α, β, and λ.

2.12 Further Reading

Throughout recent history human survivorship has been a topic of wide interest and concern. It is not surprising, therefore, that an extensive literature on the topic has been accumulated. So extensive is this literature that it makes it impractical to even attempt to compile a comprehensive bibliography. Accordingly, the purpose of this section is to merely suggest further reading.

Relational methods for constructing model survival functions has been advocated by Brass (1974) and elsewhere. Suppose $G^{(0)}(x)$ is some "standard" life span distribution function. Then, its logit at x is defined by

$$Y^{(0)}(x) = \log_e (G^{(0)}(x)/S^{(0)}(x)), \qquad (2.12.1)$$

provided that $S^{(0)}(x) \ne 0$. Let $Y(x)$ be the logit of another distribution function $G(x)$ at x. Then, Brass proposed that the two distribution functions be related according to the equation

$$Y(x) = \alpha + \beta Y^{(0)}(x), \qquad (2.12.2)$$

where α and β are two parameters to be estimated.

Relational methods seem particularly relevant when only scanty information is available. For example, suppose the survival function $S(x)$, or equivalently the distribution function $G(x)$ is known at only two ages x_1 and x_2.

Then, given the logits $Y(x)$ and $Y^{(0)}(x)$ at $x=x_1$ and x_2, relation (2.12.2) becomes two equations that may be solved for α and β. Once α and β have been estimated, it is then possible to obtain an extrapolation of $S(x)$ to all ages x for which $Y^{(0)}(x)$ is known by using the relationship

$$S(x)=(1+\exp[\alpha+\beta Y^{(0)}(x)])^{-1}. \tag{2.12.3}$$

Evidently, either a "standard" period or cohort survival function could be used. Among other things, Hobcraft et al. (1982) discuss relational ideas in their review of age, period, and cohort effects in demography. Cox and Scott (1977) have provided a review of international studies in generation (cohort) mortality.

Facing many analysts dealing with data from developing countries is the problem of lack or inadequacy of classical demographic data sources; sources that usually include registration of vital events or periodic censuses. Brass and Coale (1968), as well as others, have developed indirect methods of mortality measurement based only on simple survey or census questions. No attempt will be made here to develop mathematical theories underlying these methods. Arthur and Stoto (1981) have provided some mathematical background to indirect methods of measurement and have also studied their robustness. Other relevant papers on indirect estimation includes those of Trussel (1975), Palloni (1979), and Feeney (1980).

A paper on the recommended list of statistical theories of mortality measurement is that of Grenander (1956). Wegman et al. (1980) have provided an intersting time series approach to life table construction.

Problems and Miscellaneous Complements

1. (a) In reference to (2.2.9) and (2.2.10) show that

$$\int_0^a xg(x)dx=aG(a)-\int_0^a G(x)dx$$

$$=-aS(a)+\int_0^a S(x)dx, \quad a>0.$$

(b) By assumption,

$$E[X]=\int_0^\infty xg(x)dx=\lim_{a\uparrow\infty}\int_0^a xg(x)dx$$

is finite. Therefore,

$$\lim_{a\uparrow\infty}\int_a^\infty xg(x)dx=0.$$

But,

$$aS(a) \leq \int_a^\infty xg(x)dx$$

so that $aS(a) \to 0$ as $a \uparrow \infty$. Applying this result in part (a), yields (2.2.10).

2. (a) If X is a discrete random variable taking values in the nonnegative integers $x = 0, 1, 2, \ldots$, then the relationship

$$E[X] = \sum_{x=0}^\infty P[X > x]$$

is easy to verify. For let

$$P[X = x] = g(x), \quad x = 0, 1, 2, \ldots$$

and let

$$S(x) = P[X > x] = g(x+1) + g(x+2) + \ldots.$$

By summing on $x = 0, 1, 2, \ldots$, it is easy to see that

$$\sum_{x=0}^\infty xg(x) = \sum_{x=0}^\infty S(x).$$

Because all terms are nonnegative, the series on the left and right either converge or diverge together.

(b) The moments of the Weibull distribution may be computed in terms of the gamma function. Let the random variable X have a Weibull distribution with parameters α and β. Then for $n > 0$,

$$E[X^n] = \int_0^\infty x^n \mu(x) S(x) dx$$

$$= \alpha \beta \int_0^\infty x^{n+\beta-1} e^{-\alpha x^\beta} dx$$

$$= \frac{1}{\alpha^{n/\beta}} \int_0^\infty y^{n/\beta} e^{-y} dy = \frac{1}{\alpha^{n/\beta}} \Gamma\left(\frac{\beta+n}{\beta}\right).$$

Write down formulas for the mean and variance of X. In the programming language APL, the gamma function is implemented as a primitive.

(c) Unlike the Weibull, the expected value of the Gompertz distribution cannot be expressed in terms of an elementary function. However, with time measured in years,

$$E[X] \simeq \sum_{x=0}^N S(x)$$

will provide a rough approximation to $E[X]$ for some N, depending on the values of the parameters α and β. In many cases, $N = 100$ will suffice.

3. The conditional density of the random variable X, given that $x < X \leq x + h$, is $g(x+s)/d(x, h)$, $0 \leq s \leq h$. Therefore,

$$P[X > x + y \mid x < X \leq x + h] = \frac{1}{d(x, h)} \int_y^h g(x+s) ds$$

$$= \frac{G(x+h) - G(x+y)}{d(x, h)},$$

which is equivalent to (2.5.6).

4. (a) Show that if one assumes that either the survival function is a straight line between the points $\langle x, S(x) \rangle$ and $\langle x+h, S(x+h) \rangle$ or the conditional density in Problem 3 is uniform on $(x, x+h]$, then, see (2.5.8) and (2.5.9),

$$L(x, h) = h[S(x+h) + \tfrac{1}{2}d(x, h)]$$

and

$$q(x, h) = \frac{2hm(x, h)}{2 + hm(x, h)}.$$

(b) Under the assumptions of part (a), show that

$$L(x, 1) = \frac{S(x) + S(x+1)}{2},$$

for the case of a complete life table.

5. Let $f(t)$ be the logistic function defined in (2.8.3) with $\theta_1 < \theta_2$.
   (a) Show that if $\theta_3 > 0$, then

$$\lim_{t \to \infty} f(t) = \theta_2$$

and

$$\lim_{t \to -\infty} f(t) = \theta_1.$$

   (b) Show that

$$f'(t) = \frac{df(t)}{dt} = \frac{\theta_3(f(t) - \theta_1)(\theta_2 - f(t))}{\theta_2 - \theta_1}.$$

Thus, if $\theta_3 > 0$, then $f(t)$ increases with $t \in R$.
   (c) Show that

$$f(\tau) = \frac{\theta_1 + \theta_2}{2}.$$

   (d) Show that

$$f''(t) = \frac{f'(t)\theta_3}{\theta_2 - \theta_1}[\theta_1 + \theta_2 - 2f(t)].$$

Then, conclude that $f''(\tau) = 0$.

6. Initial estimates of the parameters in the logistic function are of critical importance in iterative nonlinear least squares procedures designed to estimate them. The results of Problem 5 are useful in obtaining initial estimates. Suppose the values $\hat{f}(t_i)$ of the function $f(t)$ at the points t_i, $1 \leq i \leq n$, are observed.

(a) Then, the initial estimates of the parameters θ_1 and θ_2 may be chosen as

$$\hat{\theta}_1(0) = \min_{1 \leq i \leq n} \hat{f}(t_i)$$

and

$$\hat{\theta}_2(0) = \max_{1 \leq i \leq n} \hat{f}(t_i).$$

(b) From part (c) of Problem 5, a number $\hat{\tau}(0)$ such that the equation

$$\hat{f}(\hat{\tau}(0)) = \frac{\hat{\theta}_1(0) + \hat{\theta}_2(0)}{2}$$

holds approximately is an initial estimate of the parameter τ.

(c) From parts (b) and (c) of Problem 5, it follows that

$$f'(\tau) = \theta_3 \frac{\theta_2 - \theta_1}{4}.$$

Therefore, if $\hat{m}$ is an estimate of the slope of the curve at $\hat{\tau}(0)$, then

$$\hat{\theta}_3(0) = \frac{4\hat{m}}{\hat{\theta}_2(0) - \hat{\theta}_1(0)}$$

is an initial estimate of θ_3.

7. The actual form of the force of mortality considered by Thiele (1871) was, in his notation,

$$\mu(x) = a_1 e^{-b_1 x} + a_2 e^{-1/2 b_2^2 (x-c)^2} + a_3 e^{b_3 x}.$$

The μ_0-function in (2.9.2) reduces to a form equivalent to the first term on the right when $\delta_0 = 1$ and $\gamma_0 = 0$. However, the μ_1-function in (2.9.5), introduced by Heligman and Pollard (1981), is distinctly different from the middle considered by Thiele. It is more suitable for the domain $(0, \infty)$ than the multiple of normal density considered by Thiele. The third term in $\mu(x)$ as defined above is, of course, the Gompertzian risk function. Even though Thiele's paper was published over a century ago, it is still of great interest from the point of view of methodological development.

8. Instead of working directly with the force of mortality, Heligman and Pollard (1981) considered parameterizations of the conditional probability $q(x, h)$ in (2.5.4). In the present discussion, $q(x, h)$ will be denoted by $q(x)$ for the case $h = 1$. When integrals of forces of mortality are difficult to compute, working with parameterizations of the q-probabilities and using life table algorithms can be helpful from the point of view of computer implementation. Essentially, two parameterizations were considered.

(a) Consider a function $f(x)$ of the form

$$f(x) = A^{(x+B)^C} + DE^{(\log Fx)^2} + GH^x,$$

defined for $x = 0, 1, 2, \ldots$. The eight parameters $A, B, \ldots, H$ are positive and $A, E,$ and H satisfy the conditions $0 < A < 1$, $0 < E < 1$, and $H > 1$. The conditional probability $q(x)$ is defined by

$$\frac{q(x)}{p(x)} = f(x),$$

where $p(x) = 1 - q(x)$. It is easy to show that

$$q(x) = \frac{f(x)}{1 + f(x)}, \qquad x = 0, 1, 2, \ldots.$$

Because every individual must die eventually, it is necessary that $q(x) \to 1$ as $x \to \infty$. Show that the above restrictions on the parameters imply that $q(x) \to 1$ as $x \to \infty$.

(b) An alternative parameterization of $q(x)$ considered by Heligman and Pollard was

$$q(x) = A^{(x+B)^C} + DE^{(\log Fx)^2} + \frac{GH^{x^K}}{1 + GH^{x^K}}.$$

The nine parameters $A, B, \ldots, K$ are all positive, with $A, E,$ and H satisfying the conditions $0 < A < 1$, $0 < E < 1$, and $H > 1$. Furthermore, the condition, $0 \le q(x) \le 1$, must be satisfied for all $x = 0, 1, 2, \ldots$. Observe that this condition is automatically satisfied by the q-probability defined in part (a). Show that, in this case, $q(x) \to 1$ as $x \to \infty$.

(c) To estimate the parameters in each case, Heligman and Pollard minimized the function

$$S^2 = \sum_{x=0}^{85} \left(\frac{\hat{q}(x)}{q(x)} - 1 \right)^2,$$

where $\hat{q}(x)$ was a life table estimate of the probability $q(x)$. These estimates were based on mortality rates for Australian males and females for the periods 1946–1948, 1960–1962, and 1970–1972. On the whole, these models seem to have worked quite well.

9. As a first step in demonstrating the validity of Eq. (2.9.6), show that

$$H_1(x) = e^{\alpha_1} \int_0^x e^{-\beta_1 (\log \gamma_1 s)^2} ds$$

$$= \gamma_1^{-1} e^{\alpha_1} \int_{-\infty}^{\log \gamma_1 x} e^{-\beta_1 y^2 + y} dy,$$

by putting $y = \log \gamma_1 s$ and making a change of variables in the integral on the left. Next, by completing the square, show that

$$\beta_1 \left(y^2 - \frac{y}{\beta_1} \right) = \beta_1 \left(y - \frac{1}{2\beta_1} \right)^2 - \frac{1}{4\beta_1}.$$

By substituting this relationship into the integral on the right, it can be seen, after some algebraic manipulation, that Eq. (2.9.6) is valid.

References

Acsádi, Gy. and Nemeskéri, J. (1970): *History of Human Life Span and Mortality*. Akademiai Kiado, Budapest

Arthur, W.B. and Stoto, M.A. (1981): An Analysis of Indirect Mortality Estimation. Paper presented at Annual Meeting of PAA, San Diego, California, 1982

Bennett, N.G. and Horiuchi, S. (1981): Estimating the Completeness of Death Registration in a Closed Population. *Population Index* 47:207–221

Brass, W. (1974): Perspectives in Population Prediction: Illustrated by the Statistics of England and Wales. *Jour. of Roy. Stat. Soc. A* 137 (4):532–583

Brass, W. and Coale, A.J. (1968): Methods of Analysis and Estimation in *Demography of Tropical Africa*. Princeton University Press, Princeton, N.J.

Chiang, C.L. (1968): *Introduction to Stochastic Processes in Biostatistics*. John Wiley and Sons, New York

Chiang, C.L. (1972): On Constructing Current Life Tables. *J. Amer. Statist. Assoc.* 67:538–541

Coale, A.J. and Demeny, P. (1966): *Regional Model Life Tables and Stable Populations*. Princeton University Press, Princeton, New Jersey

Cox, P.R. and Scott, W.F. (1977): International Studies in Generation Mortality. *Jour. of the Inst. of Actuaries* 104:297–333

Cramér, H. and Wold, H. (1935): Mortality Variations in Sweden – A Study in Graduation and Forecasting. *Skandinavisk Aktuarietidskrift*. 161–241

Elandt-Johnson, R.C. and Johnson, N.L. (1980): *Survival Models and Data Analysis*. John Wiley and Sons, New York

Feeney, G. (1980): Estimation of Mortality Trends from Child Survivorship Data. *Population Studies* 34:109–128

Forsythe, G.E., Malcolm, M.A., and Moler, C.B. (1977): *Computer Methods for Mathematical Computations*. Prentice-Hall, Inc., Englewood Cliffs, N.J.

Fritsch, F N. and Carlson, R.E. (1980): Monotone Piecewise Cubic Interpolation. *SIAM J. Numer. Anal.* 17:238–246

Gill, P.E. and Murray, W. (1978): Algorithms for the Solution of the Nonlinear Least-Squares Problem. *SIAM J. Numer. Anal.* 15:977–992

Gompertz, B. (1825): On the Nature of the Function Expressive of the Law of Mortality. *Philosophical Transactions* 27:513–585

Grenander, U. (1956): *On the Theory of Mortality Measurement*. Part I. *Skandinavisk Aktuarietidskrift* 71–95

Hobcraft, J., Menken, J. and Preston, S. (1982): Age, Period, and Cohort Effects in Demography – A Review. *Population Index* 48:4–43

Heligman, L. and Pollard, J.H. (1980): The Age Pattern of Mortality. *J. Inst. Actuaries* 107:49–80

Kennedy, W.J. and Gentle, J.E. (1980): *Statistical Computing*. Marcel Dekker, Inc., New York and Basel.

Keyfitz, N. (1977): *Applied Mathematical Demography*. John Wiley and Sons, New York

Keyfitz, N. (1981): Choice of Function for Mortality Analysis. *Theoretical Population Biology* 21:329–352

Keyfitz, N. and Flieger, W. (1968): *World Population – An Analysis of Vital Data*. The University of Chicago Press, Chicago and London

Makeham, W.M. (1867): On the Law of Mortality. *J. Inst. Actuaries* 13:325–367

Mode, C.J. and Busby, R.C. (1982): An Eight Parameter Model of Human Mortality – The Single Decrement Case. *The Bulletin of Mathematical Biology*. 44:647–659

Palloni, A. (1979): A New Technique to Estimate Infant Mortality with an Application for El Salvador and Colombia. *Demography* 16:455–473

Preston, S., Coale, A.J., Trussell, J. and Weinstein, M. (1980): Estimating the Completeness of Reporting of Adult Deaths in Populations That Are Approximately Stable. *Population Index* 46:179–202

Preston, S. and Hill, K. (1980): Estimating Completeness of Death Registration. *Population Studies* 34:349–366

Russel, J.C. (1958): *British Medieval Population.* American Philosophical Society, Philadelphia

Shryock, H., Siegel, J.S. and Associates (1973): *The Methods and Materials in Demography,* Chapter 14. U.S. Department of Commerce, Washington, D.C.

Smith, D. and Keyfitz, N. (1977): *Mathematical Demography – Selected Papers.* Springer-Verlag, Berlin, Heidelberg, New York

Stern, C. (1960): *Principles of Human Genetics.* W.H. Freeman and Company, San Francisco and London

Thiele, P.N. (1872): On a Mathematical Formula to Express the Rate of Mortality Throughout the Whole of Life. *J. Inst. Actuaries* 16:213–239

Trussel, T.J. (1975): A Re-estimation of the Multiplying Factors of the Brass Technique for Determining Childhood Survivorship Rates. *Population Studies* 29:97–107

Vaupel, J.W., Manton, K.G. and Stallard (1979): The Impact of Heterogeneity in Individual Frailty on the Dynamics of Mortality. *Demography* 16:439–454

Wegman, E.J., Nour, E.S. and Kukuk, C. (1980): A Time Series Approach to Life Table Construction. *Commun. in Stat.* – Theory and Methods A9(15):1587–1607

Yang, G.L. (1977): Life Table and Fraction of Last Age Interval. *Mathematical Biosciences* 33:289–295

Chapter 3. Theories of Competing Risks and Multiple Decrement Life Tables

3.1 Introduction

When death occurs, a natural human tendency is to attempt to ascribe it to some cause. A little reflection, however, leads one to the conclusion that it is not always easy to ascribe a death to a particular underlying cause, except perhaps in special circumstances of accidents or violence. Much remains to be discovered, in terms of genetic, behavioral, and environmental factors, as to why some people enjoy longer lives than others. Factors underlying individuality in human longevity will not be considered in this chapter; rather, attention will be focused on information that may gleaned from official mortality statistics.

In compiling mortality statistics, an operational cause of death is described in terms of the disease or injury initiating the train of events leading to death, or the circumstances of the accident or violence that produced the fatal injury. For purposes of statistical classification, death is assigned to one underlying cause, even if multiple causes are listed on the death certificate. This practice not only obscures the individuality of the circumstances underlying the death but also produces long lists of causes, making it necessary to group them into classes based on similar etiologies. It is beyond the scope of this chapter to provide a discussion of problems encountered in classifying causes of death. The interested reader may consult Preston, Keyfitz, and Schoen (1972) or discussions found in *Vital Statistics U.S. 1955, Supplement* (1965). Further discussion may be found in Alderson (1981).

In order to simplify the terminology, a set of causes in what follows will refer to some mutually exclusive classes of causes of death based on etiologic considerations. Such partitions of a set of causes give rise to the idea of competing risks, the notion that the causes somehow compete in a chain of events culminating in death. A primary purpose of this chapter is to present a discussion of models of competing risks. As will be seen in subsequent chapters, these models have applications beyond the analysis of mortality data.

Central to the analysis of mortality data is the question as to what survivorship pattern would result if a given cause or combination of causes were eradicated. Historically, the question arose in attempts to measure the impact on longevity of eliminating such controllable diseases as smallpox through vaccination. Another purpose of this chapter is to supply a discussion of methods used in giving answers to this question. Like many questions of scientific importance, it is much easier to ask than to give satisfactory answers.

Also included for discussion is some methodology for calculating multiple decrement life tables and associated single decrement tables from period data on causes of death. Statistically, these calculations are merely methods for the nonparametric estimation of certain functions in a simple model of competing risks. Since extensive listing of multiple decrement life tables are available in the literature, concepts will be illustrated, for the most part, by computer generated graphs. Brief attention will also be given to evolutionary models of competing risks.

3.2 Mortality in a Cohort with Competing Risks of Death

Just as in Sect. 2.2, imagine a cohort of individuals born at some epoch $t=0$ and suppose that each individual may be followed throughout his entire life span. Furthermore, suppose that with each individual there is associated a pair of numbers $\langle x, j \rangle$, where $x \in R^+$ is his age at death and j, $1 \leq j \leq m$, represents the cause of death. In its simplest form, the theory of competing risks in concerned with the distribution of the pair of random variables $\langle X, C \rangle$, representing variation in the pairs $\langle x, j \rangle$ among members of a cohort.

As a first step toward developing a theory of competing risks consider

$$P[X \leq x, \ C = j] = G_j(x), \tag{3.2.1}$$

for $x \in R^+$ and $1 \leq j \leq m$. Observe that $G_j(x)$ is the probability that death due to cause j occurs before age x. The distribution function of the random variable X is the marginal probability

$$G(x) = \sum_{j=1}^{m} P[X \leq x, \ C = j] = G_1(x) + G_2(x) + \ldots + G_m(x). \tag{3.2.2}$$

Similarly, the marginal distribution of the random variable C is

$$P[C = j] = \lim_{x \uparrow \infty} G_j(x) = \pi_j \geq 0, \tag{3.2.3}$$

the probability of dying due to cause j. Because all deaths are ascribed to some cause, it follows that the condition,

$$\pi_1 + \pi_2 + \ldots + \pi_m = 1, \tag{3.2.4}$$

is satisfied. From (3.2.2) and (3.2.4), it follows that $G(x) \to 1$ as $x \to \infty$ so that $G(\cdot)$ is a proper distribution function.

When there are $m \geq 2$ competing risks of death, the survival function in a cohort may be decomposed additively into m parts. As will be shown in a subsequent section, each part corresponds to a column in a multiple decrement life table. The decomposition in question is a consequence of the probabilistic identity

$$P[X > x] = \sum_{j=1}^{m} P[X > x, \ C = j], \tag{3.2.5}$$

which is valid for all $x \in R^+$. By definition the survival function is $S(x) = P[X > x]$ and if $H_j(x)$ is defined by

$$H_j(x) = P[X > x, C = j], \tag{3.2.6}$$

then (3.2.5) becomes

$$S(x) = H_1(x) + H_2(x) + \ldots + H_m(x), \quad x \in R^+. \tag{3.2.7}$$

The identity

$$P[X \leq x, C = j] + P[X > x, C = j] = P[C = j] \tag{3.2.8}$$

is the basis for the equations

$$G_j(x) + H_j(x) = \pi_j, \tag{3.2.9}$$

$1 \leq j \leq m$ and $x \in R^+$, which are also useful in calculations associated with multiple decrement life tables.

With the cause of death, there are associated forces of mortality $\mu_j(x)$, $1 \leq j \leq m$, defined by

$$\mu_j(x) = \lim_{h \downarrow 0} \frac{G_j(x+h) - G_j(x)}{h S(x)} = \frac{g_j(x)}{S(x)} \tag{3.2.10}$$

at those points $x \in R^+$, where the functions $G_j(x)$ have derivatives $G_j'(x) = g_j(x)$ and $S(x) \neq 0$. The total force of mortality $\mu(x)$ is defined just as in Eq. (2.2.4). Therefore, from (2.2.4), (3.2.2), and (3.2.10), it follows that

$$\mu(x) = \mu_1(x) + \mu_2(x) + \ldots + \mu_m(x). \tag{3.2.11}$$

This additive property of the forces of mortality associated with the causes of death plays a basic role in models of competing risks. It is important to observe that (3.2.11) is a logical consequence of Eq. (3.2.2).

Also implicit in Eq. (3.2.11) is the property that the survival function $S(x)$ may be factored into a product of survival functions at each $x \in R^+$. For $1 \leq j \leq m$ and $x \in R^+$, let

$$S_j(x) = \exp\left[-\int_0^x \mu_j(s) ds \right]. \tag{3.2.12}$$

Then, from (3.2.11), it follows that

$$S(x) = \exp\left[-\int_0^x \mu(s) ds \right] = S_1(x) S_2(x) \ldots S_m(x), \tag{3.2.13}$$

for $x \in R^+$. Equation (3.2.13) not only plays a basic role in models of competing risks but also yields a necessary and sufficient condition for the distribution function $G(\cdot)$ to be proper. Because $G(\cdot)$ is proper if, and only if, $S(x) \to 0$ as $x \to \infty$, it follows from (3.2.12) and (3.2.13) that $G(\cdot)$ is proper if, and only if,

$$\lim_{x \uparrow \infty} \int_0^x \mu_j(s) ds = \infty, \tag{3.2.14}$$

for some $1 \leq j \leq m$.

The last task considered in this section is that of expressing the function $G_j(x)$ defined in (3.2.1) in terms of the force of mortality $\mu_j(x)$. Given that an individual is alive at age y, $0 \leq y \leq x$, $\mu_j(y)dy$ is approximately the conditional probability of dying in a short time interval $(y, y+dy]$ due to cause j. The probability of being alive at age y is $S(y)$. Hence,

$$G_j(x) = \int_0^x S(y)\mu_j(y)dy, \qquad (3.2.15)$$

for $1 \leq j \leq m$ and $x \in R^+$.

The history of the development of ideas underlying competing risks is interesting. Some historical material may be found in David and Moeschberger (1978) and Birnbaum (1979).

3.3 Models of Competing Risks Based on Latent Life Spans

As a mathematical device for supplying answers to the question as to what survivorship pattern might result if some cause of death were eliminated, models of competing risks based on the concept of latent life spans have been considered. The basic notions underlying such models can easily be understood in terms of a physical system composed of $m \geq 1$ parts. If any part of the system fails, the whole system fails or "dies". It is also supposed that an observer can determine which part of the system failed first.

Let $Y_1, Y_2, \ldots, Y_m$ be random variables representing the latent life spans of the parts of the system. Then X, the life span of the system, is the waiting time till the first failure. Hence,

$$X = \min(Y_1, Y_2, \ldots, Y_m). \qquad (3.3.1)$$

Failure of the system is due to cause j, i.e., $C=j$, if, and only if, $X=Y_j$, $1 \leq j \leq m$. Although the analogy between human survivorship in the presence of competing causes of death and the above physical system is an oversimplification, models of competing risks based on the concept of latent life spans are useful nevertheless, because they help clarify methodological issues and provide clues for further development.

All properties of a model of competing risks based on latent life spans are derived in terms of the joint distribution of the latent random variables $Y_1, Y_2, \ldots, Y_m$. Let R_m^+ be the set of all vectors $\mathbf{y} = \langle y_1, y_2, \ldots, y_m \rangle$ with $y_i \in R^+$, $1 \leq i \leq m$. The joint survival function is defined by

$$S^*(\mathbf{y}) = P[Y_1 > y_1, Y_2 > y_2, \ldots, Y_m > y_m] \qquad (3.3.2)$$

for all points $\mathbf{y} \in R_m^+$. It will be assumed that $S^*(\mathbf{y})$ is a proper joint survival function in the sense that

$$S^*(\mathbf{0}) = P[R_m^+] = 1. \qquad (3.3.3)$$

The j-th marginal survival function is defined by

$$S_j^*(y_j) = P[Y_j > y_j] = S^*(0, \ldots, y_j, \ldots, 0) \tag{3.3.4}$$

for $y_j \in R^+$, $1 \leq j \leq m$. The latent life spans $Y_1, Y_2, \ldots, Y_m$ are independent if, and only if, the factorization

$$S^*(\mathbf{y}) = S_1^*(y_1) S_2^*(y_2) \ldots S_m^*(y_m) \tag{3.3.5}$$

holds for all points $\mathbf{y} \in R_m^+$. Further properties of joint survival functions are discussed in Problem 1.

The idea of a force of mortality, risk function, was first defined in terms of a survival function of a single variable, see Eq. (2.2.5). A natural extension of this notion to survival functions of several variables is the idea of a risk gradient. Define a function $R(\mathbf{y})$ by

$$R(\mathbf{y}) = -\log S^*(\mathbf{y}) \tag{3.3.6}$$

for all points $\mathbf{y} \in R_m^+$ such that $S^*(\mathbf{y}) \neq 0$. Let the functions $\eta_j(\mathbf{y})$, $1 \leq j \leq m$, be defined by

$$\eta_j(\mathbf{y}) = \frac{\partial R(\mathbf{y})}{\partial y_j} \tag{3.3.7}$$

for all points $\mathbf{y} \in R_m^+$ where the partial derivative on the right exists. Then, the risk gradient is the vector-valued function

$$\boldsymbol{\eta}(\mathbf{y}) = \langle \eta_1(\mathbf{y}), \eta_2(\mathbf{y}), \ldots, \eta_m(\mathbf{y}) \rangle. \tag{3.3.8}$$

Among the many uses of the risk gradient is an extention of formula (2.2.6) to a survival function of several variables. From multivariable calculus, it is known that, under rather general conditions, $R(\mathbf{y})$ may be expressed as the line integral

$$R(\mathbf{y}) = \int_0^{\mathbf{y}} \boldsymbol{\eta}(\mathbf{s}) d\mathbf{s}, \tag{3.3.9}$$

where

$$\boldsymbol{\eta}(\mathbf{s}) d\mathbf{s} = \eta_1(\mathbf{s}) ds_1 + \ldots + \eta_m(\mathbf{s}) ds_m$$

and the integral is taken along a path connecting the points $\mathbf{0}$ and $\mathbf{y}$. A discussion of properties and line integrals and their probabilistic implications is contained in Problems 2 and 3. From (3.3.6) and (3.3.9), it can be seen that the desired extention of formula (2.2.6) is

$$S^*(\mathbf{y}) = \exp\left[-\int_0^{\mathbf{y}} \boldsymbol{\eta}(\mathbf{s}) d\mathbf{s} \right]. \tag{3.3.10}$$

As one would expect, there is a connection between the forces of mortality $\mu_j(x)$, $1 \leq j \leq m$, defined (3.2.10) and the components of the risk gradient $\boldsymbol{\eta}(\mathbf{y})$ when the joint distribution of the random variables $\langle X, C \rangle$ is induced by the latent random variables $Y_1, Y_2, \ldots, Y_m$. As a first step in deriving this connection, for every $1 \leq j \leq m$, $\mathbf{y} \in R_m^+$, and $h > 0$, let $A_j(h, \mathbf{y})$ be the event that

$y_j < Y_j \leq y_j + h$ and $Y_k > y_k$ for all $k \neq j$, where $y_1, y_2, \ldots, y_m$ are the components of $\mathbf{y}$. Let $B(\mathbf{y})$ be the event that $Y_j > y_j$ for all j. Then, an alternative way of viewing Eq. (3.3.7) is

$$\eta_j(\mathbf{y}) = \lim_{h \downarrow 0} \frac{1}{h} P[A_j(h, \mathbf{y}) | B(\mathbf{y})]. \tag{3.3.11}$$

When X is defined as in (3.3.1), then $X > x$ if, and only if, $Y_j > x$ for all j, $1 \leq j \leq m$. Therefore, the survival function of X is

$$S(x) = P[X > x] = S^*(x, x, \ldots, x)$$
$$= P[B(x\mathbf{1})], \quad x \in R^+, \tag{3.3.12}$$

where $x\mathbf{1} = \langle x, x, \ldots, x \rangle$. With X so-defined, it can also be seen that

$$\frac{G_j(x+h) - G_j(x)}{S(x)} = P[A_j(h, x\mathbf{1}) | B(x\mathbf{1})]. \tag{3.3.13}$$

Therefore, from this equation, (3.2.10), and (3.3.11), it follows that

$$\mu_j(x) = \eta_j(x\mathbf{1}), \quad 1 \leq j \leq m. \tag{3.3.14}$$

Formula (3.2.15) for the function $G_j(x)$ continues to hold when the functions $S(x)$ and $\mu_j(x)$ are given as in (3.3.12) and (3.3.14).

An extensive theory of the reliability of physical systems made up of many parts has been reported in the monograph by Barlow and Proschan (1975). Prentice et al. (1978) seem to have been the first to use the term, latent, to describe the random variables $Y_1, Y_2, \ldots, Y_m$.

3.4 Simple Parametric Models of Competing Risks

As an aid to understanding the concepts introduced in Sects. 3.2 and 3.3, two illustrative parametric examples will be given in this section.

Example 3.4.1. *Constant Forces of Mortality*

Suppose the risk functions $\mu_j(x)$ defined in (3.2.10) are constant. Then, there are nonnegative parameters λ_j such that

$$\mu_j(x) = \lambda_j \tag{3.4.1}$$

for all $x \in R^+$ and $1 \leq j \leq m$. In this case, the survival function on the left in (3.2.7) takes the form

$$S(x) = \exp[-\lambda x], \quad x \in R^+, \tag{3.4.2}$$

where $\lambda = \lambda_1 + \lambda_2 + \ldots + \lambda_m > 0$. From (3.2.15), it follows that

$$G_j(x) = \int_0^x e^{-\lambda y} \lambda_j \, dy = \frac{\lambda_j}{\lambda}(1 - \exp[-\lambda x]), \tag{3.4.3}$$

for $x \in R^+$. Hence,

$$\lim_{x \uparrow \infty} G_j(x) = \pi_j = \lambda_j/\lambda, \qquad 1 \le j \le m. \tag{3.4.4}$$

The function $H_j(x)$ defined in (3.2.6) has the simple form

$$H_j(x) = \pi_j S(x), \qquad x \in R^+. \tag{3.4.5}$$

Example 3.4.2. *A Survival Function on* R_3^+

Given the theory of latent life spans developed in Sect. 3.3, it is easy to construct illustrative examples of survival functions on R_m^+. At least two approaches may be followed. Either the function $R(\mathbf{y})$ defined in (3.3.6) or the risk gradient $\boldsymbol{\eta}(\mathbf{y})$ in (3.3.8) may be specified. In many cases, the simplest approach is to specify the function $R(\mathbf{y})$.

Consider, for example, a function $R(\mathbf{y})$ defined on R_3^+ by

$$R(\mathbf{y}) = \lambda_1 y_1 + \lambda_2 y_2 + \lambda_3 y_3 + \lambda_{12} y_1 y_2$$
$$+ \lambda_{13} y_1 y_3 + \lambda_{23} y_2 y_3 + \lambda_{123} y_1 y_2 y_3, \tag{3.4.6}$$

with the λ's being nonnegative parameters. In this case, the joint survival function of the latent random variables $Y_1, Y_2, \dots, Y_m$ is

$$S^*(\mathbf{y}) = \exp[-R(\mathbf{y})], \qquad \mathbf{y} \in R_3^+. \tag{3.4.7}$$

The first component of the risk gradient is

$$\eta_1(\mathbf{y}) = \lambda_1 + \lambda_{12} y_2 + \lambda_{13} y_3 + \lambda_{123} y_2 y_3. \tag{3.4.8}$$

The survival function of the random variable $X = \min(Y_1, Y_2, Y_3)$ is

$$S(x) = S^*(x\mathbf{1}) = \exp[-\delta_1 x + \delta_2 x^2 + \delta_3 x^3)], \tag{3.4.9}$$

with $x \in R^+$, $\delta_1 = \lambda_1 + \lambda_2 + \lambda_3$, $\delta_2 = \lambda_{12} + \lambda_{13} + \lambda_{23}$, and $\delta_3 = \lambda_{123}$.

Other properties of the survival function $S^*(\mathbf{y})$ in (3.4.7) are developed in Problem 4. When $\delta_2 = \delta_3 = 0$, the survival function in (3.4.9) reduces to that in (3.4.2) for $m = 3$. Observe that a model of competing risks with constant forces of mortality is equivalent to one based on independent latent life spans with exponential distributions. Another example of a joint survival function with marginal exponential distributions is given in Problem 5.

3.5 Equivalent Models of Competing Risks

An interesting equivalence phenomenon arises when models of competing risks based on latent life spans are considered. Let $\langle X, C \rangle$ be the pair of random variables in a competing risk model induced by the latent random variables $Y_1, Y_2, \dots, Y_m$ as in Sect. 3.3. Similarly, let the pair of random variables be induced by the latent random variables $Z_1, Z_2, \dots, Z_m$. Two models of competing risks are said to be equivalent if the vector random variables $\langle X, C \rangle$ and $\langle X^*, C^* \rangle$ have the same distribution.

A more formal definition of the idea of equivalence is the following. Let

$$G_j(x) = P[X \leq x, C = j]$$

and

$$G_j^*(x) = P(X^* \leq x, C^* = j). \tag{3.5.1}$$

Then, two competing risk models are equivalent if, and only if,

$$G_j(x) = G_j^*(x) \tag{3.5.2}$$

for all $x \in R^+$ and $1 \leq j \leq m$.

The following theorem is of basic importance in the theory of models of competing risks based on latent life spans. As will be seen, it is a direct consequence of the factorization property, see (3.2.13).

Theorem 3.5.1. Every model of competing risks based on dependent latent life spans $Y_1, Y_2, \ldots, Y_m$ is equivalent to one based on independent latent life spans $Z_1, Z_2, \ldots, Z_m$.

Proof: Let $S^*(\mathbf{y})$ be the joint survival function of the latent random variables $Y_1, Y_2, \ldots, Y_m$. Let $Z_1, Z_2, \ldots, Z_m$ be independent latent random variables with the survival functions

$$S_j(z) = \exp\left[-\int_0^z \mu_j(x)dx\right], \tag{3.5.3}$$

where $1 \leq j \leq m$, $\mu_j(x)$ is defined as in (3.3.14), and $\eta_j(\mathbf{y})$ is the j-th component of the risk gradient associated with $S^*(\mathbf{y})$. By independence, the joint survival function of the random variables $Z_1, Z_2, \ldots, Z_m$ is

$$S^{**}(\mathbf{z}) = S_1(z_1)S_2(z_2) \ldots S_m(z_m), \tag{3.5.4}$$

with $\mathbf{z} = \langle z_1, z_2, \ldots, z_m \rangle$.

As above, let $\langle X, C \rangle$ and $\langle X^*, C^* \rangle$ be the pairs of latent random variables induced, respectively, by the latent random variables $Y_1, Y_2, \ldots, Y_m$ and $Z_1, Z_2, \ldots, Z_m$. Then, the survival functions of X and X^* are

$$S(x) = P[X > x] = S^*(x\mathbf{1})$$

and $\hspace{10cm} (3.5.5)$

$$S^0(x) = P[X^* > x] = S^{**}(x\mathbf{1}).$$

But, by the factorization property, see (3.2.13), and (3.5.4), it follows that

$$S(x) = S^0(x) \tag{3.5.6}$$

for all $x \in R^+$. Since the two models of competing risks have the same forces of mortality $\mu_j(x)$ associated with the j-th cause of death, it follows from (3.5.6) and formula (3.2.15) that Eq. (3.4.2) is satisfied for all $x \in R^+$ and $1 \leq j \leq m$. Hence, the pairs of random variables $\langle X, C \rangle$ and $\langle X^*, C^* \rangle$ have the same distribution. $\square$

The material presented in this section represents a simplification and a reworking of ideas contributed by several other workers. Further details may be found in Cox (1962), Chap. 10, Tsiatis (1975), and Elandt-Johnson and Johnson (1980), Chap. 9. Langberg et al. (1978) have developed a general theory for converting dependent models into independent ones.

3.6 Eliminating Causes of Death and Nonidentifiability

Having outlined some basic concepts underlying models of competing risks based on latent life spans, answers to the question of what survivorship pattern would result if one or more causes of death were eliminated can be formulated. Actually, in view of the many factors affecting human mortality, factors whose change in time seem to be governed by an unpredictable evolutionary process, it is beyond the scope and power of the simple mathematical models considered so far to give completely satisfactory answers to the question. What seems to be needed is a deeper basic understanding of the many factors affecting human survivorship so that steps could be taken toward the development and implementation of more realistic models. Despite the oversimplifications embodied in models of competing risks based on latent life spans, formulating answers to the question within the framework of these models brings the methodological issues into sharper focus.

As in (3.3.2), let $S^*(\mathbf{y})$ be the joint survival function of the latent random variables $Y_1, Y_2, \ldots, Y_m$. Rather than considering a general case, the concepts will be illustrated for the case $m=3$. Within the present framework, answers to the question under consideration are formulated in terms of the marginal survival functions of the joint survival function $S^*(y_1, y_2, y_3)$. Suppose, for example, the third cause of death were eliminated. Then, it is assumed that the marginal survival function,

$$S_{12}^*(y_1, y_2) = S^*(y_1, y_2, 0),\qquad (3.6.1)$$

governs survivorship. Proceding as in (3.3.12), $S(3, x)$, the survival function for a cohort if the third cause of death were eliminated, is given by

$$S(3, x) = S_{12}^*(x, x).\qquad (3.6.2)$$

The marginal survival functions $S_{13}^*(y_1, y_2)$ and $S_{23}^*(y_2, y_2)$ with the corresponding cohort survival functions $S(2, x)$ and $S(1, x)$ are defined similarly. If the second and third causes of death were eliminated so that the first would be the only cause, then the cohort survival function, $S(23, x)$, would be

$$S(23, x) = S^*(x, 0, 0) = S_1^*(x).\qquad (3.6.3)$$

The second and third causes acting alone would be treated similarly.

Although the above answers to the question under consideration are easily formulated, it is difficult to express them in numerical form, because the functional form of the joint survival function $S^*(\mathbf{y})$, $\mathbf{y} \in R_m^+$, cannot usually be determined from available information. What an investigator usually has available are multiple decrement life table estimates, based on period data, of the functions $H_j(x)$, or equivalently, the functions $G_j(x)$, for $1 \le j \le m$, see Eq. (3.2.9). In principle, given estimates of the functions $G_j(x)$, the force of mortality $\mu_j(x)$ associated with the j-th cause of death may be estimated, see Eq. (3.2.10). Further details will be developed in subsequent sections of this chapter.

Armed with estimates of the forces of mortality, an investigator can then estimate the survival functions in (3.2.12). Thus, in principle, the joint survival function $S^{**}(\mathbf{z})$ in (3.5.4) for the independent latent random variables $Z_1, Z_2, ..., Z_m$ in Theorem 3.5.1 can be estimated. In general, however, the joint survival function $S^*(\mathbf{y})$ of the dependent latent random variables $Y_1, Y_2, ..., Y_m$ cannot be uniquely determined from a knowledge of the functions $G_j(x)$, $1 \le j \le m$, i.e., from the distribution of the vector random variable $\langle X, C \rangle$. The inability to make this determination is known as the problem of nonidentifiability.

Nádas (1971) and Basu and Ghosh (1978) have studied this problem from a theoretical point of view. Nádas showed that for the case $m = 2$ and the latent life spans follow a bivariate normal distribution, the distribution of $\langle X, C \rangle$ completely determines the five parameters. Of course, the same result holds if the latent life spans follow a bivariate lognormal distribution. Basu and Ghosh showed that the result of Nádas cannot be extended to the case $m = 3$ and gave results for other joint distributions, including a two-dimensional version of the one considered in Example 3.4.2.

Due to the fact that the joint survival function of the independent latent random variables $Z_1, Z_2, ..., Z_m$ can be estimated from available information, it is frequently used to estimate survival patterns that would result if a given cause or a combination of causes of death were eliminated. Consider, for example, the case $m = 3$ and let $S^0(3, x)$ be the survival function for a cohort if the third cause of death were eliminated under the assumption of independent latent life spans. Then,

$$S^0(3, x) = S^{**}(x, x, 0) = S_1(x) S_2(x), \qquad x \in R^+. \tag{3.6.4}$$

It can be shown by example that the functions in (3.6.2) and (3.6.4) are not identical. As one would expect, the forms of the functions $S(3, x)$ and $S^0(3, x)$ depend very much on whether one assumes the latent life spans are dependent or independent, even though the models from which these functions are derived are equivalent in the sense of Theorem 3.5.1.

Before presenting a concrete example, it should be pointed out that, although a knowledge of the risk functions $\mu_j(x)$, $1 \le j \le m$, does not determine the joint survival function of the latent random variables $Y_1, Y_2, ..., Y_m$ in Theorem 3.5.1, it at least supplies partial information on the risk gradient $\boldsymbol{\eta}(\mathbf{y})$ in (3.3.8), because of Eq. (3.3.14). Suppose, for example, an investigator specifies the joint survival function $S^*(\mathbf{y})$ in terms of a risk gradient, using Eq. (3.3.10). Given estimates of the forces of mortality $\mu_j(x)$, $1 \le j \le m$, Eq. (3.3.14) could be used, as a partial check on the validity of a proposed model based on dependent latent life spans, to see whether exploratory graphs of the functions

$\eta_j(x\mathbf{1})$, $x \in R^+$, conformed roughly to the empirical graphs of those of the functions $\mu_j(x)$, $1 \leq j \leq m$.

It should also be mentioned that the methods of formulating answers to the question under consideration discussed in this section is not exhaustive. Other methods have been proposed. Elandt-Johnson (1976), for example, proposed a scheme based on certain limiting conditional distributions, see Sect. 5 of her paper. Kimball (1958, 1969), has proposed methods for eliminating causes of death based on conditioning the multinomial distribution. For a review on critique of models used in competing risk analysis, see Gail (1975).

Example 3.6.1. *A Case where the Functions $S(3, x)$ and $S^0(3, x)$ Differ*

Consider again the survival function on R_3^+ discussed in Example 3.4.2. By applying formula (3.6.2) and using (3.4.6), it is easy to see that

$$S(3, x) = \exp[-(\delta_{31}x + \delta_{32}x^2)], \qquad x \in R^+, \tag{3.6.5}$$

where $\delta_{31} = \lambda_1 + \lambda_2$ and $\delta_{32} = \lambda_{12}$.

By computing the components of the risk gradient and using (3.3.14), it can also be shown that

$$\mu_1(x) = \lambda_1 + (\lambda_{12} + \lambda_{13})x + \lambda_{123}x^2$$

and   (3.6.6)

$$\mu_2(x) = \lambda_2 + (\lambda_{12} + \lambda_{23})x + \lambda_{123}x^2.$$

Therefore, from (3.6.4), it follows that

$$S^0(3, x) = \exp\left[-\int_0^x (\mu_1(y) + \mu_2(y))dy\right]$$

$$= \exp[-(\xi_{31}x + \xi_{32}x^2 + \xi_{33}x^3)], \qquad x \in R^+, \tag{3.6.7}$$

where $\xi_{31} = \lambda_1 + \lambda_2$, $\xi_{32} = \lambda_{12} + (\lambda_{13} + \lambda_{23})/2$, and $\xi_{33} = \frac{2}{3}\lambda_{123}$. The functions in (3.6.5) and (3.6.7) are manifestly different.

Example 3.6.2. *A Graphical Example Comparing the Survival Functions $S(x)$, $S(3, x)$, and $S^0(3, x)$*

Although the functions in (3.6.5) and (3.6.7) are clearly different from an analytical point of view, it is difficult to get insights into the magnitudes of the differences for all values of $x \in R^+$ unless numerical examples are studied. Consequently, a graphical example will be considered by assigning hypothetical values to the nine parameters. As a basis for measuring the impact of eliminating the third cause of death, the graph of the survival function $S(x)$ defined by Eq. (3.4.9), with all three causes operational, will also be plotted.

A simple hypothetical scheme was used to reduce the dimension of the parameter space from nine to three by expressing the interaction parameters λ_{12}, λ_{13}, λ_{23}, and λ_{123} as functions of the base parameters λ_1, λ_2, and λ_3. The particular functional relationships chosen were $\lambda_{12} = (\lambda_1\lambda_2)/2$, $\lambda_{13} = (\lambda_1\lambda_3)/2$, $\lambda_{23} = (\lambda_2\lambda_3)/2$, and $\lambda_{123} = (\lambda_1\lambda_2\lambda_3)/3$. The rationale for making these choices con-

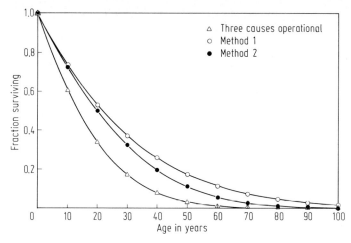

Fig. 3.6.2. A comparison of two methods for eliminating a cause of death

sisted of merely wanting to consider a case of weak interaction among the causes that could be succinctly described. Interaction is weak in the sense that if the parameters λ_1, λ_2, and λ_3 have values in the interval $(0, 1)$, then all interaction parameters are smaller than the base parameters. The values chosen for the base parameters were $\lambda_1 = \frac{1}{70}$, $\lambda_2 = \frac{1}{65}$, and $\lambda_3 = \frac{1}{60}$. If either cause 1, 2, or 3 were the only one operational, then these choices would correspond to expectations of life at birth of 70, 65, and 60 years, respectively.

Presented in Fig. 3.6.2 are the computer generated graphs of the functions (3.4.9), (3.6.5), and (3.6.7) for the assigned parameter values. The graph labeled – three causes operational – is that of the function $S(x)$ in (3.4.9); the graph with the designation – method 1 – is that of the function $S(3, x)$ in (3.6.5). Finally, the graph with the designation – method 2 – is that for the function $S^0(3, x)$ in (3.6.7).

As one would expect, the graph of $S(x)$, with all three causes operational, decreases more rapidly than those for $S(3, x)$ and $S^0(3, x)$ with cause 3 eliminated according to different assumptions. If $S(3, x)$ is viewed as the true survival function for cause 3 were eliminated, then from the graphs of $S(3, x)$ and $S^0(3, x)$ it may be seen that the assumption of independent latent life spans, when, in fact, they are dependent, leads to an underestimation of survivorship. That is, for the parameter values considered, the inequality, $S^0(3, x) \leq S(3, x)$ seems valid for all $x \in R^+$. By comparing the shapes of the graphs in Fig. 3.6.2 with those in Figs. 2.7.1 and 2.7.2, it can be seen that the model under consideration is not a realistic one for human survivorship.

3.7 Estimating a Multiple Decrement Life Table from Period Data

As an aid to understanding basic concepts, the theory of competing risks has been expressed in terms of a hypothetical cohort. In practice, however, the functions $H_j(x)$, $1 \le j \le m$, defined in (3.2.6), are estimated from period data, using extensions of the algorithms discussed in Sects. 2.4 and 2.5. It will become apparent in what follows that these extensions will not depend on any assumptions regarding an underlying model of competing risks based on latent life spans.

As a first step in understanding the basic ideas used in deriving these extensions to multiple decrement life table estimation, suppose r is the maximum age in a population; and, let $0 = x_0 < x_1 < \ldots < x_n = r$ be $n \ge 1$ numbers determining a partition $I_v = (x_{v-1}, x_v]$, $1 \le v \le n$, of the interval $(0, r]$. Let q_{vj} be the conditional probability of dying due to cause j at age $x \in I_v$, given survival to age x_{v-1}. Then,

$$q_{vj} = \frac{G_j(x_v) - G_j(x_{v-1})}{S(x_{v-1})}, \tag{3.7.1}$$

for $1 \le j \le m$ and $1 \le v \le n$. Given survival to age x_{v-1},

$$q_v = q_{v1} + q_{v2} + \ldots + q_{vm} \tag{3.7.2}$$

is the conditional probability of dying at some age $x \in I_v$. The conditional probability in (3.7.2) is, in fact, that defined in (2.4.1) and used in a single decrement life table. Just as in Eq. (2.4.3), $S(x_0) = S(0) = 1$ and $S(x_v) = p_1 p_2 \ldots p_v$, $1 \le v \le n$, where $p_v = 1 - q_v$. Therefore, if we let $P_0 = 1$ and $P_v = p_1 p_2 \ldots p_v$ for $1 \le v \le n$, then, from (3.7.1) it follows that

$$S(x_{v-1}) q_{vj} = P_{v-1} q_{vj} = G_j(x_v) - G_j(x_{v-1}) \tag{3.7.3}$$

is the probability of dying at some age $x \in I_v$ due to cause j.

The basic ideas underlying multiple decrement life table procedures center around getting estimates of the conditional probabilities in (3.7.1) and (3.7.2) from period data. By using these estimates to estimate $P_{v-1} q_{vj}$ in (3.7.3), estimates of the functions $H_j(x) = P[X > x, C = j]$ at the points x_v, $1 \le v \le n$, may be obtained from the relationship

$$H_j(x_{v-1}) = G_j(x_v) - G_j(x_{v-1}) + \ldots + G_j(x_n) - G_j(x_{n-1}), \tag{3.7.4}$$

which is valid for $1 \le j \le m$ and $1 \le v \le n$. At this point, it should be noted that no assumptions concerning latent life spans have been used in the derivation of (3.7.4).

Usually, the calculations suggested by the above formulas are carried out for abridged life tables of the type described in Sect. 2.5. Thus, age intervals of the general form $(x, x+h]$, $h > 0$, are objects of attention. More specifically, the age intervals $(0, 1]$, $(1, 5]$, $(5, 10]$, $\ldots$, $(80, 85]$, $(85, \infty)$ are considered and it is supposed that an investigator has counts $D_j(x, h)$ of the number of deaths due to the j-th cause occurring in the age interval $(x, x+h]$ during a given period. The total number $D(x, h)$ of deaths at ages in the interval $(x, x+h]$ occurring during the period is

$$D(x, h) = D_1(x, h) + D_2(x, h) + \ldots + D_m(x, h). \tag{3.7.5}$$

In addition to this information, it is also supposed that an investigator has estimates of $K(x, h)$, the number of persons in the age interval $(x, x+h]$

exposed to risk of death during the period, as well as estimates of the fractions $f(x, h)$ in formula (2.5.9).

A first step in calculating a multiple decrement life table is frequently that of calculating the associated single decrement life table, using procedures outlined in Sect. 2.5. Let $\hat{q}(x, h)$ be an estimate of the conditional probability in (2.5.4) obtained by substituting $\hat{m}(x, h)$ in (2.5.10) into (2.5.9). Then, $\hat{q}(x, h)$ may be represented in the form

$$\hat{q}(x, h) = \frac{D(x, h)}{N(x)}, \tag{3.7.6}$$

where $N(x) = h^{-1}[K(x, h) + h(1 - f(x, h))D(x, h)]$. Sometimes $N(x)$ is called the effective number of lives at age x. In analogy with (3.7.1), let $q_j(x, h)$ be the conditional probability of dying from the j-th cause during the age interval $(x, x+h]$, given that an individual is alive at age x. Then, $q_j(x, h)$ is estimated by

$$\hat{q}_j(x, h) = \frac{D_j(x, h)}{N(x)}. \tag{3.7.7}$$

From (3.7.6) and (3.7.7), it follows that

$$\hat{q}_j(x, h) = \frac{D_j(x, h)}{D(x, h)} \hat{q}(x, h), \tag{3.7.8}$$

for $1 \leq j \leq m$.

For the case of abridged multiple decrement life tables, the above formulas may be applied as follows in estimating the probabilities in (3.7.4). A first step could consist of calculating the estimates $\hat{q}(x, h)$ and $\hat{q}_j(x, h)$ in (3.7.6) and (3.7.8). By using the estimates $\hat{q}(x, h)$ and the procedures outlined in Sect. 2.5, estimates $\hat{S}(x)$ of the values of the survival function at the ages $x = 0, 1, 5, 10, \ldots, 85$ could be obtained. Then, in view of (3.7.3),

$$\hat{d}_j(x, h) = \hat{S}(x)\hat{q}_j(x, h) \tag{3.7.9}$$

is an estimate of $d_j(x, h) = G_j(x+h) - G_j(x)$, the probability of dying due to the j-th cause during the age interval $(x, x+h]$. Thus, from (3.7.4) and (3.7.9), it may be seen that

$$\hat{H}_j(x) = \hat{d}_j(x, h) + \hat{d}_j(x+h, h) + \ldots + \hat{d}_j(85, \infty) \tag{3.7.10}$$

are the desired period estimates of the functions $H_j(x) = P[X > x, \ C = j]$, $1 \leq j \leq m$, at the ages $x = 0, 1, 5, 10, \ldots, 85$.

Mention should be made of the special procedure that is followed in calculating the estimate $\hat{d}_j(85, \infty)$ on the right in (3.7.10). Having survived till age 85, death at some age past 85 is certain. It makes sense, therefore, to let $\hat{q}(85, \infty) = 1$ in (3.7.8) so that $\hat{q}_j(85, \infty) = D_j(85, \infty)/D(85, \infty)$, $1 \leq j \leq m$. Observe that these estimates do not require an estimate of the number of persons past 85 at risk of death during a given period.

A detailed numerical example, taken from the mortality experience of U.S. white males in 1970, has been provided in Chap. 10 of Elandt-Johnson and

Johnson (1980). These authors use an actuarial notation and also provide estimates of functions other than that in (3.7.10). An extensive compilation of multiple decrement life tables covering many countries over several periods have been published by Preston, Keyfitz, and Schoen (1972). Selected multiple decrement life tables will be presented in graphical form in a subsequent section of this chapter.

3.8 Estimating Single Decrement Life Tables from Multiple Decrement Life Tables

From the point of view of the theoretical framework developed in this chapter, the topic to be discussed in this section is that of estimating marginal survival functions associated with the independent latent random life spans $Z_1, Z_2, \ldots, Z_m$ in Theorem 3.5.1. Consider, for example, the survival function $S_j(z)$ of the latent random variable Z_j in (3.5.3). From (3.2.10) it follows that

$$S_j(z) = \exp\left[-\int_0^z \frac{g_j(x)}{S(x)}\, dx\right] = \exp\left[-\int_0^z \frac{G_j(dx)}{S(x)}\right], \tag{3.8.1}$$

for $z \in R^+$. The Stieltjes integral in the exponent on the right is defined as follows. For any $z \in R^+$, let $0 = x_0 < x_1 < \ldots < x_n = z$ be numbers such that $\Delta_n = \max_{1 \le v \le n} (x_v - x_{v-1}) \to 0$ as $n \to \infty$. Then, provided the limit exists,

$$\int_0^z \frac{G_j(dx)}{S(x)} = \lim_{n \to \infty} \sum_{v=1}^n \frac{G_j(x_v) - G_j(x_{v-1})}{S(x_{v-1})}$$

$$= \lim_{n \to \infty} \sum_{v=1}^n q_{vj}, \tag{3.8.2}$$

see (3.7.1). This result suggests that if an investigator had life table estimates $\hat{q}_{vj} = (\hat{G}_j(x_v) - \hat{G}_j(x_{v-1})/S(x_{v-1})$ for a sufficiently fine mesh of points $0 = x_0 < x_1 < \ldots, x_n = r$, for some $n > 1$, then the survival function in (3.8.1) could be estimated by the approximations

$$\hat{S}_j(x_k) = \exp\left[-\sum_{v=1}^k \hat{q}_{vj}\right] \tag{3.8.3}$$

for $1 \le k \le n$.

Next suppose an investigator has estimates $\hat{S}(x_k)$ of the survival function in (3.3.12) at the points (x_k), $1 \le k \le n$, with all m causes operational, i.e., acting simultaneously. Under the assumption of independent latent life spans, it follows that

$$\hat{S}^0(j, x_k) = \frac{\hat{S}(x_k)}{\hat{S}_j(x_k)}, \qquad 1 \le k \le n, \tag{3.8.4}$$

would be an estimate of the survival function at these points if the j-th cause of death, $1 \le j \le m$, were eliminated. In some compilations of multiple decrement life tables, the estimates in (3.8.3) are listed.

In some data sets, information is available on a sufficiently fine mesh of points to make the above estimation techniques useful. But, when the mesh of points is coarse, as in abridged life tables, these techniques would not be satisfactory. The next problem to be discussed is that of extending to foregoing ideas to abridged multiple decrement life tables.

To make the desired extension, it will be convenient to have a succinct notation. Let $\mathbf{A} = [i : 1 \le i \le m]$ be the set of the first $m \ge 1$ positive integers. Subsets of $\mathbf{A}$ will be denoted by $\mathbf{a}$. If $m = 2$, then the subsets of $\mathbf{A}$ are φ, the empty set, plus the sets (1), (2), and (1, 2). For $m = 3$, the number of subsets of $\mathbf{A}$ is $2^3 = 8$. For any nonempty subset $\mathbf{a} \in \mathbf{A}$, let $q(\mathbf{a}, x, h)$ be the conditional probability that an individual alive at age x dies in the age interval $(x, x+h]$, under the assumptions of independent latent life spans and only the causes in the set $\mathbf{a}$ are operational. If $\mathbf{a} = \mathbf{A}$, then $q(\mathbf{a}, x, h) = q(x, h)$, the conditional probability estimated in (3.7.6) under no assumptions on the latent life spans. When the set $\mathbf{a}$ is the singleton $\mathbf{a} = (j)$, the conditional probability $q(\mathbf{a}, x, h)$ will be denoted by $q(j, x, h)$.

In general terms, the problem of estimating single decrement life tables from a multiple decrement one may be described as follows. For any nonempty set $\mathbf{a} \in \mathbf{A}$, let $S^0(\mathbf{a}, x)$ be the survival function if the causes in the set $\mathbf{a}$ were eliminated, under the assumption of independent latent life spans. Given life table estimates $\hat{q}(x, h)$ and $\hat{q}_j(x, h)$, $1 \le j \le m$, see (3.7.6), (3.7.7), and (3.7.8), the problem is that of deriving estimates $\hat{S}^0(\mathbf{a}, x)$ of the survival function $S^0(\mathbf{a}, x)$ at the points $x = 0, 1, 5, 10, \ldots, 85$.

The algorithm for estimating the survival function in a single decrement life table will again be useful. For any nonempty set $\mathbf{a} \in \mathbf{A}$, let $p(\mathbf{a}, x, h) = 1 - q(\mathbf{a}, x, h)$ be the conditional probability that an individual alive at age x is still alive at age $x + h$, under the assumptions of independent latent life spans and only the causes in the set $\mathbf{a}$ are operational. If the causes in the nonempty set $\mathbf{a} \in \mathbf{A}$ were eliminated, then only those causes in the complementary set $\mathbf{a}^c$ would be operational. Therefore, $S^0(\mathbf{a}, 0) = 1$, $S^0(\mathbf{a}, 1) = p(\mathbf{a}^c, 1, 1)$, $S^0(\mathbf{a}, 5) = S^0(\mathbf{a}, 1)p(\mathbf{a}^c, 1, 4)$, and in general the recursion relation

$$S^0(\mathbf{a}, x) = S^0(\mathbf{a}, x - 5)p(\mathbf{a}^c, x - 5, 5), \qquad (3.8.5)$$

is valid for $x = 10, 15, \ldots, 85$. From (3.8.5) it is clear that the problem reduces to deriving estimates of the conditional probabilities $p(\mathbf{a}, x, h)$ for any nonempty set $\mathbf{a} \in \mathbf{A}$.

Several methods of estimating the conditional probabilities $p(\mathbf{a}, x, h)$ have been proposed. Only one of these methods, which has also been used by Chiang (1968) and others, will be described here. A basic assumption characterizing this method is that of local proportionality of the forces of mortality $\mu_j(x)$ defined in (3.2.10). According to this assumption, for every age interval $(x, x+h]$, there are constants $c_j(x, h)$ such that

$$\frac{\mu_j(y)}{\mu(y)} = c_j(x, h) \qquad (3.8.6)$$

for all $y \in (x, x+h]$ and $1 \leq j \leq n$. In (3.8.6), $\mu(y)$ is the total force of mortality, see (3.2.11). It is natural to doubt whether this assumption is a sound one, particularly if h is large. Experience has shown, however, that the assumption seems to lead to reasonable results for standard abridged life tables.

To derive an equation connecting $p(\mathbf{a}, x, h)$ with $p(x, h)$ the case in which the set $\mathbf{a}$ is the singleton (j) will be considered first. With only the j-th cause of death operational,

$$p(j, x, h) = P[Z_j > x+h \mid Z_j > x]$$

$$= \exp \left[- \int_x^{x+h} \mu_j(y) dy \right]$$

$$= \exp \left[-c_j(x, h) \int_x^{x+h} \mu(y) dy \right], \tag{3.8.7}$$

the last step being an implication of (3.8.6). But,

$$p(x, h) = \exp \left[- \int_x^{x+h} \mu(y) dy \right]. \tag{3.8.8}$$

Therefore, from (3.8.7) and (3.8.8) it may be seen that

$$p(j, x, h) = [p(x, h)]^{c_j(x, h)}, \tag{3.8.9}$$

for $1 \leq j \leq m$.

Equation (3.8.9) is useful because the constants $c_j(x, h)$ may be expressed in terms of the conditional probabilities $q_j(x, h)$ and $q(x, h)$, which are estimated during the course of calculating an abridged multiple decrement life table with all causes of death operational. The desired expression may be derived by first observing that

$$G_j(x+h) - G_j(x) = \int_x^{x+h} S(y) \mu_j(y) dy$$

$$= c_j(x, h) \int_x^{x+h} S(y) \mu(y) dy$$

$$= c_j(x, h)(G(x+h) - G(x)). \tag{3.8.10}$$

Then, by dividing this equation by $S(x) \neq 0$ and solving for $c(x, h)$, it can be seen that

$$c_j(x, h) = \frac{q_j(x, h)}{q(x, h)}, \quad 1 \leq j \leq m. \tag{3.8.11}$$

This result is not surprising in view of the assumption that the functions $\mu_j(y)$, $1 \leq j \leq m$, are assumed to be in constant proportion on the interval $(x, x+h]$.

Equation (3.8.9) may easily be extended to any nonempty subset $\mathbf{a} \in A$. For let

$$c(\mathbf{a}, x, h) = \sum_j c_j(x, h), \quad j \in \mathbf{a}. \tag{3.8.12}$$

Then, by using arguments similar to those above, it may be shown that
Eq. (3.8.9) becomes

$$p(\mathbf{a}, x, h) = [p(x, h)]^{c(\mathbf{a}, x, h)}. \tag{3.8.13}$$

The method just outlined is appealing because it is simple to apply. Suppose, for example, that the life table estimates $\hat{q}_j(x, h)$, $1 \leq j \leq m$, and $\hat{q}(x, h)$ are available with all causes operational and no assumptions on the latent life spans. Then,

$$\hat{c}_j(x, h) = \frac{\hat{q}_j(x, h)}{\hat{q}(x, h)}, \quad 1 \leq j \leq m, \tag{3.8.14}$$

are the estimates of the constants in (3.8.11). If $\mathbf{a}$ is the nonempty set of causes to be eliminated, then $\hat{c}(\mathbf{a}^c, x, h)$ is computed by substituting the estimates in (3.8.14) into (3.8.12). This estimate and $\hat{p}(x, h) = 1 - \hat{q}(x, h)$ is then substituted into (3.8.13) to calculate the estimates $p(\mathbf{a}^c, x, h)$ at $x = 0, 1, 5, 10, \ldots, 80$. Finally, these estimates are used recursively in (3.8.5) to estimate $S^0(\mathbf{a}, x)$ at $x = 0, 1, 5, 10, \ldots, 85$.

Elandt-Johnson and Johnson (1980) in their Chap. 11 have described four other methods, in addition to that just described, for calculating a single decrement life table associated with a multiple decrement table. They also gave an example comparing the five methods. In their example, the five methods of calculation produced very similar results. A simple example, applying formula (3.8.13) for one value of x, is given in Problem 6.

3.9 Evolutionary Changes in the Structure of Causes of Death

In their extensive compilation of national multiple decrement life tables, Preston, Keyfitz, and Schoen (1972) used twelve broad classifications of causes of death. Included in their classification was the cause, other and unknown. It would be of interest to study actual evolutionary changes in the structure of causes of death as represented by changes in the period estimates of the parameters $\pi_1, \pi_2, \ldots, \pi_m$ in (3.2.3), the probability that a newborn will eventually die of the various causes. In many national period life tables, however, the estimated fraction of death assigned to other and unknown causes constitutes a substantial part of the whole. Consequently, time series representations of period estimates of the π-probabilities may not be informative with respect to the actual evolutionary changes in the structure of the causes of death that have occurred in a population.

Despite substantial numbers of deaths being assigned to other and unknown causes, interesting relationship between the expectation of life at birth and the structure of the π-probabilities may be demonstrated, see for example Tables I-1a and I-1b in Preston et al. Among the national tables compiled by these authors, those for United States had some of the lowest estimates of the percentage of deaths attributable to other and unknown causes for the period considered, making the study of evolutionary changes in the π-probabilities meaningful. To reduce the tables to a manageable size, a four-way classification of causes of death was used. The classification, Neoplasms and Cardiovascular, were included because they were among the largest probabilities of death in 1964, the last period reported; the classification, Influenza-Pneumonia-Bronchitis, was included, because the diseases in this class were among the leading causes of death in 1900, the earliest period reported. Finally, a class called, Residual, representing all other causes of death, was obtained by subtraction.

Before discussing Tables 3.9.1 and 3.9.2, it should be mentioned that Preston et al. used a method of their own devising for calculating multiple decrement life tables rather than the one outlined in Sect. 3.7. There are good reasons for believing, however, that if the method of Sect. 3.7 had been used, the resulting tables would not have differed substantially.

From an inspection of Tables 3.9.1 and 3.9.2 several interesting trends become apparent. As the expectation of life at birth increased for both males

Table 3.9.1. Evolutionary changes in the structure of causes of death – U.S. males 1900–1964

Period	Expectation of life at birth	Neo-plasms	Cardio-vascular	Influenza-Pneumonia-Bronchitis	Residual
1900	45.65[a]	3.37[b]	19.06[b]	13.64[b]	63.93[b]
1910	49.49	4.88	24.94	11.84	58.34
1920	54.01	6.70	29.01	14.61	49.68
1930	57.58	8.46	34.78	8.82	47.94
1940	61.02	10.54	41.89	6.53	41.04
1950	65.46	13.34	56.16	3.31	27.19
1960	66.65	15.05	57.46	4.13	23.36
1964	66.91	15.70	56.88	3.68	23.74

[a] Expressed in years.
[b] Period estimate of probability that a newborn eventually dies of a given cause expressed in percent.

Table 3.9.2. Evolutionary changes in the structure of causes of death – U.S. females 1900–1964

Period	Expectation of life at birth	Neo-plasms	Cardio-vascular	Influenza-Pneumonia-Bronchitis	Residual
1900	48.35[a]	6.27[b]	19.74[b]	15.30[b]	58.69[b]
1910	53.13	8.73	26.93	13.05	51.29
1920	55.92	9.74	30.86	15.17	44.23
1930	61.01	11.94	36.98	8.71	42.37
1940	65.43	13.61	43.73	6.75	35.91
1950	71.03	15.33	59.78	3.47	21.42
1960	73.26	15.36	62.91	4.02	17.71
1964	73.78	15.34	63.04	3.55	18.07

[a] Expressed in years.
[b] Period estimate of probability that a newborn eventually dies of a given cause expressed in percent.

and females, the percentages of deaths falling in the classes, Neoplasms and Cardiovascular, representing diseases associated with aging, steadily increased. Accompanying this increase was a decrease in the percentages of deaths attributable to causes in the class, Influenza-Pneumonia-Bronchitis, a decrease due in large part to effective methods of treatment such as antibiotics. Another interesting trend exhibited in the tables was the steady decline in deaths assigned to the class, Residual. This decline was due in part to increasing accuracy in identifying causes of death. From Tables 3.9.1 and 3.9.2, it can also be seen that the greater the expectation of life at birth, the higher the percentage of deaths due to Cardiovascular diseases. This observation suggests that significant future gains in human longevity, in populations such as that of the U.S., should come from the prevention and effective control of cardiovascular diseases.

3.10 Graphs of Multiple Decrement Life Tables – A Study of Proportional Forces of Mortality

As indicated in Sect. 3.7, the columns of an abridged multiple decrement life table consist, in part, of period estimates $\hat{H}_j(x)$ of the functions $H_j(x)$, defined in (3.2.6), at the ages $x = 0, 1, 5, 10, \ldots, 85$. These estimates are useful and interesting in their own right, but comparisons among tables are difficult, because of the voluminous information they contain. Visual comparisons can be greatly facilitated if multiple decrement life tables are represented in graphical form. Graphical forms of tables are also very useful in exploring the plausibility of simplifying assumptions underlying attempts at formulating parsimonious parametric models of competing risks. The purpose of this section is to display partial results from two experiments in computer graphics. In one experiment, two selected multiple decrement life tables were represented in graphical form; in a second, the plausibility of a model of competing risks with proportional forces of mortality was explored.

The two multiple decrement life tables that were selected for graphical representation were those for U.S. males and females in 1964 as listed by Preston, Keyfitz, and Schoen (1972). Presented in Fig. 3.10.1 is a graph of the table for males; while that for females is presented in Fig. 3.10.2. The markings on the graph represent the actual listed values taken from the tables. A monotone interpolation procedure, due to Fritsch and Carlson (1980), was again used to insert eye-pleasing points between the actual values. In both graphs, the classes, Influenza-Pneumonia-Bronchitis and Residual, in Tables 2.9.1 and 2.9.2 have been combined to form a new class, which is again labeled, Residual. Corresponding to the label, All Causes, in Figs. 3.10.1 and 3.10.2 are estimates $\hat{S}(x)$ of the survival function $S(x)$ on the left in (3.2.7). Therefore, at each data point, the additive property displayed in Eq. (3.2.7) holds for the estimates.

Sex differentials in U.S. multiple decrement life tables for 1964 may easily be seen by viewing Figs. 3.10.1 and 3.10.2. According to Tables 3.9.1 and 3.9.2, the expectation of life at birth was 66.91 and 73.78, respectively, for U.S. males and females in 1964. A reflection of this difference in expectations is that the graphs of the estimate $\hat{S}(x)$ of the survival function $S(x)$ for females lies above that for males, especially at ages past 40. The graphs of the estimates of the H-functions corresponding to the classes, Cardiovascular and Residual, are also uniformly higher for females than for males. One of the most striking properties of the curves in Figs. 3.10.1 and 3.10.2 is the lack of sex differentials in the

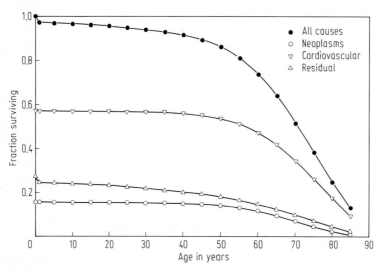

Fig. 3.10.1. Graph of multiple decrement life table – U.S. males 1964

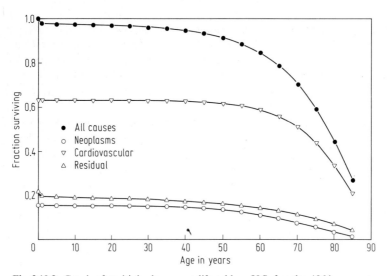

Fig. 3.10.2. Graph of multiple decrement life table – U.S. females 1964

estimated graphs of the H-functions corresponding to the class, Neoplasms. Thus, under a broad classification, the U.S. experience for 1964 suggest that males and females are about equally susceptible to cancer.

A model of competing risks is said to have proportional forces of mortality if there are nonnegative constants $c_1, c_2, \ldots, c_m$ such that

$$\mu_j(x) = c_j \mu(x) \tag{3.10.1}$$

for all $x \in R^+$ and $1 \leq j \leq m$. If the implications of (3.10.1) have empirical validity, the task of finding useful parametric models of competing risks would be greatly simplified. For if condition (3.10.1) is satisfied, then

$$G_j(x) = \int_0^x S(y)\,\mu_j(y)\,dy = c_j \int_0^x S(y)\,\mu(y)\,dy = c_j\,G(x), \tag{3.10.2}$$

for all $x \in R^+$ and $1 \leq j \leq m$. Moreover, if $G(x)$ is a proper distribution function so that $G(x) \to 1$ as $x \to \infty$, it follows from (3.2.3) that $c_j = \pi_j$, $1 \leq j \leq m$. Hence, the constants of proportionality in (3.10.1) may be identified with the probabilities of eventually dying from the various causes.

Furthermore, the equation $G_j(x) = \pi_j\,G(x)$ together with (3.2.9) imply that

$$H_j(x) = \pi_j\,S(x) \tag{3.10.3}$$

for all $x \in R^+$ and $1 \leq j \leq m$. Another implication of (3.10.1) is that the survival functions in (3.2.12), corresponding to the independent latent life spans in models of competing risks discussed in Theorem 3.5.1, take the form

$$S_j(x) = [S(x)]^{\pi_j}, \tag{3.10.4}$$

for all $x \in R^+$ and $1 \leq j \leq m$.

Equations (3.10.3) and (3.10.4) make clear why the task of finding useful parametric models of competing risks would be greatly simplified if the implications of condition (3.10.1) have empirical validity. For, in principle, if good parameterized forms of a survival function $S(x)$ were available, such as those discussed in Sect. 2.9, then a parametric model of competing risks could be completely determined by specifying the parameters $\pi_1, \pi_2, \ldots, \pi_m$. A useful application of such parameterized models would be that of computing model multiple decrement life tables and the associated single decrement tables.

Given a set of multiple decrement life tables stored in a computer, Eq. (3.10.3) may be used to derive experimental checks on the validity of proportionality assumption (3.10.1). For example, to compute estimates of the H-function corresponding to the class, Neoplasms, for U.S. males in 1964 under assumption (3.10.1), the survival function labeled, All Causes, in Fig. 3.10.1, an estimate of $S(x)$, was multiplied by 0.157, the 1964 period estimate of the probability that a newborn will eventually die from a disease belonging to the class, Neoplasms; see Table 3.9.1. In Fig. 3.10.3, the resulting curve is labeled, Neo Pro; the matching observed curve, based on values taken from the life table, is labeled, Neo Obs. The curves labeled, Car Pro, Car Obs, Res Pro, and Res Obs in Table 3.10.3 have corresponding interpretations for the classifications, Cardiovascular and Residual. It should be recalled, however, that to compute the estimates for the curve labeled, Res Pro, the curve labeled,

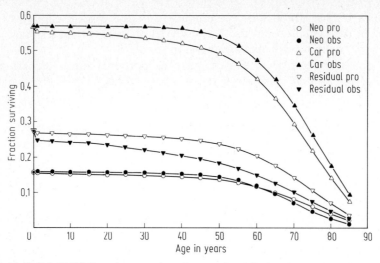

Fig. 3.10.3. Validation of assumption of proportional forces of mortality – U.S. males 1964

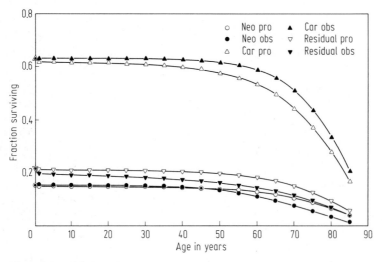

Fig. 3.10.4. Validation of assumption of proportional forces of mortality – U.S. females 1964

All Causes, in Fig. 3.10.1 was multiplied by $0.2742 = 0.0368 + 0.2374$, representing the combination of two classifications; see the last row of Table 3.9.1. The corresponding curves for U.S. females in 1964 are presented in Fig. 3.10.4. In order to produce eye-pleasing curves, the monotone interpolation procedure was again used to insert points between observed values.

From inspection of Figs. 3.10.3 and 3.10.4, it can be seen that the curves calculated under the assumption of proportional forces of mortality do not always closely match the corresponding observed curves. In both sexes, the match is closest for the class, Neoplasms, especially at ages less than 40. It is also apparent that the values of the H-function, estimated under the assumption of proportionality for the class, Cardiovascular, lie below the observed curve in both sexes. For the class, Residual, the values of the H-function estimated under condition (3.10.1) lie above the observed curve in both sexes. A possibly significant observation, for the two classes just discussed, is that the curves calculated under the proportionality assumption would match the observed curves more closely if the discrepancies in the first year were removed. This observation suggests that if condition (3.10.1) were assumed to hold only for ages $x \geq 1$, then the calculated and observed curves would match more closely.

In conclusion, according to U.S. mortality experience in 1964, the proportionality assumption in (3.10.1) will lead to a rough approximation of reported multiple decrement life tables, with respect to the classification of causes of death in Figs. 3.10.3 and 3.10.4. Whether model life tables calculated under assumption (3.10.1) prove useful will depend, in part, on the extent to which an investigator is willing to trade off sacrifices in accuracy for simplicity in conceptualization.

3.11 Graphs of Single Decrement Life Tables Associated with Multiple Decrement Tables

In terms of the discussion in Sect. 3.6, single decrement life tables associated with multiple decrement tables refer to listings of values of survival functions that would result when a given cause, or combination of causes, is eliminated, under the assumption of independent latent life spans. The purpose of this section is to display computer generated graphs of the survival functions that would result if causes corresponding to the classifications, Neoplasm, Cardiovascular, and Residual, as used in Sect. 3.10, were eliminated. Specifically, if these classes are labeled 1, 2, and 3, respectively, and $S_1(x)$, $S_2(x)$, and $S_3(x)$ are the latent survival functions defined in (3.2.12), then the functions plotted were

$$S^0(1, x) = S_2(x) S_3(x),$$
$$S^0(2, x) = S_1(x) S_3(x), \tag{3.11.1}$$

and

$$S^0(3, x) = S_1(x) S_2(x).$$

As a basis for comparison, the survival function

$$S(x) = S_1(x) S_2(x) S_3(x), \tag{3.11.2}$$

with all causes operational, was also plotted.

The portions of the single decrement life tables chosen for graphical display were those for U.S. males and females in 1964 as listed by Preston, Keyfitz, and Schoen (1972). Taken directly from these authors were the survival functions, $S^0(1, x)$ and $S^0(2, x)$, that would result if the classes, Neoplasms and Cardiovascular, were eliminated. The survival function, $S^0(3, x)$, resulting from the elimination of the class, Residual, was calculated according to the formula

$$S^0(3, x) = \frac{[S(x)]^2}{S^0(1, x)\, S^0(2, x)}.$$ (3.11.3)

Plotted in Fig. 3.11.1 are the graphs of the survival functions in question for males; those for females are plotted in Fig. 3.11.2. The techniques described in Sect. 3.10 were again used to generate these graphs. As expected, the estimated graphs of the survival functions corresponding to the classes, Neoplasms, Cardiovascular, and Residual, eliminated lie above that of non-eliminated, i.e., the function in (3.11.2). Beyond age 60, the estimated graph of the function $S^0(2, x)$ lies above all others in both sexes, clearly illustrating the impact of eliminating the class, Cardiovascular, the leading causes of death. At ages less than 50, the estimated graph of the function $S^0(3, x)$ lies above all others in both sexes, indicating that the causes of death in the class, Residual, have their greatest impact among younger people. Included in this class are deaths due to violence and motor vehicle accidents.

Estimated years of life added to life expectancy at birth if a given cause of death were eliminated are always of interest. According to Tables 3.9.1 and 3.9.2, the expectations of life at birth for U.S. males and females in 1964 were 66.91 and 73.78 years, respectively. According to calculations listed by Preston et al., if the causes in the class, Neoplasms, were eliminated, these expectations would increase to 69.18 and 76.34 for the two sexes; if the causes in the class, Cardiovascular, were eliminated, the corresponding expectations would increase to 80.21 and 90.85. As yet, there seems to be no national population with an expectation of life at birth equal to 90.85 years for females.

The value of the graphs in Figs. 3.11.1 and 3.11.2 lies not so much in their being an actual representation of a reality that would exist if a given class of diseases were eliminated as in exploring hypothetical forms of reality with

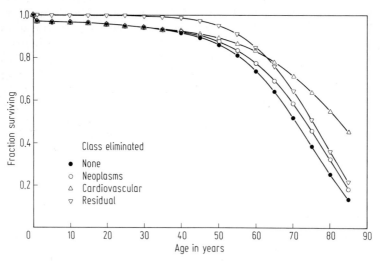

Fig. 3.11.1. Survival functions resulting from elimination of classes of diseases – U.S. males 1964

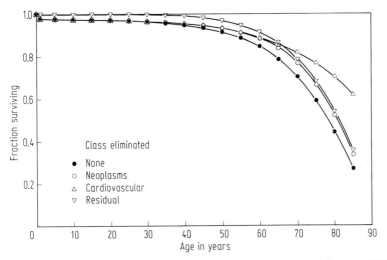

Fig. 3.11.2. Survival functions resulting from elimination of classes of diseases – U.S. females 1964

some plausibility. Unlike deaths due to infectious agents that can be averted by administering antibiotics, vaccinations, innoculations, or other control measures, discussions about eliminating deaths assigned to the classes, Neoplasms and Cardiovascular, deaths that are associated with a complex of genetic, environmental, and behavioral factors, must be tempered with caution.

3.12 Graphs of Latent Survival Functions and Forces of Mortality

When estimates of the survival functions $S^0(1, x)$, $S^0(2, x)$, $S^0(3, x)$, and $S(x)$, discussed in the preceding section, are available, it is a simple matter to calculate estimates of the latent survival functions, $S_1(x)$, $S_2(x)$, and $S_3(x)$, appearing on the right in (3.11.2). These estimates are of interest for at least two reasons. Firstly, keeping in mind the limited plausibility of statements about only one class of causes of death being operational in the real world, graphs of latent survival functions do provide some insights into the impact of each classification on longevity. Secondly, from the point of view of constructing parametric models of competing risks, it is of interest to isolate the survival functions associated with the independent latent life spans described in Theorem 3.5.1.

Presented in Fig. 3.12.1 are graphs of the latent survival functions for U.S. males in 1964; those for U.S. females in 1964 are presented in Fig. 3.12.2. In both figures, the survival function $S(x)$ in (3.11.2), with all causes operational, was plotted as a basis for comparisons. Each graph of a latent survival function is labeled by a specific class of causes of death; while the graph of that in (3.11.2) is labeled, All Causes. The techniques described in Sect. 3.10 were again used to generate these figures.

Of the three latent survival functions plotted in Figs. 3.12.1 and 3.12.2, that for the class, Cardiovascular, being the only causes of death is the most plausible relative to real-world experience. For in both sexes, the graph of this latent survival function has about the same shape as that for all causes, although it lies uniformly above it as expected. According to the latent survival functions for the classes, Neoplasms and Residual, if only the causes of death in these classes were operational, as much as 90% of a female birth cohort would still be living at age 85. Rates of survivorship this high are beyond the recorded mortality experience of national populations.

As part of a preparatory and exploratory process leading to the construction of parametric models of competing risks, graphs of forces of mortality are also of interest. An attempt was made, therefore, to graph the forces of mortality associated with the classes of causes of death under consideration, utilizing information available in abridged multiple decrement life tables for U.S. males and females in 1964. Mathematically, the functions under consideration are those defined in (3.2.10), having the additive property displayed in (3.2.11). Since estimates of the H-functions defined in (3.2.6) were available only at the ages $x=0,1,5,10,\ldots,85$, crude estimates of these functions were the only ones possible.

As a first step in computing estimates of the forces of mortality on a finer age scale, the monotone interpolation procedure, previously mentioned, was used to interpolate the H-functions, as well as the survival function $S(x)$ for all causes, down to a yearly age scale $x=0,1,2,\ldots,85$. Then, the conditional probabilities,

$$q_j(x)=\frac{H_j(x)-H_j(x+1)}{S(x)} \tag{3.12.1}$$

and

$$q(x)=\frac{S(x)-S(x+1)}{S(x)}, \tag{3.12.2}$$

were calculated for $1\leq j\leq 3$ and $x=0,1,2,\ldots,80$. It is actually these conditional probabilities that were plotted in Figs. 3.12.3 and 3.12.4, respectively, for U.S. males and females in 1964. After interpolation, plots of these conditional probabilities provide a crude picture of the graphs of the forces of mortality. Observe that the q-probabilities in (3.12.2) play the role of the total force of mortality on the left in (3.2.11).

If an investigator chooses to specify a parametric model of competing risks in terms of forces of mortality, some tentative clues regarding their construction may be gleaned from the exploratory graphs presented in (3.12.3) and (3.12.4). One of these clues is that a force of mortality for deaths in the early years should be associated with the class, Residual, since the force of mortality for this class was the only one in both sexes that was essentially nonzero at age zero. A second clue is that the force of mortality associated with the class, Cardiovascular, appears to be Gompertzian, since it rises steeply after age 60 in both sexes. The forces of mortality associated with the classes, Neoplasms and Residual, may not be Gompertzian, however, because they rise rather

slowly in both sexes after age 60. This observation suggests that non-Gompertz-
ian forces of mortality, other than those discussed in Sect. 2.9, may have to
be introduced when attempting to associate a parametric force of mortality
with each class of causes of death. Even if parametric models of competing
risks based on dependent latent life spans were introduced, graphs, such as
those in Figs. 3.12.3 and 3.12.4, will provide clues, through Eqs. (3.3.14), as to
what necessary conditions the functions in the risk gradient must satisfy.

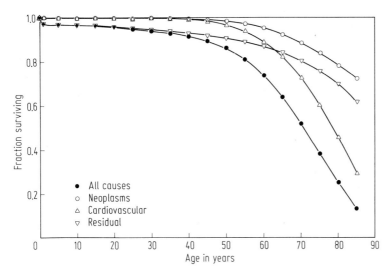

Fig. 3.12.1. Latent survival functions – U.S. males 1964

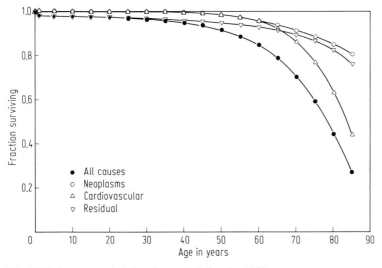

Fig. 3.12.2. Latent survival functions – U.S. females 1964

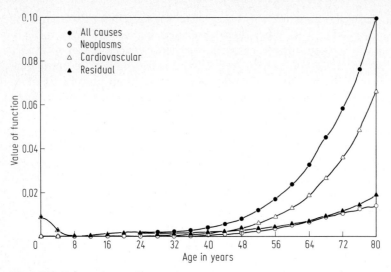

Fig. 3.12.3. Forces of mortality – U.S. males 1964

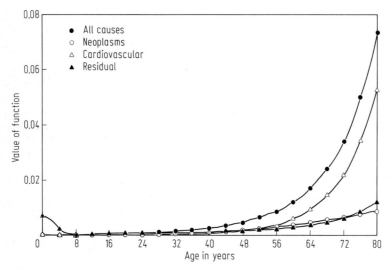

Fig. 3.12.4. Forces of mortality – U.S. females 1964

3.13 An Evolutionary Model of Competing Risks

Just as in the single decrement case, forces of mortality may change with time in models of competing risks. Accordingly, the purpose of this section is to provide a brief and tentative outline of an evolutionary model of competing risks that is a generalization of the structure discussed in Sect. 2.7.

Suppose that for each epoch $t \in R$ there is a force of mortality $\mu_j(t, x)$, $x \in R^+$, associated with the j-th cause of death, $1 \leq j \leq m$. At each epoch $t \in R$, it will be supposed that the additive property

$$\mu(t, x) = \mu_1(t, x) + \mu_2(t, x) + \ldots + \mu_m(t, x), \tag{3.13.1}$$

holds for all $x \in R^+$. The function $\mu(t, x)$ on the left in (3.13.1) may be identified as the force of mortality defined in Sect. 2.7.

A period survival function $S_p(t, x)$ may again be defined as in (2.7.4) at each epoch $t \in R$. Furthermore, the period functions $G_{pj}(t, x)$ and $H_{pj}(t, x)$, the analogues of those defined in (3.2.1) and (3.2.6), take the forms

$$G_{pj}(t, x) = \int_0^x S_p(t, y) \mu_j(t, y) \, dy$$

and   (3.13.2)

$$H_{pj}(t, x) = \int_x^\infty S_p(t, y) \mu_j(t, y) \, dy,$$

for $t \in R$, $x \in R^+$, and $1 \leq j \leq m$.

For a cohort born at epoch $t \in R$, a survival function $S_c(t, x)$ may be defined as in (2.7.3). The analogues of the formulas in (3.13.2) for cohorts take the forms

$$G_{cj}(t, x) = \int_0^x S_c(t, y) \mu_j(t + y, y) \, dy$$

and   (3.13.3)

$$H_{cj}(t, x) = \int_x^\infty S_c(t, y) \mu_j(t + y, y) \, dy,$$

for $t \in R$, $x \in R^+$, and $1 \leq j \leq m$.

In an evolutionary model of competing risks, it becomes necessary to distinguish between period and cohort probabilities of dying due to a given cause of death. At epoch $t \in R$, the period probability of dying due to the j-th cause is

$$\pi_{pj}(t) = \lim_{x \to \infty} G_{pj}(t, x), \tag{3.13.4}$$

for $1 \leq j \leq m$. For a cohort born at epoch $t \in R$, the probability of dying due to the j-th cause is

$$\pi_{cj}(t) = \lim_{x \to \infty} G_{cj}(t, x). \tag{3.13.5}$$

When the forces of mortality in (3.13.1) are changing in time, the probabilities in (3.13.4) and (3.13.5) could be substantially different. It is also of interest to observe that the equations

$$\pi_{p1}(t) + \pi_{p2}(t) + \ldots + \pi_{pm}(t) = 1$$

and   (3.13.6)

$$\pi_{c1}(t) + \pi_{c2}(t) + \ldots + \pi_{cm}(t) = 1$$

hold for all $t \in R$, if, and only if, conditions (2.7.5) are satisfied.

Problems and Miscellaneous Complements

1. Equations of type (3.3.4), expressing marginal survival functions in terms of a joint survival function, can easily be understood in terms of special cases.

(a) Suppose, for example, the random variables Y_1 and Y_2, taking values in R_2^+, have the joint probability density function $f(y_1, y_2)$. Then, the joint survival function of Y_1 and Y_2 is

$$S^*(y_1, y_2) = \int_{y_1}^{\infty} \int_{y_2}^{\infty} f(s_1, s_2)\, ds_2\, ds_1, \quad \langle y_1, y_2 \rangle \in R_2^+.$$

(b) For $y_1 \in R^+$, the marginal density function of Y_1 is

$$f_1(y_1) = \int_0^{\infty} f(y_1, s_2)\, ds_2$$

and the survival function of Y_1 is

$$S(y_1) = \int_{y_1}^{\infty} f_1(s_1)\, ds.$$

(c) From (a) and (b), it follows that

$$S^*(y_1, 0) = \int_{y_1}^{\infty} \int_0^{\infty} f(s_1, s_2)\, ds_2\, ds_1 = S_1(y_1), \quad y_1 \in R^+.$$

(d) If the joint survival function of the nonnegative random variables $Y_1, Y_2, \ldots, Y_m$ is considered for any $m \geq 2$, then the above ideas can be easily extended to any marginal survival function corresponding to any subset of these random variables. In general, a marginal survival function of a subset of random variables is determined from the joint survival function by substituting zeros in the complementary set. The assumption that these random variables have a density is not necessary for this procedure to be valid.

2. Line integrals of the type occurring in (3.3.9) are discussed in many text books on calculus in terms of applications in physics. Kolman and Trench (1971), in their Chap. 4, have provided a lucid discussion of these integrals. Such standard text books as Thomas (1972) also contain pertinent discussions. In this complement, the basic concepts will be illustrated in terms of simple examples.

(a) Line integrals of the form appearing in (3.3.9) are evaluated by expressing a curve connecting the points $\mathbf{0}$ and $\mathbf{y}$ in parametric form. For example, suppose $m = 2$, $\mathbf{y} = \langle 1, 1 \rangle$, and the risk gradient is $\eta_1(y_1, y_2) = y_1$ and $\eta_2(y_1, y_2) = y_2$. Then, formally, the line integral in (3.3.9) is

$$R(1, 1) = \int_{\langle 0, 0 \rangle}^{\langle 1, 1 \rangle} (s_1\, ds_1 + s_2\, ds_2),$$

where the integral is taken along a path connecting the points $\langle 0, 0 \rangle$ and $\langle 1, 1 \rangle$. If this path is a straight line, then the parametric equations of this line are

$$s_1 = \varphi_1(t) = t \quad \text{and} \quad s_2 = \varphi_2(t) = t, \quad \text{for } 0 \le t \le 1.$$

Thus, $ds_1 = dt = ds_2$, and the above integral becomes

$$R(1,1) = \int_0^1 (t \, dt + t \, dt) = 2 \int_0^1 t \, dt = 1.$$

More generally, if the integral is taken along a straight line connecting the points $\langle 0, 0 \rangle$ and $\langle y_1, y_2 \rangle$, then the parametric equations of this line are $s_1 = \varphi_1(t) = y_1 t$, $s_2 = \varphi_2(t) = y_2 t$, $0 \le t \le 1$. Hence, $ds_1 = y_1 \, dt$, $ds_2 = y_2 \, dt$, and the integral in (3.3.9) becomes

$$R(y_1, y_2) = \int_0^1 (y_1^2 t \, ds + y_2^2 t \, dt)$$

$$= (y_1^2 + y_2^2) \int_0^1 t \, dt = \frac{y_1^2 + y_2^2}{2}.$$

As it should, this function, defined for all $\langle y_1, y_2 \rangle \in R_2^+$, has a gradient of the form assumed at the beginning of the discussion.

(b) A question that naturally arises is whether the integral in (3.3.9) depends on the path connecting the points $\mathbf{0}$ and $\mathbf{y}$. It turns out that the integral is independent of the path if the integral is an exact differential of a function, a condition that always holds for the risk gradient. For, by (3.3.7), it follows from the definition of the differential of $R(\mathbf{y})$ that

$$dR(\mathbf{y}) = \eta_1(\mathbf{y}) \, dy_1 + \eta_2(\mathbf{y}) \, dy_2 + \ldots + \eta_m(\mathbf{y}) \, dy_m.$$

To see why the integral in (3.3.9) is independent of the path connecting the points $\mathbf{0}$ and $\mathbf{y}$ suppose $m = 2$ and let $s_1 = \varphi_1(t)$ and $s_2 = \varphi_2(t)$, $0 \le t \le 1$, be any path connecting the points $\langle 0, 0 \rangle$ and $\langle y_1, y_2 \rangle$ such that $\varphi_1(0) = \varphi_2(0) = 0$, $\varphi_1(1) = y_1$, and $\varphi_2(1) = y_2$. Then,

$$\int_{\langle 0, 0 \rangle}^{\langle y_1, y_2 \rangle} \boldsymbol{\eta}(\mathbf{s}) \, d\mathbf{s} = \int_0^1 (\eta_1(\varphi_1(t), \varphi_2(t)) \, \varphi_1'(t) + \eta_2(\varphi_1(t), \varphi_2(t)) \, \varphi_2'(t)) \, dt$$

$$= \int_0^1 \left(\frac{dR(\varphi_1(t), \varphi_2(t))}{dt} \right) dt$$

$$= R(\varphi_1(1), \varphi_2(1)) - R(\varphi_1(0), \varphi_2(0))$$

$$= R(y_1, y_2) - R(0, 0) = R(y_1, y_2),$$

since $R(0,0) = 0$ for the case under consideration. The argument may be easily generalized to any $m > 2$. Hence, the integral in (3.3.9) is independent of the path.

(c) In constructing a model of competing risks based on dependent latent life spans, an investigator may wish to specify the model in terms of a risk gradient vector $\boldsymbol{\eta}(\mathbf{y})$ in (3.3.8). This raises the question as to what conditions this vector must satisfy in order that it be the gradient vector of some function $R(\mathbf{y})$ defined on R_m^+.

When a function $R(\mathbf{y})$ has continuous second order partial derivatives at all points $\mathbf{y}$ in some set C, then the condition

$$\frac{\partial^2 R(\mathbf{y})}{\partial y_i \partial y_j} = \frac{\partial^2 R(\mathbf{y})}{\partial y_j \partial y_i}, \qquad 1 \le i, \ j \le m,$$

is satisfied for all $\mathbf{y} \in C$. Therefore, in order that an arbitrarily specified vector-valued function $\boldsymbol{\eta}(\mathbf{y})$ be the gradient of some function $R(\mathbf{y})$ with continuous second order derivatives on a set C, it is necessary that the condition

$$\frac{\partial \eta_j(\mathbf{y})}{\partial y_i} = \frac{\partial \eta_i(\mathbf{y})}{\partial y_j}, \qquad 1 \le i, \ j \le m,$$

hold for all $\mathbf{y} \in C$. It will be useful to check whether this condition is satisfied when a vector-valued function $\boldsymbol{\eta}(\mathbf{y})$ is considered as a possible choice for a risk gradient.

(d) Show that the vector-valued function $\boldsymbol{\eta}(\mathbf{y}) = \langle \eta_1(\mathbf{y}), \eta_2(\mathbf{y}) \rangle = \langle y_1 y_2, y_1 \rangle$, defined for $\mathbf{y} \in R_2^+$ cannot be a risk gradient. Construct a simple example of a risk gradient on R_2^+.

3. A question that naturally arises is whether the path of integration from the point $\mathbf{0}$ to $\mathbf{y}$ may be chosen in such a way that some properties of the joint distribution of the latent life spans becomes apparent. Such a pathway, yielding explicit formulas for certain conditional survival functions, will be exhibited.

Let C be a path connecting the points $\mathbf{0}$ and $\mathbf{y}$. Then, the integral in (3.3.9) along C may be represented in the symbolic form

$$\int_C \boldsymbol{\eta}(\mathbf{s}) \, d\mathbf{s}.$$

Let C_1 be a path connecting the points $\mathbf{0}$ and $\mathbf{x}$, and let C_2 be a path connecting the points $\mathbf{x}$ and $\mathbf{y}$. Then, $C = C_1 + C_2$ symbolizes a path from $\mathbf{0}$ to $\mathbf{y}$. A useful property of line integrals is that

$$\int_C \boldsymbol{\eta}(\mathbf{s}) \, d\mathbf{s} = \int_{C_1} \boldsymbol{\eta}(\mathbf{s}) \, d\mathbf{s} + \int_{C_2} \boldsymbol{\eta}(\mathbf{s}) \, d\mathbf{s}.$$

This additivity property may be generalized to any number $n \ge 2$ of sub-pathways C_i, $1 \le i \le n$, such that the terminal point of C_i is the initial point of C_{i+1}. A pathway C from $\mathbf{0}$ to $\mathbf{y}$ may be represented in the symbolic form $C = C_1 + C_2 + \ldots + C_n$.

(a) For the case $m = 2$, consider the problem of choosing a path C from $\mathbf{0}$ to $\mathbf{y}$ in such a way that the line integral

$$R(y_1, y_2) = \int_C \eta_1(s_1, s_2) \, ds_1 + \eta_2(s_1, s_2) \, ds_2$$

may be expressed in a useful computational form. One way of choosing a path C from the point $\langle 0, 0 \rangle$ to the point $\langle y_1, y_2 \rangle$ is to move along the y_1-axis from $\langle 0, 0 \rangle$ to the point $\langle y_1, 0 \rangle$. Let this path be C_1. Then, move along the verticle line $L = y_1$ from the point $\langle y_1, 0 \rangle$ to the point $\langle y_1, y_2 \rangle$. Let this path be C_2. Observe that $C = C_1 + C_2$.

The parametric equations for the path C_1 are $s_1 = \varphi_1(t) = t$ and $s_2 = \varphi_2(t) = 0$ for $0 \leq t \leq y_1$. Therefore,

$$\int_{C_1} \eta(s)\, ds = \int_0^{y_1} \eta(s_1, 0)\, ds_1.$$

The parametric equations for the path C_2 are $s_1 = \varphi_1(t) = y_1$, a fixed number, and $s_2 = \varphi_2(t) = t$ for $0 \leq t \leq y_2$. Thus,

$$\int_{C_2} \eta(s)\, ds = \int_0^{y_2} \eta_2(y_1, s_2)\, ds_2.$$

From the additive property of pathways, it follows that

$$R(y_1, y_2) = \int_0^{y_1} \eta_1(s_1, 0)\, ds_1 + \int_0^{y_2} \eta_2(y_1, s_2)\, ds_2$$

for all $\langle y_1, y_2 \rangle \in R_2^+$.

(b) For the case $m = 3$, the result in (a) may be extended to

$$R(y_1, y_2, y_3) = \int_0^{y_1} \eta_1(s_1, 0, 0)\, ds_1$$

$$+ \int_0^{y_2} \eta_2(y_1, s_2, 0)\, ds_2 + \int_0^{y_3} \eta_3(y_1, y_2, s_3)\, ds_3$$

for all $\langle y_1, y_2, y_3 \rangle \in R_3^+$.

(c) The particular pathways just discussed are very useful for setting down formulas for conditional survival functions. For example, when $m = 3$ and the joint survival function of the latent variables Y_1, Y_2, and Y_3 is given by $\exp[-R \cdot (y_1, y_1, y_3)]$, where the function R is given in (b), then marginal survival function of the random variables Y_1 and Y_2 is

$$P[Y_1 > y_1, Y_2 > y_2] = \exp\left[-\int_0^{y_1} \eta_1(s_1, 0, 0)\, ds_1 - \int_0^{y_2} \eta_2(y_1, s_2, 0)\, ds_2 \right].$$

Using this expression and the result in (b), show that

$$P[Y_3 > y_3 | Y_1 > y_1, Y_2 > y_2] = \exp\left[-\int_0^{y_3} \eta_3(y_1, y_2, s_3)\, ds_3 \right].$$

4. Let $R(y_1, y_2, y_3)$ be the function defined in (3.4.6) for $\mathbf{y} = \langle y_1, y_2, y_3 \rangle \in R_3^+$.

(a) Show that the components of the risk gradient $\boldsymbol{\eta}(\mathbf{y}) = \langle \eta_1(\mathbf{y}), \eta_2(\mathbf{y}), \eta_3(\mathbf{y}) \rangle$ are

$$\eta_1(\mathbf{y}) = \lambda_1 + \lambda_{12} y_2 + \lambda_{13} y_3 + \lambda_{123} y_2 y_3,$$
$$\eta_2(\mathbf{y}) = \lambda_2 + \lambda_{12} y_1 + \lambda_{23} y_3 + \lambda_{123} y_1 y_3,$$

and

$$\eta_3(\mathbf{y}) = \lambda_3 + \lambda_{13} y_1 + \lambda_{23} y_2 + \lambda_{123} y_1 y_2.$$

(b) Let $\mu(x)$ be the force of mortality associated with survival function $S(x)$ in (3.4.9). Then,

$$\mu(x) = \delta_1 + 2\delta_2 x + 3\delta_3 x^2.$$

The force of mortality $\mu_j(x)$, associated with the j-th cause of death, is $\mu_j(x)$ $=\eta_j(x, x, x)$, $1 \leq j \leq 3$. Verify that the additivity property

$$\mu(x) = \mu_1(x) + \mu_2(x) + \mu_3(x)$$

holds for the special case under consideration.

(c) Let $G_1(x)$, $G_2(x)$, and $G_3(x)$ be the functions defined in (3.2.15). Show that for $x \in R^+$

$$G_1(x) = \int_0^x \exp[-(\delta_1 y + \delta_2 y^2 + \delta_3 y^3)](\lambda_1 + (\lambda_{12} + \lambda_{13}) y + \lambda_{123} y^2) \, dy,$$

where the parameters δ_1, δ_2, and δ_3 are defined as in (3.4.9). The integral formulas for $G_2(x)$ and $G_3(x)$ are similar. Integrals of the above type would have to be evaluated by numerical methods for given values of the λ-parameters.

(d) Show that

$$\int_0^{y_1} \eta_1(s_1, 0, 0) \, ds_1 \quad = \lambda_1 y_1,$$

$$\int_0^{y_2} \eta_2(y_1, s_2, 0) \, ds_2 = \lambda_2 y_2 + \lambda_{12} y_1 y_2,$$

$$\int_0^{y_3} \eta_3(y_1, y_2, s_3) \, ds_3 = \lambda_3 y_3 + \lambda_{13} y_1 y_3 + \lambda_{23} y_2 y_3 + \lambda_{123} y_1 y_2 y_3.$$

(e) Show that

$$P[Y_3 > y_3 \mid Y_1 > y_1, Y_2 > y_2] = \exp[-\gamma(y_1, y_2) y_3],$$

where $\gamma(y_1, y_2) = \lambda_3 + \lambda_{13} y_1 + \lambda_{23} y_2 + \lambda_{123} y_1 y_2$.

5. Marshall and Olkin (1967a, 1967b) proposed a joint survival function of the form

$$P[Y_1 > y_1, Y_2 > y_2] = \exp[-\lambda_1 y_1 - \lambda_2 y_2 - \lambda_{12} \max(y_1, y_2)],$$

where $\langle y_1, y_2 \rangle \in R_2^+$ and the λ's are positive parameters.

(a) Show that the marginal survival functions are

$$P[Y_1 > y_1] = \exp[-(\lambda_1 + \lambda_{12}) y_1]$$

and

$$P[Y_2 > y_2] = \exp[-(\lambda_2 + \lambda_{12}) y_2]$$

for $y_1 > 0$ and $y_2 > 0$. This is another example of a bivariate survival function with marginal exponential survival functions. There are, in fact, infinitely many bivariate survival functions with marginal exponential survival functions.

(b) In this example,

$$R(y_1, y_2) = \lambda_1 y_1 + \lambda_2 y_2 + \lambda_{12} \max(y_1, y_2).$$

Show that for $y_1 < y_2$, the risk gradient is $\boldsymbol{\eta}(\mathbf{y}) = \langle \lambda_1, \lambda_2 + \lambda_{12} \rangle$; but, for $y_1 > y_2$, the risk gradient is $\boldsymbol{\eta}(\mathbf{y}) = \langle \lambda_1 + \lambda_{12}, \lambda_2 \rangle$. In this case the risk gradient does not exist along the line $y_1 = y_2$.

6. To illustrate the relative ease with which formula (3.8.13) may be applied, suppose $m=4$ and that the life table estimates are $\hat{q}_1(x,h)=0.006$, $\hat{q}_2(x,h)=0.015$, $\hat{q}_3(x,h)=0.005$, and $\hat{q}_4(x,h)=0.007$ for a given x and h. Then, $\hat{q}(x,h)=\hat{q}_1(x,h)+\ldots+\hat{q}_4(x,h)=0.033$ and $\hat{p}(x,h)=1-\hat{q}(x,h)=0.967$. Furthermore, $\hat{c}_1(x,h)=\hat{q}_1(x,h)/\hat{q}(x,h)=0.182$ so that $\hat{p}(1,x,h)=(\hat{p}(x,h))^{\hat{c}_1(x,h)}=0.994$. Calculate $\hat{p}(2,s,h)$, $\hat{p}(3,x,h)$, $\hat{p}(4,x,h)$, and $\hat{p}(\mathbf{a},x,h)$ for $\mathbf{a}=(1,2)$.

References

Alderson, M. (1981): *International Mortality Statistics*. Facts on File, Inc., New York, N.Y.

Barlow, R.E. and Proschan, F. (1975): *Statistical Theory of Reliability and Life Testing*. Holt, Rinehart, and Winston, Inc.

Basu, A.P. and Ghosh, J.K. (1978): Identifiability of the Multinormal and Other Distributions Under Competing Risks Model. *Journal of Multivariate Analysis* 8:413–429

Birnbaum, Z.W. (1979): *On the Mathematics of Competing Risks*. DHEW Publication No. (PHS) 79–1351

Chiang, C.L. (1968): *Introduction to Stochastic Processes in Biostatistics*. John Wiley and Sons, Inc., New York

Cox, D.R. (1962): *Renewal Theory*. Methuen, London

David, H.A. and Moeschberger, M.L. (1978): *The Theory of Competing Risks*. Griffin's Statistical Monographs and Courses No. 39, MacMillan, New York

Elandt-Johnson, R.C. (1976): Conditional Failure Time Distributions Under Competing Risk Theory with Dependent Failure Times and Proportional Hazard Rates. *Scand. Actuarial J.* 37–51

Elandt-Johnson, R.C. and Johnson, N.L. (1980): *Survival Models and Data Analysis*. John Wiley and Sons, New York

Fritsch, F.N. and Carlson, R.E. (1980): Monotone Piecewise Cubic Interpolation. *SIAM. J. Numer. Anal.* 17:238–246

Gail, M. (1975): A Review and Critique of Some Models Used in Competing Risk Analysis. *Biometrics* 31:209–222

Kimball, A.W. (1958): Disease Incidence Estimation in Population Subject to Multiple Causes of Death. *Inter. Statistical Institute Bulletin* 36:193–204

Kimball, A.W. (1969): Models for the Estimation of Competing Risks From Grouped Data. *Biometrics* 25:329–37

Kolman, B. and Trench, W.F. (1971): *Elementary Multivariable Calculus*. Academic Press, New York

Langberg, N., Proschan, F. and Quinzi, A.J. (1978): Converting Dependent Models Into Independent Ones, Preserving Essential Features. *Ann. of Prob.* 6:174–181

Marshall, A.W. and Olkin, I. (1967a): A Multivariate Exponential Distribution. *J. Amer. Stat. Assoc.* 62:30–44

Marshall, A.W. and Olkin, I. (1967b): A Generalized Bivariate Exponential Distribution. *J. Appl. Prob.* 4:291–302

Nádas, A. (1971): The Distribution of the Identified Minimum of a Normal Pair Determines the Distribution of the Pair. *Technometrics* 13:201–202

Prentice, R.L., Kalbfleisch, J.D., Peterson, A.V., Flournoy, N., Farewell, V.T., and Breslow, N.E. (1978): The Analysis of Failure Times in the Presence of Competing Risks. *Biometrics* 34:541–554

Preston, S.H., Keyfitz, N. and Schoen, R. (1972): *Causes of Death – Life Tables for National Populations*. Seminar Press, New York

Thomas, G.B. (1972): *Calculus and Analytic Geometry*. Addison-Wesley, Reading, Mass.

Tsiatis, A. (1975): A Nonidentifiability Aspect of the Problem of Competing Risks. *Proc. Natl. Acad. Sci. USA* 72:20–22

Vital Statistics Instruction Manual, Part II. Procedures for Coding Multiple Causes of Death in 1955 (1962): National Office of Vital Statistics, Washington, D.C.

Chapter 4. Models of Maternity Histories and Age-Specific Birth Rates

4.1 Introduction

At any time point, the size of a population closed to in-and-out migration will increase or decrease, depending on the balance between births and deaths. Consequently, along with the study of mortality discussed in the previous two chapters, the study of fertility, the factors and circumstances underlying the flow of births into a population, has long been of central interest to students of population. In countries in which births are registered according to the age of the mother and estimates of the size of the female population are available by age groups, age-specific birth (fertility) rates are routinely calculated. During the past two decades, however, the investigation of fertility has also been approached through the study of maternity histories, a study that is sometimes referred to as birth interval analysis. With respect to data requirements, the two approaches to investigating fertility are distinct. Period age-specific birth rates provide summary measures of the birth experiences of all women in the childbearing ages during short time periods, consisting of one or more calendar years. Analyzing data on maternity histories, on the other hand, entails following cohorts of women over longer time periods, which may, particularly in the analysis of historical data, include entire reproductive careers, starting with marriage to the onset of physiological sterility.

Among the problems of central interest to mathematical demography is that of developing structures with a capability for handling the cohort and period points of view within a single unifying framework. In discussing the problem of dealing with these two points of view within the context of making population projections, Keyfitz (1968), page 88, wrote: "A thoroughgoing incorporation of this cohort approach is part of the mathematical demography of the future." Investigators will differ in their approaches to developing structures for dealing with cohort and period effects within a unifying framework. What framework, or perhaps combination of frameworks, becomes widely accepted among populationists in some future time period cannot be foreseen. In this chapter, solutions to the problem of developing structures which treat cohort and period effects from a unified point of view will be approached by extending renewal theory, a branch of applied probability having roots involving many investigators and going back several decades, see, for example, Lotka (1939) and Feller (1968).

4.2 A Potential Birth Process

Consider a cohort of females born at some epoch $t=0$ and assume these individuals are followed throughout their reproductive careers. To simplify an initial organization of ideas, it will be assumed that all events are conditioned on survival to the end of the childbearing ages, say age 50. If X is a random variable representing age at death in the cohort, then this conditioning event may be represented by $X > 50$.

Reproductive careers of females in a cohort begin when they mature and become sexually active. Observed social conventions surrounding the beginning of sexual activity in a cohort can vary greatly within and among cultures; but, in this exposition, the initiation of this activity will be referred to simply as marriage. Let T_0 be a random variable with distribution function $G_0(x)$ $=P[T_0 \leq x]$, representing the age of females in a cohort at marriage. If a female never marries, then this event is symbolized by $T_0 = \infty$; if she marries, then T_0 is finite, i.e., $T_0 < \infty$. In most societies, marriage is not universal so that the probability, $P[T_0 = \infty]$, is positive. Usually, random variables encountered in probabilistic formulations are finite with probability one. In this connection, it is interesting to observe that infinite-valued random variables may be introduced in a meaningful way when considering natural phenomena.

A first step in modeling reproductive histories in a cohort of females is to consider the ages following marriage that live births occur. Let T_k, $k \geq 1$, be a sequence of random variables representing these ages in a cohort of women. Then, $Z_k = T_k - T_{k-1}$, $k \geq 1$, is a sequence of random variables representing waiting times among live births. Like T_0, the Z-random variables may be infinite-valued. For if a female experiences no live births, then $Z_1 = \infty$; if she experiences one live birth, then $Z_1 < \infty$ but $Z_2 = \infty$. In general, if she experiences $k \geq 1$ live births, then $Z_j < \infty$ for all $1 \leq j \leq k$ and $Z_{k+1} = \infty$. Before proceding further, the temporary simplifying assumption that once a female marries she remains married will also be introduced.

The next task considered in the formulation of a model of reproductive histories will be that of assigning joint distributions of the random variables in the sequence $T_0, Z_1, Z_2, \dots$. A simple case arises when it is assumed that the random variables in this sequence are independent. Under this assumption, the joint distributions of the random variables in question are completely determined once the distribution functions of the random variables $Z_1, Z_2, \dots$ are specified. It is, of course, easy to think of reasons why the Z-random variables may not be independent. However, because the assumption of independence leads to the mathematical operation of convolution, which may be efficiently implemented on a computer using fast Fouier transforms, it is worthwhile to consider models based on this simplifying assumption.

A female is said to be of parity $v \geq 0$ if she has experienced v live births. When more complicated models are introduced, it will be useful to think of the waiting time to a live birth as conditioned on parity. Accordingly, for $v \geq 0$, let $H(v, t)$ be the distribution function of the random variable Z_{v+1}, the waiting time to birth $v+1$ following the v-th birth. Let p_v be the probability the v-th birth occurs, $v \geq 1$. Then,

$$p_v = P[Z_v < \infty] = \lim_{t \uparrow \infty} H(v-1, t). \tag{4.2.1}$$

An alternative way of specifying the model is to suppose the probabilities in (4.2.1) are given and $G(v-1, t)$ is a proper distribution function. Then, for $v \geq 1$,

$$H(v-1, t) = p_v G(v-1, t), \tag{4.2.2}$$

where $G(v-1, t) \uparrow 1$ as $t \uparrow \infty$. From now on it will also be assumed that $\lim_{t \downarrow 0} H(v, t) = H(v, 0) = 0$ for all $v \geq 0$.

Several functions of basic interest can be derived in terms of a random function $K(t)$, representing the number of live births occurring in a cohort after t time units of marriage. It will also be assumed momentarily that the time origin is the age of marriage so that $K(t)$ represents the number of live births occurring during the time interval $(0, t]$. The random function $K(t)$, which is defined in terms of the Z-random variables, will be called the potential birth process.

As a first step in analyzing the potential birth process, the task of expressing the probabilities $P[K(t) = n]$, $n = 0, 1, 2, ...,$ in terms of the distribution functions $H(v, t)$, $v \geq 0$, will be considered. Let $h(v, t)$ be the p.d.f. of $H(v, t)$. First observe that $K(t) = 0$ if, and only if, $Z_1 > t$. Therefore,

$$P[K(t) = 0] = 1 - H(0, t). \tag{4.2.3}$$

Next observe that $K(t) = 1$ if, and only if, $Z_1 = s$ for some $s \in (0, t]$ and $Z_2 > t - s$. By integrating over the possible values of s and using the assumption of independence, it can be seen that

$$P[K(t) = 1] = \int_0^t h(0, s)(1 - H(1, t-s)) ds$$
$$= H^{(1)}(t) - H^{(2)}(t), \tag{4.2.4}$$

where $H^{(1)}(t) = H(0, t)$ and $H^{(2)}(t)$ is the convolution integral

$$H^{(2)}(t) = \int_0^t h(0, s) H(1, t-s) ds. \tag{4.2.5}$$

To extend formula (4.2.4) to the case $P[K(t) = n]$ for any $n \geq 1$, it will be useful to identify the functions on the right in (4.2.4) with the distribution functions of the partial sums of the random variables $Z_1, Z_2, ...$ For $n \geq 1$, let $H^{(n)}(t)$ be the distribution function of the random variable $Z_1 + Z_2 + ... + Z_n$, the waiting time to the n-th birth. Then $H^{(1)}(t) = H(0, t)$. To see that $H^{(2)}(t)$ in (4.2.5) is the distribution function of the random variable $Z_1 + Z_2$, observe that $Z_1 + Z_2 \leq t$ if, and only if, $Z_1 = s$ for some $s \in (0, t]$ and $Z_2 \leq t - s$. Integrating over s and using the assumption of independence leads to formula (4.2.5).

More generally, if $H^{(n-1)}(t)$ is the distribution function of the random variable $Z_1 + Z_2 + ... + Z_{n-1}$ for $n \geq 2$ and $h^{(n-1)}(t)$ is its p.d.f., then, by using an induction argument, it can be shown that

$$H^{(n)}(t) = \int_0^t h^{(n-1)}(s) H(n-1, t-s) \, ds. \tag{4.2.6}$$

Furthermore, a straight-forward extention of the argument used to derive (4.2.4) leads to the formula

$$P[K(t) = n] = H^{(n)}(t) = H^{(n+1)}(t), \tag{4.2.7}$$

which is valid for all $n \geq 1$ and $t > 0$. If the function $H^{(0)}(t)$ is defined by $H^{(0)}(t) = 1$ for all $t > 0$, then, in view of (4.2.3), (4.2.7) is valid for all $n \geq 0$ and $t > 0$.

It is shown in Problem 1, that in human populations $K(t)$ is a finite-valued random variable for every $t > 0$. Consequently, it is meaningful to consider its mean. By definition,

$$E[K(t)] = \sum_{n=0}^{\infty} n P[K(t) = n] = \sum_{n=1}^{\infty} P[K(t) \geq n]. \tag{4.2.8}$$

But, $K(t) \geq n$ if, and only if, $Z_1 + Z_2 + \ldots + Z_n \leq t$. Hence,

$$P[K(t) \geq n] = H^{(n)}(t), \tag{4.2.9}$$

and it follows that

$$E[K(t)] = \sum_{n=1}^{\infty} H^{(n)}(t). \tag{4.2.10}$$

The last task considered in this section is that of extending the foregoing to the case marriage occurs at an arbitrary age x. From now the birth of the cohort will be taken as the time origin. Let

$$q_n(x, t) = P[K(t) = n \mid T_0 = x], \tag{4.2.11}$$

and let

$$p_n(t) = P[K(t) = n] \tag{4.2.12}$$

for $n \geq 0$. Now the potential birth process starts at a random time T_0 with distribution function $G_0(x)$ and p.d.f. $g_0(x)$. If $T_0 > t$, then $K(t) = 0$ with probability one; if $T_0 \leq t$, then the potential birth process started at some $x \in (0, t)$. Therefore,

$$p_n(t) = (1 - G_0(t)) \delta_0(n) + \int_0^t g_0(x) q_n(x, t-x) \, dx, \tag{4.2.13}$$

where $\delta_0(n) = 1$ if $n = 0$ and $\delta_0(n) = 0$ if $n \neq 0$ and $n > 0$.

Given that marriage occurs at age x,

$$v(x, t) = E[K(t) \mid T_0 = x] \tag{4.2.14}$$

is the mean number of live births experienced in a cohort during the time interval $(x, x+t]$, $t > 0$. Viewing a cohort from the epoch of birth,

$$v(t) = E[K(t)] \tag{4.2.15}$$

is the mean number of live births experienced during the time interval $(0, t]$. Equation (4.2.13) implies that

$$v(t) = \int_0^t g_0(x) v(x, t-x) \, dx. \tag{4.2.16}$$

Under the assumption $T_0, Z_1, Z_2, \ldots$ is a sequence of independent random variables, the age-dependent convolutions in (4.2.13) and (4.2.16) reduce to ordinary convolutions. For in this case, it follows from (4.2.7) and (4.2.10) that

$$q_n(x, t-x) = H^{(n)}(t-x) - H^{(n+1)}(t-x) \tag{4.2.17}$$

and

$$v(x, t-x) = E[K(t-x)] = \sum_{n=1}^{\infty} H^{(n)}(t-x) \tag{4.2.18}$$

for $0 < x \leq t$.

4.3 Cohort Net and Gross Maternity Functions

Two widely used concepts in mathematical demography are the gross and net maternity functions. Although these concepts are usually used in connection with female age-specific birth rates in a given period, they will be formulated in this section in terms of the potential birth process for a cohort introduced in Sect. 4.2.

A simplifying assumption underlying the potential birth process was that it started with marriage and continued through the childbearing years. In reality, it may be stopped prior to the end of the childbearing years. There may be many reasons for stoppage of the potential birth process, including dissolution of marriage and death of the female. In this section, however, it will be assumed that death of the female is the only event that stops the potential birth process.

As before, let X be a random variable with distribution function $G(x)$, representing lifespan in a birth cohort of females. Let $N(t)$ be a random function representing the number of live births experienced during the time interval $(0, t]$. Here, 0 represents the epoch at which the cohort was born. The random function $N(t)$ is determined by stopping the potential birth process $K(t)$ at the age of death. It will be supposed that the random variable X is independent of the random variables $T_0, Z_1, Z_2, \ldots$. Let $t > 0$ be fixed. Then, if $X > t$, so that death occurs after time t, $N(t)$ is just the value of the potential birth process at t. Hence,

$$N(t) = K(t) \quad \text{if } X > t. \tag{4.3.1}$$

But, if $X \leq t$, so that death occurs prior to t, $N(t)$ is the value of the potential birth process at the age of death. Therefore,

$$N(t) = K(X) \quad \text{if } X \leq t. \tag{4.3.2}$$

For a cohort born at epoch 0,

$$m(t) = E[N(t)] \tag{4.3.3}$$

is the mean number of live births experienced during the time interval $(0, t]$, $t > 0$. Equations (4.3.1) and (4.3.2) may be used to express $m(t)$ in terms of the function $v(t)$, the mean function of the potential birth process, see (4.2.15), and the lifespan distribution function $G(x)$ with p.d.f. $g(x)$. A little reflection leads to the formula

$$m(t) = \int_0^t v(x)g(x)dx + S(t)v(t), \tag{4.3.4}$$

for $t > 0$. Observe that the first term on the right in (4.3.4) arises from (4.3.2); while the second term is derived from (4.3.1).

A function that plays an important subsequent role is

$$b(t) = \frac{dv(t)}{dt}, \quad t > 0, \tag{4.3.5}$$

the birth rate function of the potential birth process. By differentiating (4.3.4) with respect to t, it can be shown that

$$\frac{dm(t)}{dt} = v(t)g(t) + S'(t)v(t) + S(t)b(t). \tag{4.3.6}$$

But, $S'(t) = -g(t)$. Therefore, from (4.3.6), it follows that the density of $m(t)$ is

$$\frac{dm(t)}{dt} = S(t)b(t), \quad t > 0. \tag{4.3.7}$$

Hence,

$$m(t) = \int_0^t S(x)b(x)dx, \quad t > 0. \tag{4.3.8}$$

The density in (4.3.7) will be referred to as the cohort net maternity function; while the density $b(t)$ in (4.3.5) will be called the gross maternity function. As $t \uparrow \infty$, the random function $K(t)$ and $N(t)$ converge, with probability one, to random variables K and N, representing the potential and actual number of births experienced in a cohort of females by the end of the childbearing ages. The total fertility rate, TFR, in a cohort is defined by

$$TFR = E[K]; \tag{4.3.9}$$

the net fertility rate, NFR, is defined by

$$NFR = E[N]. \tag{4.3.10}$$

In terms of the densities $b(t)$ and $S(t)b(t)$, (4.3.9) and (4.3.10) may be expressed in the form

$$TFR = \int_0^\infty b(t)dt \tag{4.3.11}$$

and

$$\text{NFR} = \int_0^\infty S(t)b(t)\,dt. \tag{4.3.12}$$

Concepts involving one-sex models are widely used in the analysis of population data. Usually, only the female population is considered. In such circumstances, names of rates sometimes have the word gross attached to them to indicate only female offspring are being considered. For example, Coale (1972) uses the term, gross reproductive rate, for the expression in (4.3.11) when only female births are considered. To convert TFR to the gross reproductive rate, one would merely multiply by q, the probability that an offspring is female at birth. A commonly used value for q is $\frac{100}{205}$.

Period age-specific birth rates are frequently calculated for the age intervals $(x, x+h]$, where $x = 15, 20, \ldots, 45$ and $h = 5$. Suppose, for example, that $B(x, h)$ is the number of births experienced during a given period by mothers in the age interval $(x, x+h]$. Let $K(x, h)$ be an estimate of the number of live females with ages in this interval. Then,

$$\hat{R}_B(x, h) = \frac{B(x, h)}{K(x, h)} \tag{4.3.13}$$

is the period age-specific birth rate associated with the age interval $(x, x+h]$. A period estimate of the total fertility rate is usually calculated according to the formula

$$\widehat{\text{TFR}} = 5 \sum_{x=15}^{45} \hat{R}_B(x, 5). \tag{4.3.14}$$

It is of interest to make some connection between the rates in (4.3.13) and (4.3.14) and the potential birth process discussed in Chap. 2. Under the assumption that the law governing the productive careers of cohort is stationary in time, $\hat{R}_B(x, h)$ is evidently an estimate of

$$R_B(x, h) = \frac{v(x+h) - v(x)}{h} = \frac{1}{h} \int_x^{x+h} b(s)\,ds. \tag{4.3.15}$$

Although the support of the birth rate function $b(x)$ is formally represented as the infinite interval $[0, \infty)$, it is actually a finite interval of childbearing ages, say $[15, 50]$. That is, $b(x) \neq 0$ if $x \in [15, 50]$ and $b(x) = 0$ if $x \notin [15, 50]$. Therefore, from (4.3.11) and (4.3.15), it follows that

$$\text{TFR} = 5 \sum_{x=15}^{45} R_B(x, 5) = \int_{15}^{50} b(x)\,dx. \tag{4.3.16}$$

Equation (4.3.16) supplies a justification for using (4.3.14) as an estimate of TFR.

Among the items of interest not considered in this section is that of deducing a formula for $P[N(t) = n]$, $n \geq 0$, the distribution of the actual birth process $N(t)$ at $t > 0$. Problem 2 may be consulted for further details. It should also be mentioned that conventions for selecting the age interval used in estimating the age-specific birth rate (4.3.13) are not, evidently, standarized. When age at last birthday is used as an indicator of age, intervals $[x, x+h)$ closed on the left and open on the right are frequently used in classifying individuals with respect to age.

4.4 Parity Progression Ratios

A frequently used concept in the analysis of data on maternity histories is that of parity progression ratios, see Leridon (1977), Chap. 2. These ratios seem to have several meanings, depending, among other things, on the sample of women an investigator is considering. The purpose of this section is to explore the concept of parity progression ratios in relation to the structures discussed in Sects. 4.2 and 4.3.

As a first step towards defining these ratios, consider an idealized cohort of women who marry at some epoch $t=0$ and survive to the end of the childbearing ages. Also make the further simplifying assumption that the maternity histories of women in a cohort are realizations of the potential birth process introduced in Sect. 4.2. The distribution of the number of live births experienced by women in the cohort as a function of duration $t>0$ of marriage is given by

$$P[K(t)=n]=H^{(n)}(t)-H^{(n+1)}(t), \tag{4.4.1}$$

for $n=0, 1, 2, \ldots$, see Eq. (4.2.7). When the sequence of distribution functions $H(v, t)$, $v\geq 0$, are defined as in (4.2.2) it can be shown that

$$\lim_{t\uparrow\infty} H^{(n)}(t)=p_1 p_2 \cdots p_n \tag{4.4.2}$$

for $n\geq 1$; Problem 3 may be consulted for details. Therefore, the distribution of the random variable K, representing final parity in the cohort of women, is given by

$$P[K=n]=\lim_{t\uparrow\infty} P[K(t)=n]=p_1 p_2 \cdots p_n(1-p_{n+1}), \tag{4.4.3}$$

for $n\geq 1$ and $P[K=0]=1-p_1$.

For the cohort of women under consideration, the sequence π_n of parity progression ratios is defined as the conditional probabilities

$$\pi_n=P[K\geq n|K\geq n-1]=\frac{P[K\geq n]}{P[K\geq n-1]}, \tag{4.4.4}$$

where $n\geq 1$. It is natural to ask whether there is a simple relationship connecting the p-probabilities on the right in (4.4.2) and the parity progression ratios as defined in (4.4.4). From (4.4.3), it follows that

$$P[K\geq n]=p_1 p_2 \cdots p_n \tag{4.4.5}$$

for $n \geq 1$. By substituting (4.4.5) into (4.4.4) and using the condition $P[K \geq 0]$ $= 1$, it can be seen that $\pi_n = p_n$, for all $n \geq 1$. Hence, the p-probabilities on the right in (4.2.1) are actually the parity progression ratios for the cohort under consideration. In terms of parity progression ratios, the mean of the random variable K is given by

$$E[K] = p_1 + p_1 p_2 + p_1 p_2 p_3 + \cdots, \tag{4.4.6}$$

as can be seen by summing (4.4.5) on $n \geq 1$. There will be only finitely many nonzero terms in the series on the right, because, in human populations, final parity in a cohort is a bounded random variable so that $p_n = 0$ for all $n \geq n_0$, some positive integer.

Parity progression ratios in the potential birth process considered above may also reflect family planning intentions. A simple but concrete example of this situation arises when it is decided that m male and f female offspring are an ideal family. Although progress has been made in controlling the sex of children, in most cases a child's sex is indeterminate. In such circumstances, the actual family size attained by a couple may exceed $d = m + f$, the desired family size. For if couples continue the childbearing process until at least m males and f females have been born, then the range of the random variable K is $d + x$, for $x = 0, 1, 2, \ldots$. Furthermore, the p.d.f. of the random variable K is

$$P[K = d + x] = \binom{d + x - 1}{m - 1} p^m q^{f+x} + \binom{d + x - 1}{f - 1} p^{m+x} q^f, \tag{4.4.7}$$

where $x = 0, 1, 2, \ldots$. Outlined in Problem 4 is the argument underlying the derivation of the p.d.f. in (4.4.7).

Models of a potential birth process may be specified in many ways. If, for example, the parity progression ratios are specified by $p_n = P[K \geq n]/P[K \geq n - 1]$, $n \geq 1$, using the p.d.f. in (4.4.7), then $p_1 = p_2 = \ldots = p_d = 1$, reflecting the desires of couples to continue having children until the desired family size d is reached. For all $n > d$, $0 < p_n < 1$, reflecting inabilities to control fertility. Sheps (1963) has given a detailed analysis of sex preferences based on the p.d.f. in (4.4.7). These is a rather extensive literature on sex preferences; the paper of McClelland (1979) is a recent contribution.

Only rarely, if ever, would an investigator have a sample of women from which the pure forms of the parity progression ratios discussed above could be estimated. Among the factors complicating the specification and estimation of parity progression ratios in the potential birth process is that of heterogeniety among members of a cohort regarding sex preferences. One way of treating such heterogeniety is to suppose the p.d.f. of K is a mixture of densities of form (4.4.7), with $p(m, f)$ representing the distribution (p.d.f.) of sex preferences $\langle m, f \rangle$ among members of a cohort. Further discussion of this approach may be found in Mode (1975).

Factors other than sex preferences also contribute to heterogeniety in family planning intentions among members of a cohort. Another approach to treating such heterogeniety is to express parity progression ratios in terms of covariates, reflecting such variables as age of mother when attaining a given parity and perhaps socioeconomic factors. Curtin et al. (1979) have proposed a general scheme, related to analysis of variance procedures, for handling qualitative variables pertaining to heterogeniety in parity progression ratios. Like many demographic variables of interest, parity progression ratios are sometimes estimated from sample surveys, containing samples of women in various stages of the childbearing process during a given period. Among other things, Pullum (1979) has suggested that period parity progression ratios may be estimated by the fraction of women at each parity who state at the interview that they would like to continue childbearing. Given such estimates, one can then estimate a period distribution of desired family size, using formula (4.4.3).

In formulating and implementing computer simulation models, careful attention must be given to the problem of expressing p_n in such a way that it is a nonnegative function on the domain of covariates being considered. A widely used method for constructing such a function is to consider the so-called logit transformation

$$\log_e \frac{p_n}{1-p_n} = f_n(\mathbf{X}), \qquad n \geq 1, \tag{4.4.8}$$

where $f_n(\mathbf{X})$ is a function of a vector $\mathbf{X}$ of covariates. From (4.4.8), it follows that

$$p_n = \frac{\exp[f_n(\mathbf{X})]}{1+\exp[f_n(\mathbf{X})]}, \tag{4.4.9}$$

so that p_n is nonnegative for all values of the covariate vector $\mathbf{X}$. As a first approximation, the function $f_n(\mathbf{X})$ is frequently assumed to be linear. Thus, if $\mathbf{X} = \langle X_1, X_2, \ldots, X_k \rangle$ is a k-dimensional vector of covariates, then

$$f_n(\mathbf{X}) = \beta_{n1} X_1 + \ldots + \beta_{nk} X_k, \tag{4.4.10}$$

where the β's are unknown parameters. Problems centering around the definition of covariates and parameter estimation will not be considered here.

4.5 Parametric Distributions of Waiting Times Among Live Births

Once the distribution functions $H(v-1, t)$, $v \geq 1$, of waiting times among live births have been specified, see (4.2.2), the potential birth process defined in Sect. 4.2 is essentially determined, except for a distribution of age at first marriage. The purpose of this section is to provide a brief overview of the literature in which parametric forms of these distributions have been investigated. Before discussing a sample of these forms, however, it is appropriate to note that there is a rather extensive demographic literature on the reproductive histories of women. Sometimes, the term, birth intervals, is used in describing these histories. Among the papers in this literature published during the period, 1955–1970, are Tuan (1958), Dandekar (1963), Potter (1963), and Wolfers (1968). Sheps (1965) was a pioneer in using life table methods to obtain nonparametric estimates of the distribution functions of waiting times among live births.

Example 4.5.1. *Applications of the Exponential Distribution*

Singh and his collaborators have applied the exponential distribution extensively in models of waiting times among live births, see, for example, Singh (1968) and Singh et al. (1974). The approach followed by these workers consisted of partitioning these waiting times among births into independent random variables Y_v, $v \geq 1$, representing waiting times till conception, and constants h_1 and h_2 for the lengths of the gestation and postpartum sterile periods. Thus, following marriage, among those women in a cohort experiencing at least one live birth, waiting times till the first live birth is given by the random variable $Z_1 = Y_1 + h_1$. Among women experiencing at least two live births, waiting times among births are given by the random variables $Z_v = h_2 + Y_v + h_1 = Y_v + h$ for $v \geq 2$.

In Singh et al. (1974), the constants h_1 and h_2 were chosen as 0.75 and 0.5 years, respectively. That is, nine months were allowed for gestation and six for postpartum sterile periods. Moreover, the random variables Y_v were assumed to follow exponential distributions with positive parameters λ_v, $v \geq 1$. Under this assumption, the conditional distribution functions $G(v-1, t) = P[Z_v \leq t | Z_v < \infty]$ on the right in (4.2.2) take the form

$$G(0, t) = 1 - \exp[-\lambda_1(t-h_1)], \qquad t > h_1$$

and   (4.5.1)

$$G(v-1, t) = 1 - \exp[-\lambda_v(t-h)],$$

for $t \geq h$ and $v \geq 2$.

Basic to the computer implementation of the potential birth process defined in Sect. 4.2 is the compution of $H^{(n)}(t)$, the distribution function of the random variable $Z_1 + Z_2 + \ldots + Z_n$, $n \geq 1$, see (4.2.6) and (4.2.7). Let $G^{(n)}(t)$ be the convolution of the distribution functions $G(0, t), \ldots, G(n-1, t)$. Then, from (4.2.2), it follows that

$$H^{(n)}(t) = \alpha(n) G^{(n)}(t),$$ (4.5.2)

where $\alpha(n) = p_1 p_2 \ldots p_n$. But $Z_1 + Z_2 + \ldots + Z_n = Y_1 + Y_2 + \ldots + Y_n + nh_1 + (n-1)h_2$. Therefore, apart from the translation constant $h_n = nh_1 + (n-1)h_2$, the distribution function $G^{(n)}(t)$ is completely determined by the distribution function $F^{(n)}(t)$ of the random variable $Y_1 + Y_2 + \ldots + Y_n$.

Let $f^{(n)}(t)$ be the p.d.f. of the distribution function $F^{(n)}(t)$. Then,

$$f^{(1)}(t) = \lambda_1 \exp[-\lambda_1 t], \qquad t > 0,$$ (4.5.3)

when $\lambda_1 \neq \lambda_2$,

$$f^{(2)}(t) = \lambda_1 \lambda_2 (\psi(1, 2) \exp[-\lambda_1 t] + \psi(2, 2) \exp[-\lambda_2 t])$$ (4.5.4)

for $t > 0$, where $\psi(1, 2) = (\lambda_2 - \lambda_1)^{-1}$ and $\psi(2, 2) = (\lambda_1 - \lambda_2)^{-1}$. In general, when all the λ-parameters are unequal, it can be shown by induction that

$$f^{(n)}(t) = \beta(n) \sum_{k=1}^{n} \psi(k, n) \exp[-\lambda_k t], \qquad t > 0,$$ (4.5.5)

where $\beta(n)=\lambda_1\lambda_2\dots\lambda_n$. The ψ-parameters are determined according to the recursive relations

$$\psi(k,n)=\frac{\psi(k,n-1)}{\lambda_n-\lambda_k}, \qquad 1\le k\le n-1$$

and   (4.5.6)

$$\psi(n,n)=\sum_{k=1}^{n-1}\frac{\psi(k,n-1)}{\lambda_k-\lambda_n}$$

for $n\ge2$. Problem 5 may be consulted for details.

If the constants $\gamma(k,n)$ are defined by $\gamma(k,n)=\beta(n)\lambda_k^{-1}$, then an integration of (4.5.5) yields the formula

$$F^{(n)}(t)=\sum_{k=1}^{n}\gamma(k,n)\psi(k,n)(1-\exp[-\lambda_k t])$$ (4.5.7)

for the desired distribution function, a formula that is valid for $t>0$ and $n\ge2$. To complete the derivation of formula (4.5.2) for the model under consideration, it suffices to observe that $G^{(1)}(t)=G(0,t)$ and

$$G^{(n)}(t)=F^{(n)}(t-h_n)$$ (4.5.8)

for $n\ge2$ and $t>h_n$. An advantage of assuming that waiting times till conception follow exponential distributions is that the distribution function $H^{(n)}(t)$ in (4.5.2) may be expressed in terms of elementary functions of the λ-parameters and the exponential function. Feller (1966), page 40, has given explicit formulas for the ψ-parameters; but in an age of computers there is less of a need for explicit formulas, since the desired values may be easily computed, using the recursion relations in (4.5.6).

A limitation of the model under discussion is that no pregnancy wastage is taken into account, i.e., all conceptions are supposed to lead to live births. A major difficulty from the point of view of statistical estimation is that the model contains too many parameters. Despite these disadvantages, the model is of interest, since it can easily be implemented on the computer for arbitrary assignments of the λ-parameters.

Example 4.5.2. *A Simplified Model Based on the Exponential Distribution*

An analysis of birth interval data by life table methods sometimes shows that, apart from the first birth following marriage, mean times among subsequent births are approximately equal. This observation has led Singh et al. (1974) and others to consider a simplification of the preceding model by supposing that $\lambda_i=\lambda_2$ for $i\ge2$ but $\lambda_1\ne\lambda_2$. Under this assumption, the distribution function $F^{(n)}(t)$ of the random variable $Y_1+Y_2+\dots+Y_n$ is that of a sum of $n-1$ independent and identically distributed exponential random variables with a common parameter λ_2 convolved with that of an exponential random variable Y_1 with parameter λ_1.

As is well-known, for $n \geq 2$ the distribution of the sum $Y_2 + \ldots + Y_n$ is a gamma whose distribution function is

$$\Gamma_{n-1}(\lambda_2, t) = \frac{\lambda_2^{n-1}}{\Gamma(n-1)} \int_0^t s^{n-2} \exp[-\lambda_2 s] ds, \quad t > 0. \tag{4.5.9}$$

Consequently, the sequence of distribution functions $F^{(n)}(t)$ take the form,

$$F^{(1)}(t) = 1 - \exp[-\lambda_1 t],$$

and for $n \geq 2$

$$F^{(n)}(t) = \frac{\lambda_2^{n-1}}{\Gamma(n-1)} \int_0^t F^{(1)}(t-s) s^{n-2} \exp[-\lambda_2 s] ds$$

$$= \Gamma_{n-1}(\lambda_2, t) - \left(\frac{\lambda_2}{\delta}\right)^{n-1} \exp[-\lambda_1 t] \Gamma_{n-1}(\delta, t), \tag{4.5.10}$$

where $t > 0$ and $\delta = \lambda_2 - \lambda_1 > 0$, by assumption.

Formula (4.5.2) is, of course, still valid but the F's on the right in (4.5.8) are now given by the sequence of distribution functions defined in (4.5.10). The formulas in (4.5.10) seem adequate for computational purposes. Readers interested in more explicit formulas may consult Singh et al. Additional technical details regarding (4.5.9) and (4.5.10) are given in Problem 6.

Singh et al. applied the above model to a sample of 494 Indian women married between the ages 13 and 16 years. Because no women in the sample attempted to control their fertility, the parity progression ratios were assigned the values $p_i = 1$ for $i \geq 2$. In this application, $1 - p_1$ was interpreted as the fraction of sterile women. Under these further simplifying assumptions, there were three parameters to estimate; namely, p_1, λ_1, and $\lambda_2 = b\lambda_1$, $b > 0$. Estimates of the three parameters were obtained by the method of moments. Used as the empirical distribution in this exercise was the observed distribution of the number of births experienced by the sample of women after ten years of marriage. The estimates reported by the authors were $\hat{p}_1 = 0.977$, $\hat{\lambda}_1 = 0.34$, and $\hat{b} = 1.53$. As can be seen from Table 1 in Singh et al., there was a close fit between the empirical and observed theoretical distributions.

Example 4.5.3. *A Double Exponential Distribution*

Instead of working with a distribution of waiting times till conception, an alternative approach is to specify a parametric family for the distribution functions $H(v-1, t)$, $v \geq 1$, governing waiting times among live births. An interesting example of this approach has been given by Ross and Madavan (1981). Chosen as a model for birth intervals by these authors was the parametric form $H(v-1, t) = p_v G(v-1, t)$, see (4.2.2), where $G(v-1, t)$ is the two-parameter double exponential family, $v \geq 1$, defined by

$$G(v-1, t) = \exp[-\exp(\alpha_v - \beta_v t)] \tag{4.5.11}$$

for $t > 0$ and by $G(v-1, t) = 0$ for $t \leq 0$. In (4.5.11), α_v and β_v are positive parameters. From an inspection of the formula in (4.5.11), it can be seen that the parameter α_v essentially determines the magnitude of the values of $G(v-1, t)$ in a right-hand neighborhood of zero; the parameter β_v governs the rate at which $G(v-1, t)$ approaches one as $t \uparrow \infty$. Some theoretical properties of the distribution function in (4.5.11) are developed in Problem 7.

Ross and Madavan refer to the formula in (4.5.11) as a Gompertz model for birth interval analysis. In this connection, it is interesting to note that the distribution function in (4.5.11) is not that of the Gompertz distribution developed in Example 2.2.3. It is, however, closely related to the double exponential distribution function $G(x) = \exp[-e^{-x}]$, defined for all $x \in R$, which arises as a limiting distribution for the largest order statistic. Galambos (1978) may be consulted for details.

A two-stage procedure was used by Ross and Madavan to estimate the three parameters in the model from birth history data for Korean women. In a first step, nonparametric estimates $\hat{H}(v-1, t)$ of the distribution functions $H(v-1, t)$ were obtained for $v = 1, \ldots, 4$ by using a life table algorithm. The parity progression ratios were then estimated by taking the maximum values

$$\hat{p}_v = \max_t \hat{H}(v-1, t), \tag{4.5.12}$$

for $v = 1, \ldots, 4$. Given these estimates, the parameters α_v and β_v were estimated by a linear least squares procedure based on the relationship

$$\log_e\left[-\log_e\left(\frac{H(v-1, t)}{p_v}\right)\right] = \alpha_v - \beta_v t, \tag{4.5.13}$$

where $t > 0$ and $v = 1, \ldots, 4$.

An advantage of the above two-stage procedure is that it can be implemented with a minimum of computer software. When, however, library programs for finding extreme values of a function of several variables are available, it would seem desirable to use either a nonlinear least squares procedure combined with a nonparametric estimate of the distribution function, or a maximum likelihood procedure to estimate the three parameters p_v, α_v, and β_v simultaneously. According to the authors, the nonparametric procedure combined with linear least squares, yielded good fits between the theoretical and empirical distribution functions for Korean cohorts of women married during the periods 1941–1943, 1953–1955, and 1956–1958.

As is readily apparent, there are marked differences between the models proposed by Ross and Madavan (1981) and Singh et al. (1974). Both groups of investigators reported good fits to their data, but the groups viewed their data on birth histories from different perspectives. Singh et al. were concerned with fitting a model to the distribution of the number of births following ten years of continuous marriage; while Ross and Madavan worked directly with waiting times among births. It seems likely that if a simple exponential model were applied directly to birth interval data, a good fit between the theoretical and empirical distribution functions would not be obtained.

A possible explanation of the good fits obtained by Singh et al. is a central limit theorem for probability density functions. According to this theorem, successive convolutions of density will tend to approach a normal density under rather weak conditions. This result suggests that the parity distribution for the model of the type set forth in Example 4.5.2 will be somewhat independent of the waiting times among live births following the first when $p_v = 1$ for $v \geq 2$, as was assumed by Singh et al. Some results as well as references on central limit theorems for densities may be found in Mode (1967).

Example 4.5.4. *Distributions Based on Risk Functions*

Another approach to constructing distributions of waiting times among live births is to express them in terms of risk functions. Undoubtedly, since the idea of a risk function is one of the foundation stones of mathematical demography, this approach has been thought of by many investigators. Poole (1973), for example, used risk functions in his development of fertility measures based on birth interval data. More recently, Braun and Hoem (1979) and Hobcraft and Rodriguez (1980) have carried out extensive analysis of maternity history data based on parametric forms of the risk function.

In order to put the idea of a risk function within the notational context of this chapter, let $\theta_v(x)$ be the risk function associated with the distribution function $H(v-1, t)$, $v \geq 1$, on the left in (4.2.2). Then, for women of parity $v-1$, the distribution function of the waiting times till the v-th birth takes the form

$$H(v-1, t) = 1 - \exp\left[-\int_0^t \theta_v(x)dx\right], \qquad t > 0. \tag{4.5.14}$$

Unlike the risk functions considered heretofore, the integral

$$\delta_v = \lim_{t \uparrow \infty} \int_0^t \theta_v(x)dx \tag{4.5.15}$$

is assumed to converge to a finite number. Letting $t \uparrow \infty$ in (4.5.14), it is easy to see that the parity progression ratio p_v takes the form

$$p_v = 1 - \exp[-\delta_v]. \tag{4.5.16}$$

Hobcraft and Rodriguez (1980) considered several parametric risk functions of the form $\theta_v(x) = \delta_v f_v(x)$, where $f_v(x)$ was a probability density function. In what follows, the subscript v will be dropped and $f_v(x)$ will be denoted by $f(x)$ to lighten the notation. One of the forms of $f(x)$ considered by these investigators was a gamma type p.d.f. defined by

$$f(x) = \frac{\beta^\alpha}{\Gamma(\alpha)}(x-\gamma)^{\alpha-1}\exp[-\beta(x-\gamma)], \tag{4.5.17}$$

for $x > \gamma$ and by $f(x) = 0$ for $x \leq \gamma$.

In the data considered by these investigators, the index and scale parameters α and β were estimated by an age and parity classification; but, the location parameter γ was held constant at 0.75 years, corresponding to a gestation period of nine months.

A second form considered by Hobcraft and Rodriguez was a piece-wise linear p.d.f. defined by $f(x)=0$ for $x<\gamma$,

$$f(x)=\begin{cases} \dfrac{1}{c}\left(\dfrac{x-\gamma}{\tau_1-\gamma}\right), & \text{for } \gamma\le x\le\tau_1, \\[2mm] \dfrac{1}{c}, & \text{for } \tau_1<x\le\tau_2, \\[2mm] \dfrac{1}{c}\left(\dfrac{\tau_3-x}{\tau_3-\tau_2}\right), & \text{for } \tau_2<x\le\tau_3, \end{cases} \qquad (4.5.18)$$

where c is a normalizing constant. Probability density functions of the form displayed in (4.5.18) are of methodological interest, because they are simple and depend solely on location parameters. According to (4.5.18), the p.d.f. $f(x)$ rises linearly for $\gamma\le x\le\tau_1$, remains constant for $\tau_1<x\le\tau_2$, and decreases linearly for $\tau_2<x\le\tau_3$. Observe that when $\tau_1=\tau_2$, the p.d.f. $f(x)$ reduces to a triangular density. The parameters γ and τ_3 represent the times of entry into and exits from the risks of childbearing. In the data analyses reported by the investigators, the parameters γ and τ_3 had the values 9 and 72 months.

A third p.d.f., in which a piece-wise linear form was combined with a gamma p.d.f., was also considered by Hobcraft and Rodriguez. In this form, $f(x)$ is defined by $f(x)=0$ for $x\le\gamma$,

$$f(x)=\begin{cases} \dfrac{1}{c}\left(\dfrac{x-\gamma}{\tau_1-\gamma}\right)x^{\alpha-1}\exp[-\beta x], & \text{for } \gamma<x\le\tau_1 \\[2mm] \dfrac{1}{c}x^{\alpha-1}\exp[-\beta x], & \text{for } \tau_1<x<\infty. \end{cases} \qquad (4.5.19)$$

Again $\gamma=0.75$ years and c is a normalizing constant.

Using data on maternity histories for Columbian women, collected as part of the World Fertility Survey, Hobcraft and Rodriguez estimated the parameters in the risk functions just described according to a parity-age classification. Within each classification, discretized, nonparametric, empirical estimates of the risk functions, the q-probabilities, were obtained by life table methods. The parameters in the risk functions were then, apparently, estimated by fitting the parametric forms to the empirical risk functions by nonlinear least squares. As judged by measuring the distances of empirical from theoretical distribution functions though an error mean square criterion, when compared with its competitors, the model determined by the p.d.f. in (4.5.19) received high marks. Only parities greater than or equal to two were reported by the investigators. A brief description of the model considered by Braun and Hoem (1979) is given in Problem 8.

The foregoing examples are merely a sampling of the literature on the development of parametric models for distributions of waiting times among live births. Further insights into the evolution of ideas concerning the analysis of maternity histories and the development of parametric models may be obtained by consulting the papers Pathak (1971), Svinivasan (1968), Suchindran and Lachanbruch (1974), Venkatacharya (1972), Braun (1980), Lee and Lin

(1976), Ruzicka (1976), and Suchindran and Linger (1977). George and Mathai (1975) have developed a model for birth intervals based on a generalization of the gamma distribution and expressed it in terms of hypergeometric functions. A seminar on the analysis of maternity histories, which was sponsored by the International Union for the Scientific Study of Population, was held at the London School of Hygiene and Tropical Medicine during April 9–11, 1980.

4.6 Parametric Forms of the Distribution of Age at First Marriage in a Cohort

A final ingredient needed to specify the potential birth process described in Sect. 4.2 is the distribution function $G_0(x)$ of age at first marriage in a cohort of females. When compared to the literature on parametric birth interval models, that on parametric forms for the distribution function under consideration is more limited. In this section, two parametric families of distribution functions will be considered.

Example 4.6.1. *A Model Based on a Double Exponential Risk Function*

After extensive analyses of historical data on those who eventually married, Coale (1971) proposed a distribution for age at first marriage in a cohort determined by a three-parameter double exponential risk function defined by

$$\eta(x) = \beta \exp[-\gamma \exp(-\delta x)] \qquad (4.6.1)$$

for $x > 0$, where β, γ, and δ are positive parameters. It is of interest to note the similarity between the risk function in (4.6.1) and the distribution function proposed by Ross and Madavan (1981) in Example 4.5.3.

Presented in the paper of Coale is evidence for the proposition that there exists standard parameter values, $\beta_s = 0.174$, $\gamma_s = 4.411$, and $\delta_s = 0.309$. Given these standard values, further evidence is presented to the effect that the distribution of age at first marriage in many populations is governed by a family of risk functions depending on two parameters α and k, which vary among populations. The parameter α is the earliest age at marriage; while the parameter k governs the speed at which marriage occurs. According to Coale's proposal, in an i-th population, there exist parameter values α_i and k_i such that the risk function $\eta_i(x)$ determining age at marriage is given by

$$\eta_i(x) = \eta(x - \alpha_i), \qquad (4.6.2)$$

for $x > \alpha_i$ and $\eta_i(x) = 0$ for $x \leq \alpha_i$, where the parameter values β_i, γ_i and δ_i are determined by $\beta_i = \beta_s/k_i$, $\gamma_i = \gamma_s$, and $\delta_i = \delta_s/k_i$.

Let p be the fraction of individuals in a cohort who eventually marry and define the function $S(x)$ by

$$S(x) = \exp\left[-\int_0^x \eta(s)\,ds\right], \qquad (4.6.3)$$

for $x > 0$. Then, when age at first marriage is governed by the risk function in (4.6.1), the distribution function $G_0(x)$ in a cohort is given by $G_0(x) = 0$ for $x \leq \alpha$ and by

$$G_0(x) = p(1 - S(x - \alpha)), \tag{4.6.4}$$

for $x > \alpha$, whatever be the values of the parameters β, γ, and δ. Some further properties of the risk function in (4.6.1) and suggestions for computer implementation are given in Problem 9. A distribution for age at marriage, proposed by Coale and McNeil (1972), is discussed in Problem 10.

Example 4.6.2. *A Model Based on the Lognormal Distribution*

Among the widely used distributions in statistics is the lognormal. Its origins go back at least a century to Galton (1879) and McAlister (1879). Aitchison and Brown (1963) have given an account of its history, properties, and applications in economics. As a model for age at first marriage in a birth cohort of males or females, it is easy to describe. To fix ideas, consider the conditional distribution of age at first marriage among those in a cohort who eventually marry. With reference to the random variable T_0 introduced in Sect. 4.2, the conditional distribution under consideration is that of T_0, given that $T_0 < \infty$. As above, let α be the smallest value of T_0, the youngest age at marriage. Age at first marriage in a cohort is said to follow a lognormal distribution if the conditional distribution of the random variable $\log_e(T_0 - \alpha)$, given that $T_0 < \infty$, is normal with mean μ and variance σ^2.

A convenient property of the lognormal distribution is that the conditional distribution function $P[T_0 \leq x | T_0 < \infty]$ may be computed in terms of $\Phi(z)$, the standard normal (cumulative) distribution function. For, by definition,

$$P[T_0 \leq x | T_0 < \infty] = \Phi\left(\frac{\log_e(x - \alpha) - \mu}{\sigma}\right), \tag{4.6.5}$$

for $x > \alpha$. For many investigators with access to well-developed computing facilities, with statistical libraries stored in the computer, the computer implementation of formula (4.6.5) will be easy. There are computer programs, for example, such that if an array of values $z = (\log_e(x - \alpha) - \mu)/\sigma$ is entered into the computer, the computer will return the values $\Phi(z)$.

Let $P[T_0 = \infty] = p$, the fraction in a cohort who eventually marry. Then, the distribution function of the random variable T_0 is given by

$$G_0(x) = p\, \Phi\left(\frac{\log_e(x - \alpha) - \mu}{\sigma}\right), \tag{4.6.6}$$

for $x > \alpha$ and $G_0(x) = 0$ for $x \leq \alpha$. By differentiating with respect to x, it follows that

$$g_0(x) = \frac{p}{\sqrt{2\pi}(x - \alpha)\sigma} \exp\left[-\frac{1}{2}\left(\frac{\log_e(x - \alpha) - \mu}{\sigma}\right)^2\right] \tag{4.6.7}$$

for $x > \alpha$ and $g_0(x) = 0$ for $x \leq \alpha$, is the p.d.f. of $G_0(x)$. The risk function $\eta(x)$ $= g_0(x)/p(1 - G_0(x))$ of the conditional distribution of T_0, given that $T_0 < \infty$, has no simple analytic form but it can be easily computed.

Although the parameters μ and σ have a straight-forward mathematical interpretation, this interpretation by itself is not informative in describing the pace of marriage in a cohort. It turns out, however, that the parameters μ and σ are related to the quantiles of the lognormal distribution in a simple way. Suppose, for the moment, that $p=1$ so that all members of the cohort eventually marry. For any fraction q, $0<q<1$, let x_q be a number such that $G_0(x_q) = q$. Observe that x_q is merely the age by which a fraction q of the cohort have married. Let z_q be the q-th quantile of the standard normal distribution, i.e., $\Phi(z_q) = q$, $0<q<1$. There exist computer programs such that given the fraction q, the computer will return the value z_q. Because the random variable $Z = [\log_e(T_0 - \alpha) - \mu]/\sigma$ has a normal distribution with mean zero and variance one, it follows that the quantiles x_q and z_q are related according to the equation

$$x_q = \alpha + \exp[\mu + \sigma z_q]. \tag{4.6.8}$$

In particular, if $q=0.5$, then $z_q=0$; hence, the median of the lognormal distribution is $x_q = \alpha + \exp[\mu]$. Consequently, given a value of α, the median age at marriage is completely determined by the parameter μ. From (4.6.8), it can also be seen that the parameter σ is directly related to the pace of marriage: the greater the value of σ, the larger the value of x_q and hence the slower the pace of marriage in a cohort. To relax the condition that all members of the cohort eventually marry, for any p, $0<p<1$, and q, $0<q\leq p$, let x_q be a number such that $G_0(x_q)=q$. Then, formula (4.6.8) continues to hold if the z-value on the right is replaced by z_s, where $s=q/p$. Some additional properties of the lognormal distribution are developed in Problem 11.

Example 4.6.3. *Validation of Lognormal*

A widely used method for validating a proposed parametric model for a distribution is to determine whether the parameter values can be chosen in such a way that the theoretical distribution fits an empirical one satisfactorily. Used as a criterion for goodness of fit was the error mean square. Two sources of the data on ages at marriage were utilized in the validation exercise. One source was a nineteenth century Belgium commune; a second was the U.S. 1973 National Survey of Family Growth (NSFG), which contained information not only on the age of the bride at first marriage but also on the age of her first husband.

Watkins and McCarthy (1980) have studied the female life cycle in La Hulpe, a Belgian commune, during the period 1847–1866. As an event of interest, they considered age at first marriage and estimated its distribution function, using life table methods. A listing of the resulting empirical distribution function was given in Table 2, column 6, of their paper. After division of this function by 0.765, the estimated cohort probability of marriage by age 50, the parameters α, μ, and σ in the lognormal distribution were estimated by fitting the theoretical distribution function in (4.6.5) to the empirical one by a nonlinear least squares procedure.

Among the questions asked married women aged 15–44 years who were interviewed in NSFG, a survey conducted by the National Center for Health

Statistics, was their ages and that of their husbands at first marriage. To minimize truncation biases due to incomplete marriage experience in a cohort, only data from women aged 40–44 at the time of the interview were used. The sample was further restricted to white women. After screening the sample to eliminate obvious errors in reporting, there remained 718 data points, giving the ages of the bride and groom in the marriages considered. Even if it were assumed that the ages of the brides were reported correctly, there is at least one possible source of error in the reported ages of the grooms; namely, when interviewed, the respondent may not have remembered the age of her first husband correctly. It should also be pointed out that the marriage may not have been the first for the groom. All ages recorded from the NSFG data were those at last birthday. Consequently, the distribution functions for age at marriage could be estimated only at integral values, i.e., whole years. A nonlinear least squares procedure was again used to estimate the three parameters in the lognormal distribution by fitting the theoretical distribution function in (4.6.5) to the empirical ones.

In implementing the nonlinear least square procedure, all calculations were carried out with the programming language APL and a library of statistical programs issued by IBM.[1] An algorithm due to Marquardt (1963) formed the mathematical basis for the nonlinear least squares, derivative free programs in this library. The estimates of the parameters α, μ, and σ obtained by using these programs will be referred to as the La Hulpe females, U.S. white females, and U.S. white males.

Presented in Table 4.6.1 are three sets of estimates of the parameters α, μ, and σ for the empirical distribution functions under consideration along with the corresponding values of the error mean square. As judged by these values, the fits of the theoretical distribution function in (4.6.5) to the empirical ones appear satisfactory. For, as can be seen from Table 4.6.1, the values of the error mean square were 1.0116×10^{-4}, 3.1966×10^{-5}, and 9.2841×10^{-5}, respectively, for the La Hulpe females, U.S. white females, and U.S. white males.

Given the parameter estimates in Table 4.6.1, it is also of interest to use formula (4.6.8) to estimate the median age at marriage among those in a cohort

Table 4.6.1. Nonlinear least squares estimates of parameters in lognormal distribution

Empirical distributions	Parameter			Mean square error
	α	μ	σ	
La Hulpe females	15.971	2.487	0.472	1.0116×10^{-4}
U.S. white females	12.982	2.040	0.398	3.1966×10^{-5}
U.S. white males	16.514	1.935	0.525	9.2841×10^{-5}

[1] APL Statistical Library – Program Descriptional Operations Manual. IBM Corporation, P.O. Box 950, Poughkeepsie, N.Y. 12602.

who eventually marry. Such estimates will provide insights into the pace of marriage as reflected by the three data sets under consideration. The pace of marriage in 19-th century La Hulpe was slow, the median female age at first marriage being 27.996 years. Marriage among U.S. white females aged 40–44 years in 1973 occurred at a faster pace, the median age of first marriage being equal to 20.673. The median age of the first husbands of these famales was 23.438 years.

More vivid insights into the implications of the parameter estimates in Table 4.6.1 may be gained by inspecting graphs of density and risk functions of the lognormal distribution for each set of estimates. Contained in Fig. 4.6.1 are the graphs of the theoretical density in (4.6.7), apart from the factor p, for each set of estimates; presented in Fig. 4.6.2 are the graphs of the corresponding risk functions.

From an inspection of the graphs of the probability densities in Fig. 4.6.1, the age patterns of marriage in the three data sets can clearly be seen. Marriage of females in 19-th century La Hulpe not only commenced later in life but also occurred at a slower pace than in the modern U.S. white female cohort. The shapes of the graphs of the densities for U.S. white males and females are essentially the same, reflecting that, once started, marriage in male and female cohorts occurred at approximately the same pace. The translation of the male density curve to the right of that for females merely reflects that the marriage process began at a later age in the male cohort.

A different perspective on patterns of age at marriage in the three data sets emerges when the graphs of the risk functions in Fig. 4.6.2 are inspected. Apart from a slight decrease after age 40, the risk function for age at marriage in La Hulpe is essentially monotone increasing, with a relatively slow rate of rise. The graphs of the risk functions for U.S. white males and females, on the other hand, present quite a different picture. Following the earliest age at marriage in both male and female cohorts, there is a steep rise in the risk functions, indicating a rapid pace of marriage. After this steep rise, both risk functions decline slowly, pointing to fewer marriages as the cohorts aged. Unlike the monotone

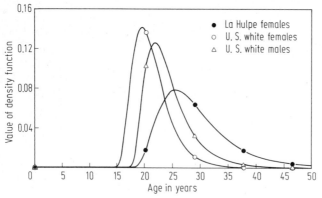

Fig. 4.6.1. Theoretical probability density functions for La Hulpe females and U.S. white males and females based on lognormal distribution

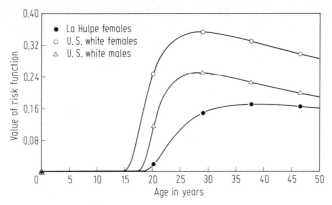

Fig. 4.6.2. Theoretical risk functions for La Hulpe females and U.S. white males and females based on lognormal distribution

increasing risk function in (4.6.1) proposed by Coale (1971), that for the lognormal is not always monotone increasing, as can be seen from the graphs in Fig. 4.6.2. The lognormal, therefore, is theoretically distinct from the distribution described in Example 4.6.1.

Example 4.6.4. *On the Joint Distribution of the Ages of Brides and Grooms – The Bivariate Lognormal*

A univariate lognormal distribution is useful for describing variation in ages at marriage in both male and female cohorts, because it not only has empirical validity but the parameters μ and σ are also easily interpreted as indicators of the pace of marriage in cohorts, see (4.6.8). Another advantage of the lognormal distribution is that it can be extended to the bivariate case, using the theory of the bivariate normal, and tested as a possible joint distribution of the ages of brides and grooms. As an aid to formulating ideas, think of the pairs of ages of brides and grooms as reported by white women aged 40–44 in the 1973 NSFG and the distribution from whence this sample may have come. Let T_1 and T_2 be random variables representing the ages of the brides and grooms, respectively, so that each data point may be viewed as a realization of the random vector $\mathbf{T}' = \langle T_1, T_2 \rangle$. Let α_1 and α_2 be the youngest ages at marriage for the brides and grooms.

Then, the random vector $\mathbf{T}' = \langle T_1, T_2 \rangle$ is said to follow a bivariate lognormal distribution with location vector $\langle \alpha_1, \alpha_2 \rangle$ if the random vector $\mathbf{X}' = \langle X_1, X_2 \rangle = \langle \log_e(T_1 - \alpha_1), \ \log_e(T_2 - \alpha_2) \rangle$ follows a bivariate normal distribution with mean vector $\mu' = \langle \mu_1, \mu_2 \rangle$ and covariance matrix

$$\Sigma = \begin{bmatrix} \sigma_1^2 & \rho \sigma_1 \sigma_2 \\ \rho \sigma_1 \sigma_2 & \sigma_2^2 \end{bmatrix}. \tag{4.6.9}$$

In (4.6.9), σ_1^2 and σ_2^2 are the variances of X_1 and X_2; while ρ is the product moment correlation of X_1 and X_2. The permissible values of the parameters in (4.6.9) are $\sigma_1 > 0$, $\sigma_2 > 0$, and $-1 < \rho < 1$. These ranges of parameter values ensure that the matrix in (4.6.9) is always nonsingular.

The theory of the multivariate normal distribution has been developed in great detail in vector-matrix notation by Anderson (1958); a more elementary treatment may be found in Chap. 9 of the textbook, Mood and Graybill (1963). As part of the study of skew bivariate frequency surfaces, Pretorius (1930) used ages of brides and grooms taken from Australian data for the period 1907–1914 in illustrative numerical examples based on the bivariate lognormal distribution.

With the help of vector-matrix notation, the p.d.f. of the bivariate normal distribution may be succinctly described. From now on, a vector without a prime will represent a column vector. Let $\mathbf{x}' = \langle x_1, x_2 \rangle$, let $\mathbf{\Sigma}^{-1}$ be the inverse of the matrix in (4.6.9), and define a quadratic form $Q(x_1, x_2)$ by

$$Q(x_1, x_2) = (\mathbf{x} - \boldsymbol{\mu})' \, \mathbf{\Sigma}^{-1} (\mathbf{x} - \boldsymbol{\mu}), \tag{4.6.10}$$

for all vectors $\mathbf{x}' = \langle x_1, x_2 \rangle$ in the $x_1 x_2$-plane. Then, as is well-known, the p.d.f. $f(x_1, x_2)$ on the nonsingular bivariate normal distribution may be represented in the form

$$f(x_1, x_2) = \frac{1}{2\pi |\mathbf{\Sigma}|^{1/2}} \exp\left[-\tfrac{1}{2} Q(x_1, x_2)\right], \tag{4.6.11}$$

where $|\mathbf{\Sigma}| = \sigma_1^2 \sigma_2^2 (1 - \rho^2)$, the determinant of the covariance matrix $\mathbf{\Sigma}$. The domain of the bivariate normal p.d.f. is, of course, the whole $x_1 x_2$-plane. Let $g(t_1, t_2)$ be the p.d.f. of a vector $\mathbf{T}$ with a bivariate lognormal distribution. By using well-known techniques from mathematical statistics concerning transformations of random variables, it can be shown that

$$g(t_1, t_2) = (t_1 - \alpha_1)^{-1}(t_2 - \alpha_2)^{-1} f(\log_e(t_1 - \alpha_1), \log_e(t_2 - \alpha_2)) \tag{4.6.12}$$

for $t_1 > \alpha_1$ and $t_2 > \alpha_2$. Formula (4.6.12), expressing $g(t_1, t_2)$ in terms of $f(x_1, x_2)$, makes it possible to deduce many properties of the bivariate lognormal directly from those of the bivariate normal.

A property of the bivariate normal distribution that is useful in testing whether a set of data points may be regarded as a sample from the bivariate lognormal distribution may be described in terms of a random variable

$$Y = Q(X_1, X_2), \tag{4.6.13}$$

the random variables X_1 and X_2 being defined in the discussion preceding (4.6.9). It is well-known that when the matrix $\mathbf{\Sigma}$ is nonsingular, the random variable Y in (4.6.13) has a Chi-square distribution with two degrees of freedom. Therefore, for $y > 0$

$$H(y) = P[Y \le y] = 1 - \exp\left[-\frac{y}{2}\right]. \tag{4.6.14}$$

Formula (4.6.14) is of practical use, because it facilitates the calculation of the probability content of the set of points $\langle x_1, x_2 \rangle$ in the $x_1 x_2$-plane such that $Q(x_1, x_2) \le y$.

Based on the idea just outlined, an exercise was carried out with the 718 data points from the 1973 NSFG to test whether the bivariate lognormal was a plausible model for the joint distribution of the ages of brides and grooms. To implement the test, it was necessary to get an estimate of the parameter ρ in the bivariate normal distribution. One way of estimating ρ would be that of calculating the correlation between realizations of the random variables $X_1 = \log_e(T_1 - \alpha_1)$ and $X_2 = \log_e(T_2 - \alpha_1)$. But, because these random variables involve the unknown parameters α_1 and α_2, it was decided to base an estimate of ρ on η, the correlation between the random variables T_1 and T_2, which does not depend on the location parameters α_1 and α_2. It can be shown that the formula connecting η and ρ is

$$\eta = \frac{\exp[\rho \sigma_1 \sigma_2] - 1}{((\exp[\sigma_1^2] - 1)(\exp[\sigma_2^2] - 1))^{1/2}}. \tag{4.6.15}$$

The sample of 718 realizations of the random vector $\mathbf{T}' = \langle T_1, T_2 \rangle$ yielded $\hat{\eta} = 0.6423$ as an estimate of η. Given $\hat{\eta}$ and $\hat{\sigma}_1 = 0.398$ and $\hat{\sigma}_2 = 0.525$, the estimates of σ_1 and σ_2 for U.S. white females and males from Table 4.6.1, Eq. (4.6.15) may be solved for ρ to yield the estimate $\hat{\rho} = 0.6694$.

Let $\hat{Q}(x_1, x_2)$ be the quadratic form in (4.6.10) obtained by replacing $\boldsymbol{\mu}$ and $\boldsymbol{\Sigma}^{-1}$ by the estimates of the parameters in Table 4.6.1 for U.S. males and females and the estimate of ρ just described. Furthermore, for any pair $\langle t_1, t_2 \rangle$ of ages for a bride and groom, define x_1 and x_2 by $x_1 = \log_e(t_1 - \hat{\alpha}_1)$ and $x_2 = \log_e(t_x - \hat{\alpha}_2)$, where the estimates $\hat{\alpha}_1$ and $\hat{\alpha}_2$ for U.S. males and females are again taken from Table 4.6.1. For any $y > 0$, let $\hat{H}(y)$ be the fraction of pairs $\langle t_1, t_2 \rangle$ such that $\hat{Q}(x_1, x_2) \leq y$. If the joint distribution of the ages of the brides and grooms does indeed follow a bivariate lognormal, then the estimated fraction $\hat{H}(y)$ should approximate the value $H(y)$ of the distribution function in (4.6.14) for all $y > 0$.

A set of y-values were calculated by solving the equation $q = H(y_q)$, see (4.6.4), for y_q at the ten q-values: 0.1, 0.2, ..., 0.9, and 0.95. The resulting y-values, rounded to four decimal places, are presented in the first column of Table 4.6.2; the second column of the table contains the corresponding values of the distribution function $H(y)$ at the chosen y-values. With the help of a computer program, it was possible to estimate the fraction $\hat{H}(y)$, defined in the preceding paragraph, at the chosen y-values in the first column of Table 4.2.2, using the 718 points of the U.S. marriage data. The estimates so obtained are reported in the third column of the table.

When comparing the empirical values $\hat{H}(y_q)$ in the third column of Table 4.6.2 with the corresponding theoretical ones, $q = H(y_q)$, it can be seen that there is a close match between the empirical and theoretical values for $q = 0.10$, 0.20, ..., 0.70. The fit deteriorates, however, for $q = 0.80$, 0.90, and 0.95. These observations suggest that the bivariate lognormal works well for about seventy percent of the data points in the U.S. sample. Despite satisfactory fits of the univariate lognormal distribution to the empirical marginal distributions of the ages of brides and grooms, the results in Table 4.6.2 suggest that if a bivariate lognormal were used as the joint distribution for the ages of brides and grooms, it would tend to overestimate the fraction of points $\langle t_1, t_2 \rangle$ such that $Q(x_1, x_2) \leq y_q$ for $q \geq 0.80$. In other words, the U.S. data considered suggests

Table 4.6.2. A test of validity for the bivariate log-normal as a joint distribution for ages of brides and grooms

y-values	Distribution functions	
	theoretical $H(y)$	empirical $\hat{H}(y)$
0.2107	0.10	0.1072
0.4463	0.20	0.2173
0.7134	0.30	0.3231
1.0217	0.40	0.3858
1.3863	0.50	0.5265
1.8326	0.60	0.5933
2.4079	0.70	0.6727
3.2189	0.80	0.7451
4.6052	0.90	0.8273
5.9915	0.95	0.8747

that the scatter of points in the $t_1 t_2$-plane seems to be greater than that predicted by the bivariate lognormal distribution.

Some basic properties of the bivariate lognormal distribution are developed in Problem 12. Problem 13 contains a derivation of a conditional distribution for the bivariate lognormal distribution. When mortality is high and divorce is rare, dissolution of marriages due to deaths of husbands may be a significant factor in terminating the potential birth process. In Problem 14, treating death of husband as a stopping time for the potential birth process is discussed within the framework of the results developed in Problem 13.

4.7 Heterogeneity in Waiting Times Among Live Births

A recurring theme in the development of stochastic models of human reproduction is that of taking into account variability among individuals of a population. Among the types of variability that may be considered in a cohort is that in maternity histories. One approach to the problem of coming to grips with this variability is to view maternity histories as realizations of a mixture of terminating renewal processes. Accordingly, the purpose of this section is to extend the theory of the potential birth process introduced in Sect. 4.2 to the case of dependence on a single mixing parameter.

As a first step towards getting an overview of ideas, consider a random variable Λ, representing variability among members of a cohort, with p.d.f. $f(\lambda)$. For the sake of concreteness, it will be supposed that the support of $f(\lambda)$ is $(0, \infty)$. Given that $\Lambda = \lambda$, it will be assumed that the random variables $Z_1, Z_2, \ldots$, representing waiting times among live births, are conditionally independent. Let $h(\lambda; v-1, t)$ be the conditional p.d.f. of the random variable Z_v, given that $\Lambda = \lambda$. Then, the p.d.f. of Z_v is the mixture

$$h(v-1,t) = \int_0^\infty f(\lambda) h(\lambda; v-1, t) \, d\lambda, \tag{4.7.1}$$

defined for $t > 0$ and $v \geq 1$.

Many of the formulas derived in Sect. 4.2 continue to hold for a mixture of renewal processes. For example, let $H^{(1)}(t)$ be the distribution function associated with the p.d.f. in (4.7.1). For $n \geq 2$, let $H^{(n)}(\lambda; t)$ be the distribution function determined by formula (4.2.6) from the conditional p.d.f.'s $h(\lambda; v-1, t)$, $v \geq 1$. Then, for a mixture of renewal processes, the distribution function of the random variable $Z_1 + Z_2 + \ldots + Z_n$, representing waiting times to the n-th live birth following marriage, is given by

$$H^{(n)}(t) = \int_0^\infty f(\lambda) H^{(n)}(\lambda; t) \, d\lambda \tag{4.7.2}$$

for $t > 0$ and $n \geq 2$. Thus, formula (4.2.7) for the distribution of the potential birth process $K(t)$ as a function t of duration of marriage continues to hold.

Unlike the potential birth process introduced in Sect. 4.2, waiting times among live births are no longer independent. For let $h(t_1, t_2, \ldots, t_n)$ be the joint p.d.f. of the random variables $Z_1, Z_2, \ldots, Z_n$ in a mixture of renewal processes. Then, it follows from the assumption of conditional independence that

$$h(t_1, t_2, \ldots, t_n) = \int_0^\infty f(\lambda) \prod_{v=1}^n h(\lambda, v-1, t_v) \, d\lambda. \tag{4.7.3}$$

Because the joint p.d.f. in (4.7.3) cannot be expressed as the product of the p.d.f.'s determined by (4.7.1), the random variables $Z_1, Z_2, \ldots, Z_n$ will not, in general, be independent.

There is an extensive literature on the type of stochastic process under consideration, going back several decades. When, for example, the Z-random variables are identically distributed, given that $\Lambda = \lambda$, one is lead to the interchangeable random variables considered by de Finetti (1937) in his treatment of the concept of subjective probability, a concept that is a cornerstone of Bayesian statistics. Hewitt and Savage (1956) have given a basic and abstract treatment of symmetric measures on Cartesian products related to the interchangeable random variables of de Finetti. Interchangeable random variables are sometimes called symmetrically dependent. Many of the classical limit theorems of probability theory, i.e., the law of large numbers and the central limit theorem, have been extended to sequences of interchangeable random variables, see, for example, Chap. 6 in Re've'sz (1968).

Example 4.7.1. *Gamma Mixtures of Gamma Distributions*

Singh (1964) as well as Singh et al. (1976, 1979, and 1981) have considered mixtures of distributions in various contexts, including the timing of the first birth and selection bias in post-partum amenorrhea. Among the mixtures considered was that of a gamma and exponential. But, the distribution resulting from mixing a gamma with an exponential distribution is merely a special case of mixing a gamma distribution with a gamma. Because gamma mixtures of gamma distributions are tractable mathematically, they are useful in providing illustrative parametric examples, which may lead to further insights into the construction and computer implementation of models describing variability in maternity histories. Some properties of mixtures of this type will, therefore, be explored in the remainder of this section.

Suppose the random variable Λ introduced above follows a gamma distribution with index parameter $\alpha > 0$ and scale parameter $\beta > 0$. Then, the p.d.f. $f(\lambda)$ takes the form

$$f(\lambda) = \frac{\beta^\alpha}{\Gamma(\alpha)} \lambda^{\alpha-1} \exp[-\beta\lambda], \tag{4.7.4}$$

for $\lambda \in (0, \infty)$. It will also be supposed that the ν-th conditional parity progression ratio, given that $\Lambda = \lambda$, has the form

$$p_\nu(\lambda) = \exp[-\delta_\nu \lambda], \tag{4.7.5}$$

where δ_ν is a positive parameter, $\nu \geq 1$. Furthermore, it will be supposed that the conditional p.d.f. of the random variable Z_ν, given that $Z_\nu < \infty$ and $\Lambda = \lambda$, is that of a gamma with index parameter θ_ν and scale parameter $\lambda \beta_\nu$. Under these assumptions, see (4.2.2), the conditional p.d.f. $h(\lambda; \nu-1, t)$ in the integral on the right in (4.7.1) takes the form

$$h(\lambda; \nu-1, t) = \frac{\beta_\nu^{\theta_\nu}}{\Gamma(\theta_\nu)} t^{\theta_\nu-1} \lambda^{\theta_\nu} \exp[-(\delta_\nu + \beta_\nu t)\lambda], \tag{4.7.6}$$

for $t > 0$ and $\lambda > 0$.

Substitution of (4.7.4) and (4.7.6) into (4.7.1) leads to

$$h(\nu-1, t) = \frac{\Gamma(\alpha+\theta_\nu)}{\Gamma(\alpha)\,\Gamma(\theta_\nu)} \left(\frac{\beta}{\beta+\delta_\nu+\beta_\nu t}\right)^\alpha \left(\frac{\beta_\nu t}{\beta+\delta_\nu+\beta_\nu t}\right)^{\theta_\nu} \frac{1}{t}, \tag{4.7.7}$$

for $t > 0$, as a formula for the p.d.f. of Z_ν, a p.d.f. that is related to Snedecor's F-distribution in statistics, see Wilkes (1962). The p.d.f. in (4.7.7) may not be convenient for computational purposes. But, a case, making it possible to use computer programs in many statistical libraries, arises, if the p.d.f. of the random variable,

$$W_\nu = \frac{\beta+\delta_\nu}{\beta+\delta_\nu+\beta_\nu Z_\nu}, \tag{4.7.8}$$

is considered. By a straight-forward change of variables, it can be shown that the p.d.f. of the random variable W_ν is

$$\left(\frac{\beta}{\beta+\delta_\nu}\right)^\alpha \frac{\Gamma(\alpha+\theta_\nu)}{\Gamma(\alpha)\,\Gamma(\theta_\nu)} w^{\alpha-1}(1-w)^{\theta_\nu-1}, \tag{4.7.9}$$

for $0 < w < 1$. Moreover, (4.7.8) implies that

$$P[Z_\nu \leq t] = P\left[W_\nu \geq \frac{a_\nu}{a_\nu + \beta_\nu t}\right], \tag{4.7.10}$$

where $a_v = \beta + \delta_v$. Therefore, from (4.7.9), it follows that if $\gamma(t) = a_v (a_v + \beta_v t)^{-1}$ and $t \geq 0$, then

$$H(v-1, t) = \left(\frac{\beta}{\beta + \delta_v}\right)^\alpha \frac{\Gamma(\alpha + \theta_v)}{\Gamma(\alpha)\,\Gamma(\theta_v)} \int_{\gamma(t)}^{1} w^{\alpha-1}(1-w)^{\theta_v - 1}\,dw, \qquad (4.7.11)$$

for $v \geq 1$. The integral on the right in (4.7.11) may, of course, be evaluated in terms of the incomplete Beta function, see Kennedy and Gentle (1980), page 106, for a FORTRAN program. Problem 15 may be consulted for further details on the derivation of (4.7.7) and (4.7.9).

Substantive interpretations of the parameters of the system are always of interest in attempting to interpret the model in terms of maternity histories. Considered first are the parameters δ_v, $v \geq 1$, in the parity progression ratios defined in (4.7.5). When Λ follows a gamma distribution whose p.d.f. is that in (4.7.4), then the v-th parity progression ratio takes the form

$$p_v = \int_0^\infty f(\lambda) \exp[-\delta_v \lambda]\,d\lambda = \left(\frac{\beta}{\beta + \delta_v}\right)^\alpha. \qquad (4.7.12)$$

As one would expect, expression (4.7.12) is the multiplicative factor on the beta density in (4.7.9). From (4.7.12), it is clear that the larger the value of δ_v, the smaller is the probability of a woman of parity $v-1$, $v \geq 1$, having at least one more live birth. In a sense, then, the δ-parameters reflect intentions to have children. On the other hand, when another child is not intended but an additional live birth is experienced, these parameters may also reflect inabilities to control fertility.

Variability in waiting times among live births may be studied in terms of the conditional mean and variance of Z_v, given that $Z_v < \infty$ and $\Lambda = \lambda$. By assumption, the conditional p.d.f. of Z_v, given that $Z_v < \infty$ and $\Lambda = \lambda$, is a gamma with index parameter θ_v and scale parameter $\lambda \beta_v$. Hence,

$$E[Z_v | Z_v < \infty,\ \Lambda = \lambda] = \frac{\theta_v}{\lambda \beta_v},$$

and $\qquad (4.7.13)$

$$\mathrm{Var}[Z_v | Z_v < \infty,\ \Lambda = \lambda] = \frac{\theta_v}{\lambda^2 \beta_v^2}.$$

It is clear from these formulas that, given values of θ_v and β_v, the magnitude of the conditional mean and variance at Z_v varies inversely with λ.

Long waiting times among live births with high variability are thus associated with small values of λ. Alternatively, large values of λ are associated with shorter waiting times, of lower variability, among live births. Similarly, formula (4.7.5) for the conditional v-th parity progression ratio suggests that, given a value of δ_v, large values of λ are associated with smaller probabilities for women of parity $v-1$, $v \geq 1$, experiencing an additional live birth; while small values of λ tend to increase this probability. To the extent that small variability in waiting times among live births reflects abilities to control fertility and large values of λ tend to decrease the probability of having an additional live birth, these observations suggest that the p.d.f. of the random variable Λ corresponds roughly to a distribution of variability among members of a cohort in either abilities or intentions to control fertility. Such an interpretation seems plausible; but, it is very tentative. In considering the results of this section, their exploratory nature should be kept in mind.

Example 4.7.2. *Variances, Covariances, and Correlations of Waiting Times Among Live Births*   .

In mixtures of renewal processes, the waiting times among live births, $Z_1, Z_2, \ldots$, are no longer independent. These random variables will, therefore, have nonzero covariances among sets of women in a cohort experiencing live births. The purpose of this example is to investigate formulas for the variances, covariances, and correlations of these random variables for the gamma mixture of gamma distribution introduced above. In order to avoid a tedious argumentation required to develop formulas for more general cases, attention will be restricted to the random variables Z_1 and Z_2, conditioned on those women in a cohort who have experienced at least two live births. Women in this set will be said to have experienced at least two closed birth intervals.

Mathematically, the set of women experiencing at least two closed birth intervals is characterized by the condition (event) that $Z_1 < \infty$ and $Z_2 < \infty$. Let A_2 be the event Z_1 and Z_2 are finite. Then,

$$P[A_2 | A = \lambda] = \exp[-(\delta_1 + \delta_2) \lambda]. \tag{4.7.14}$$

Hence,

$$P[A_2] = \int_0^\infty f(\lambda) P[A_2 | A = \lambda] \, d\lambda$$

$$= \left(\frac{\beta}{\beta + \delta(2)}\right)^\alpha, \tag{4.7.15}$$

where $\delta(2) = \delta_1 + \delta_2$. From (4.7.15), it follows that the conditional expectation, $E[Z_v | A_2, A = \lambda]$, has the form

$$\left(\frac{\beta + \delta(2)}{\beta}\right)^\alpha \exp[-\delta(2) \lambda] E[Z_v | Z_v < \infty, A = \lambda]. \tag{4.7.16}$$

But, by assumption,

$$E[Z_v | Z_v < \infty, A = \lambda] = \frac{\theta_v}{\lambda \beta_v}, \tag{4.7.17}$$

see (4.7.13). Substituting (4.7.17) into (4.7.16) and evaluating the expectation, $E[A^{-1} \exp[-\delta(2) A]]$, leads to the formula

$$E[Z_v | A_2] = \frac{(\beta + \delta(2)) \theta_v}{\beta_v (\alpha - 1)}, \tag{4.7.18}$$

for $v = 1, 2$. This formula is valid provided that $\alpha > 1$.

Related, but more tedious, arguments may be used to derive formulas for the variances and covariance of the random variables Z_1 and Z_2, conditioned on the event A_2. It can, in fact, be shown that for $v=1$ or 2

$$\text{Var}[Z_v|A_2]=\frac{(\beta+\delta(2))^2\,\theta_v(\alpha+\theta_v-1)}{\beta_v^2(\alpha-1)^2(\alpha-2)}, \tag{4.7.19}$$

and

$$\text{Cov}[Z_1,Z_2|A_2]=\frac{(\beta+\delta(2))^2\,\theta_1\,\theta_2}{\beta_1\,\beta_2(\alpha-1)^2(\alpha-2)}, \tag{4.7.20}$$

provided that $\alpha>2$. From (4.7.19) and (4.7.20), it follows that, when $\alpha>2$, the conditional product moment correlation between Z_1 and Z_2, given the event A_2, is

$$\text{Corr}[Z_1,Z_2|A_2]=\sqrt{\frac{\theta_1\,\theta_2}{(\alpha+\theta_1-1)(\alpha+\theta_2-1)}}. \tag{4.7.21}$$

Observe that the random variables Z_1 and Z_2 are positively correlated.

At least two interesting properties of the model under consideration emerge from an inspection of formulas (4.7.19), (4.7.20), and (4.7.21). One of these properties is that the effect of conditioning on the event A_2 enters into the variances and covariance through the factor $\delta(2)=\delta_1+\delta_2$, the sum of the conditional parity progression ratios in (4.7.5). A detailed inspection of the derivation of formulas (4.7.19) and (4.7.20), furthermore, suggests that if conditioning were on the event A_n that $Z_1<\infty,\ldots,Z_n<\infty$ for some $n\geq2$, then $\delta(2)$ would be replaced by $\delta(n)=\delta_1+\delta_2+\ldots+\delta_n$. The model thus has the property that the variances and covariances depend on the condition event A_n of selecting those women who have experienced at least n closed birth intervals. The model, therefore, has potential usefulness in evaluating sources of bias in selecting samples of women for analyses of data on maternity histories.

Formula (4.7.21) for the product moment correlation of Z_1 and Z_2, on the other hand, has the property that it depends only on the index parameters in the underlying gamma distributions and not on the scale parameters, the β's, or the δ-parameters in the parity progression ratios. This property demonstrates that the product moment correlations among the random variables $Z_1,Z_2,\ldots,Z_n$ are invariant with respect to the condition event A_n. Such an invariance property may be useful in estimating the index parameters free of selection bias.

Some intermediate steps used in the derivation of formulas (4.7.19) and (4.7.20) are discussed in Problem 16.

Example 4.7.3. *Conditional Distributions of the Random Variable Λ*

Whatever be the interpretation of the random variable Λ, its conditional distribution, given some event characterizing a selected sample of women, is always of interest in interpreting the effects of selection. One of the simplest cases to consider is the conditional distribution of Λ, given that $Z_1=t$, the event the first live birth occurs t time units following marriage.

To derive the conditional p.d.f. for this distribution, let $f(\lambda)$ and $h(\lambda;0,t)$ be defined as in (4.7.4) and (4.7.6). Then,

$$g(\lambda,t)=f(\lambda)h(\lambda;0,t) \tag{4.7.22}$$

is the joint p.d.f. of the random variables Λ and Z_1. The marginal p.d.f. of the random variable Z_1, $h(0,t)$, is given by (4.7.7). Now $h^*(t;\lambda)$, the conditional p.d.f. of the random variable Λ, given that $Z_1 = t$, has the form $h^*(t;\lambda)$ $= g(\lambda,t)/h(0,t)$. Therefore, if $\alpha(1) = \alpha + \theta_1$ and $\beta(1,t) = \beta + \delta_1 + \beta_1 t$, then this conditional p.d.f. takes the form

$$h^*(t;\lambda) = \frac{[\beta(1,t)]^{\alpha(1)}}{\Gamma(\alpha(1))}\,\lambda^{\alpha(1)-1}\exp[-\beta(1,t)\,\lambda], \tag{4.7.23}$$

for $\lambda > 0$ and $t > 0$. But, the p.d.f. in (4.7.23) is that of a gamma distribution with index parameter $\alpha(1)$ and scale parameter $\beta(1,t)$. It is interesting to note that the regression of Λ on Z_1 has the inverse linear form

$$E[\Lambda|Z_1=t] = \frac{\alpha(1)}{\beta(1,t)} \tag{4.7.24}$$

for $t > 0$.

Formula (4.7.23) can be easily generalized to the conditional p.d.f. of Λ, given the event $Z_1 = t_1$ and $Z_2 = t_2$. Among females experiencing this event, the first birth occurs t_1 time units after marriage; while the second arrives at $t_1 + t_2$ time units following marriage. Define an index parameter $\alpha(2)$ by $\alpha(2) = \alpha + \theta_1 + \theta_2$ and a scale parameter $\beta(2,t_1,t_2)$ by $\beta(2,t_1,t_2) = \beta + \delta_1 + \delta_2 + \beta_1 t_1 + \beta_2 t_2$. Then, $h^*(t_1,t_2;\lambda)$, the conditional p.d.f. of Λ given that $Z_1 = t_1$ and $Z_2 = t_2$, is again a gamma type p.d.f. with index parameter $\alpha(2)$ and scale parameter $\beta(2,t_1,t_2)$. Problem 17 is devoted to the derivation and other properties of the joint p.d.f. of the random variables Z_1 and Z_2.

Example 4.7.4. *Distribution of Waiting Times to n-th Live Birth*

In principle, given that $\Lambda = \lambda$, the conditional p.d.f. of the random variable $Z_1 + Z_2 + \ldots + Z_n$, representing waiting times to the n-th live birth following marriage, could be worked out along the lines followed by George and Mathai (1975). Only a simple special case will be considered in this example, however. Suppose that the scale parameters in the conditional distributions governing waiting times among live births are homogeneous, i.e., $\beta_v = \beta_1$ for all $v \geq 1$. It is a well-known result that a sum of independent gamma random variables with index parameters $\theta_1, \theta_2, \ldots, \theta_n$ and a common scale parameter $\lambda\beta_1$ is again a gamma random variable with index parameter $\theta(n) = \theta_1 + \ldots + \theta_n$ and scale parameter $\lambda\beta_1$. From this result, it may be concluded that the conditional p.d.f. of $Z_1 + \ldots + Z_n$, given that $\Lambda = \lambda$, has the form

$$h^{(n)}(\lambda;t) = \frac{(\lambda\beta_1)^{\theta(n)}}{\Gamma(\theta(n))}\,t^{\theta(n)-1}\exp[-(\delta(n)+\beta_1 t)\,\lambda], \tag{4.7.25}$$

where $\lambda > 0$ and $\delta(n) = \delta_1 + \ldots + \delta_n$. Therefore, it follows that the p.d.f. of $Z_1 + \ldots + Z_n$,

$$h^{(n)}(t) = \int_0^\infty f(\lambda)h^{(n)}(\lambda;t)\,d\lambda, \quad t > 0, \tag{4.7.26}$$

has the same form as the p.d.f. in (4.7.7). All one needs to do in setting down a formula for $h^{(n)}(t)$ is to replace β_v, δ_v, and θ_v by β_1, $\delta(n)$, and $\theta(n)$, respectively in (4.7.7). With these replacements, formula (4.7.11) could also be used to compute $H^{(n)}(t)$, the distribution function of the p.d.f. $h^{(n)}(t)$ in (4.7.26). The conditional p.d.f. of Λ, given that $Z_1 + \ldots + Z_n = t$, could also be derived along the lines used in Example 4.7.4.

One advantage of being able to express the densities $h^{(n)}(t)$ in closed form is that the need for evaluating many convolutions in calculating the gross net maternity function is eliminated. For, if it is supposed that T_0, a random variable with p.d.f. $g_0(x)$ representing age at first marriage in a cohort, is independent of the random variables $Z_1, Z_2, \ldots$, then it follows that the gross maternity function defined in (4.3.5) takes the form

$$b(t) = \int_0^t g_0(x) b(x, t-x) dx, \qquad t > 0, \tag{4.7.27}$$

where

$$b(x, t-x) = \sum_{n=1}^{\infty} h^{(n)}(t-x), \qquad t > x. \tag{4.7.28}$$

To express the series on the right in a form that is easy to compute, define three sequences of constants by

$$a(n) = \frac{\beta^\alpha \beta_1^{\theta(n)} \Gamma(\alpha + \theta(n))}{\Gamma(\alpha) \Gamma(\theta(n))}, \tag{4.7.29}$$

$b(n) = \beta + \delta(n)$, and $c(n) = \alpha + \theta(n)$ for $n \geq 1$. Then, the series on the right in (4.7.28) may be represented in the compact form

$$\sum_{n=1}^{\infty} \frac{a(n) t^{\theta(n)-1}}{(b(n) + \beta_1 t)^{c(n)}} \tag{4.7.30}$$

at the argument $t > 0$. In practice, because the parity progression ratios, $p_v(\lambda)$, in (4.7.5) vanish for all $\lambda > 0$ and $n \geq n_0$, some maximum parity in a cohort, there will be only finitely many nonzero terms in (4.7.30), making the function in (4.7.28) easy to compute. More difficult problems than computer implementation are those of developing procedures to estimate the parameters and deciding whether the homogeniety assumption on the scale parameters, which underlies formula (4.7.30), leads to a useful model of maternity histories.

When the assumption that the random variable T_0 is independent of the random variables $Z_1, Z_2, \ldots$, is retained but no further restrictions are placed on the scale parameters $\beta_v > 0$, $v \geq 1$, it is still possible to write down a formula for the limiting final parity distribution $q_n(x, t)$ as defined for any $t > 0$ in (4.2.11). Thus, from a formula for $H^{(n)}(t)$ analogous to (4.7.11), it follows from (4.2.17) that for all $x > 0$

$$\lim_{t \uparrow \infty} q_n(x, t) = \left(\frac{\beta}{\beta + \delta(n)}\right)^\alpha - \left(\frac{\beta}{\beta + \delta(n+1)}\right)^\alpha, \tag{4.7.31}$$

where $n \geq 0$ and $\delta(0) = 0$. One obvious unrealistic feature of the model is that the limit on the right in (4.7.31) does not depend on x. Formula (4.7.31) suggests, however, that the variable x, at marriage, could be incorporated into the model by letting α, β, and the δ-parameters depend on x. What forms these parameters should take as functions of x is a question that needs further investigating.

Problem 18 contains a discussion relating the negative binomial distribution to the gamma mixture of gamma distributions considered in this section. Brass (1958), among others, has used the negative binomial as a distribution for family size in human populations. Sheps and Menken (1973), page 189 and elsewhere, mention mixtures of renewal processes. The approach outlined in this section, however, differs from that considered by these authors.

4.8 An Age-Dependent Potential Birth Process

Dependence on age is a natural phenomenon to consider when constructing models of the human life cycle. An example of an age-dependent phenomenon is that of the ability of the human female to conceive. Abilities to conceive vary among women in a cohort after entering the childbearing ages at about age 15 years. But, in a vast majority of cases, particularly at ages past 35 years, as a woman increases in age, her ability to conceive declines until permanent physiological sterility is reached at about age 50. Models of maternity histories accommodating some type of age-dependence are, therefore, of interest.

Just as in the preceding sections, let T_0 be a random variable, representing age at marriage in a cohort of females, and, let T_v, $v \geq 1$, be a sequence of random variables, representing ages at which these females experience live births. Successive births to women occur at increasing ages; consequently, the sequence of random variables T_v, $v \geq 0$, is monotone increasing, i.e., $T_{v-1} \leq T_v$, $v \geq 1$, with probability one. In terms of the random variables T_v, $v \geq 0$, the sequence of random variables, representing waiting times among live births, is given by

$$Z_v = T_v - T_{v-1}, \qquad v \geq 1.$$

A basic problem in constructing models of an age-dependent potential birth process is that of specifying the joint conditional distributions of the random variables $Z_1, Z_2, \ldots, Z_n$, $n \geq 2$, given that $T_0 = x$. One approach to the problem of specifying these joint distributions is through conditional distributions, given preceding events in a maternity history. For, let $\mathbf{t}^{(0)} = \langle x_0 \rangle$; and, for $v \geq 1$, let $\mathbf{t}^{(v)} = \langle x_0, t_1, \ldots, t_v \rangle$, a vector of nonnegative real numbers. Furthermore, let $A(\mathbf{t}^{(0)})$ represent marriage at age x_0; and, for $v \geq 1$, let $A(\mathbf{t}^{(v)})$ represent a maternity history such that $T_0 = x_0$, $Z_1 = t_1, \ldots, Z_v = t_v$. Then, for a woman of parity $v - 1$ and previous history $A(\mathbf{t}^{(v-1)})$, the conditional distribution function of the waiting time to the v-th live birth is defined by

$$F(v-1, \mathbf{t}^{(v-1)}, t_v) = P[Z_v \leq t_v | A(\mathbf{t}^{(v-1)})], \tag{4.8.1}$$

where $v \geq 1$. Assuming the distribution function in (4.8.1) has the p.d.f. $f(v-1, \mathbf{t}^{(v-1)}, t_v)$ and applying known laws from probability theory for multiplying conditional densities, it follows that the conditional joint p.d.f. of the random variables $Z_1, Z_2, \ldots, Z_n$, giving that $T_0 = x_0$, is

$$\prod_{v=1}^{n} f(v-1, \mathbf{t}^{(v-1)}, t_v), \qquad n \geq 1. \tag{4.8.2}$$

Basic to the computer implementation of an age-dependent potential birth process is the calculation of the conditional distribution function,

$$H^{(n)}(x, t) = P[Z_1 + \ldots + Z_n \leq t \mid T_0 = x], \tag{4.8.3}$$

of the waiting times to the n-th live birth, $n \geq 1$, following marriage at age x. In principle, once the conditional p.d.f.'s going into the product in (4.8.2) are specified, the distribution function in (4.8.3) could be computed by integrating the p.d.f. in (4.8.2) over the set of points $\langle t_1, t_2, \ldots, t_n \rangle$ such that $t_1 + t_2 + \ldots t_n \leq t$ with $x_0 = x$. But, even if the p.d.f. in (4.8.2) were completely specified, the computation of such multi-dimensional integrals is known to be a difficult numerical problem. One is, therefore, led to some scheme of simplification to make computer implementation tractable.

A computationally tractable age-dependent model of maternity histories arises if attention is focused on the joint distribution of ages of mothers when live births occur rather than waiting times among live births. The transformation connecting the ages of the mothers at the births of their children and waiting times among live births is

$$T_v = T_{v-1} + Z_v, \qquad 1 \leq v \leq n. \tag{4.8.4}$$

With respect to realizations of maternity histories, the ages of mothers when births occur are $x_v = x_{v-1} + t_v$, $1 \leq v \leq n$.

For every $v \geq 1$ and nonnegative number x_{v-1}, let $h(v-1, x_{v-1}, t_v)$ be a nonnegative p.d.f. such that

$$\int_0^\infty h(v-1, x_{v-1}, s) \, ds \leq 1. \tag{4.8.5}$$

The fundamental assumption underlying the age-dependent potential birth process to be developed in this section is stated by the equation

$$f(v-1, \mathbf{t}^{(v-1)}, t_v) = h(v-1, x_{v-1}, t_v), \tag{4.8.6}$$

which is assumed to hold for all $v \geq 1$, vectors $\mathbf{t}^{(v-1)}$, and $t_v > 0$. Given a previous history $\mathbf{t}^{(v-1)}$, equation (4.8.6) states that the conditional p.d.f. of Z_v, the waiting time to the v-th birth, $v \geq 1$, depends only on parity $v-1$ and the age x_{v-1} of the mother when this parity was attained.

Under the assumption stated in (4.8.6), the conditional joint p.d.f. of the random variables $Z_1, Z_2, \ldots, Z_n$, given that $T_0 = x_0$, may be expressed in the form

$$\prod_{v=1}^{n} h(v-1, x_{v-1}, t_v), \tag{4.8.7}$$

see (4.8.2). From Eq. (4.8.4) and the technical details given in Problem 19, it follows that the joint conditional p.d.f. of the random variables $T_1 \leq T_2 \leq \ldots \leq T_n$, given that $T_0 = x_0$, is

$$\prod_{v=1}^{n} h(v-1, x_{v-1}, x_v - x_{v-1}), \qquad n \geq 1, \tag{4.8.8}$$

where $x_0 \leq x_1 \leq x_2 \leq \ldots \leq x_n$.

By a similar argument, it can be shown that the conditional p.d.f. of the random variables $T_1, T_2, \ldots, T_{n-1}$ and Z_n, given that $T_0 = x_0$, is

$$\prod_{v=1}^{n-1} h(v-1, x_{v-1}, x_v - x_{v-1}) \, h(n-1, x_{n-1}, t_n), \tag{4.8.9}$$

for $n \geq 2$. Consequently, the condition p.d.f. of Z_n, given the past history, $T_0 = x_0$. $T_1 = x_1, \ldots, T_{n-1} = x_{n-1}$, is $h(n-1, x_{n-1}, t_n)$. Thus, under the assumption stated by Eq. (4.8.6), the potential birth process is age-dependent in the sense that the p.d.f. of the waiting time Z_n to the n-th live birth following birth $n-1$ depends only on the age x_{n-1} of the mother when parity $n-1$ was attained and not on ages at previous births. Of course, $h(0, x_0, t_1)$ is the conditional p.d.f. of Z_1, the waiting time to the first birth following marriage at age x_0. Another way of characterizing the age-dependent property is to state that the sequence of random variables T_v, $v \geq 1$, is a conditional Markov process, given that $T_0 = x$.

An immediate consequence of the age-dependent property just described is that the computation of the conditional distribution functions defined in (4.8.3) becomes numerically tractable. For let $h^{(1)}(x, t) = h(0, x, t)$ and define $H^{(1)}(x, t)$ by

$$H^{(1)}(x, t) = \int_0^t h^{(1)}(x, s) \, ds, \qquad t > 0. \tag{4.8.10}$$

Then, the sequence, $H^{(n)}(x, t)$, $n \geq 1$, may be calculated recursively by applying the formula

$$H^{(n)}(x, t) = \int_0^t h^{(n-1)}(x, s) \, H(n-1, x+s, t-s) \, ds, \tag{4.8.11}$$

where $h^{(n-1)}(x, t)$ is the density of $H^{(n-1)}(x, t)$, $H(n-1, x, t)$ is the distribution function associated with the density $h(n-1, x, t)$, $n \geq 2$, $x > 0$, and $t > 0$. Henceforth, the operation defined by (4.8.11) will be referred to as an age-dependent convolution.

Given the sequence of conditional distribution functions defined by (4.8.10) and (4.8.11), many of the formulas for the potential birth process developed in Sect. 4.2 continue to apply. For example, the conditional distribution of the potential birth process $K(t)$ at t time units following marriage at age x is

$$q_n(x, t) = P[K(t) = n \mid T_0 = x] = H^{(n)}(x, t) - H^{(n+1)}(x, t), \tag{4.8.12}$$

for $n \geq 0$, $x > 0$, and $t > 0$, provided that $H^{(0)}(x, t)$ is defined by $H^{(0)}(x, t) = 1$ for all $x > 0$ and $t > 0$. The mean of the distribution in (4.8.12), see (4.2.14), is

$$v(x, t) = \sum_{n=1}^{\infty} H^{(n)}(x, t). \tag{4.8.13}$$

Formula (4.2.16) for $v(t)$, the mean of the potential birth when a distribution of age at first marriage in a cohort is taken into account, therefore, continues to apply.

Many of the formulas for parity progression ratios developed in Sect. 4.4 are, however, no longer appropriate for an age-dependent potential birth process, because the probability of advancing an additional parity depends on the age of the mother when the last parity was attained. Thus, for $n \geq 1$,

$$p_n(x) = \lim_{t \uparrow \infty} \int_0^t h(n-1, x, s) \, ds \tag{4.8.14}$$

is the conditional probability of eventually experiencing the n-th live birth, given that birth $n - 1$ was experienced at age x. It is clear that K, a random variable representing final parity in a cohort of women surviving to the end of the childbearing years, has the property

$$P[K \geq n | T_0 = x] = \lim_{t \uparrow \infty} H^{(n)}(x, t) \tag{4.8.15}$$

for all $n \geq 0$ and $x > 0$. But, the limit on the right will not, in general, be the product of parity progression ratios as expressed in (4.4.2).

A basic step in the computer implementation of an age-dependent potential birth process is the calculation of the gross maternity function $b(x, t)$ for those women in a cohort marrying at age x. By definition, in the stochastic process under consideration, $b(x, t)$ is the density of the mean function $v(x, t)$ given by (4.8.13). Under the assumption the series in (4.8.13) can be differentiated term by term and the resulting series converges at each point $\langle x, t \rangle$; the function $b(x, t)$ takes the form

$$b(x, t) = \frac{\partial v(x, t)}{\partial x} = \sum_{n=1}^{\infty} h^{(n)}(x, t), \tag{4.8.16}$$

where, as before $h^{(n)}(x, t)$, $n \geq 1$, are the densities of the conditional distribution functions determined by (4.8.10) and (4.8.11). Actually, due to the fact that human females can experience only finitely many births during their life spans, the series in (4.8.16) will contain only finitely many nonzero terms at each point $\langle x, t \rangle$.

With $b(x, t)$ defined as in (4.8.16), the formula for $b(t)$, the gross maternity function for an age-dependent potential birth process, takes the familiar form

$$b(t) = \int_0^t g_0(x) b(x, t - x) \, dx, \tag{4.8.17}$$

for $x > 0$ and $t > 0$. Again, $g_0(x)$ is the p.d.f. of the random variable T_0. It is thus clear that the problem of finding efficient methods for calculating the gross maternity function in (4.8.17) reduces, essentially, to finding efficient ways of computing values of the series on the right in (4.8.16) on some set of $\langle x, t\rangle$-points.

Formulations based on continuous variables have aesthetic mathematical appeal when dealing with models centered on the concepts of age and time. On the other hand, when confronted with practical problems of computer implementation, it is very useful, indeed essential, to construct models closely mimicking the arithmetic done by computers. Henceforth, therefore, age x and time t will be viewed as discrete variables ranging over some finite set of points $\langle x, t\rangle$, $x, t = 0, 1, 2, \ldots$. To avoid overburdening the notation, from now on all functions defined so far in this section will have this finite set of points as their domain.

Starting from a sequence of p.d.f.'s, $h(n-1, x, t)$, $n \geq 1$, satisfying the condition $h(n-1, x, 0) = 0$ for all $n \geq 1$ and $x \geq 0$, the sequence of densities in the series on the right in (4.8.16) would be calculated according to the recursion relations determined by $h^{(1)}(x, t) = h(0, x, t)$ and the finite sums

$$h^{(n)}(x, t) = \sum_{s=0}^{t} h^{(n-1)}(x, s) h(n-1, x+s, t-s), \qquad (4.8.18)$$

defined for $n \geq 2$, $x \geq 0$, and $t \geq 0$. Even though the recursive algorithm determined by (4.8.18) can be programmed in a straight-forward manner, the computation of ten to fifteen terms of series in (4.8.16) can become prohibitive if the lattice of $\langle x, t\rangle$-points is large, unless further simplifying assumptions are introduced.

Empirical evidence, based on analyzing data on maternity histories, suggests that in some cohorts distributions of waiting times among live births become homogeneous after some number of live births. Singh et al. (1974), for example, used a model in which waiting times among live birth following the first had a common distribution, see Example 4.5.2. There is thus some justification for assuming that the potential birth process under consideration becomes only age-dependent after some number of live births. That is, there is some integer $n_0 \geq 1$ and a conditional p.d.f. $g(x, t)$ such that

$$h(n, x, t) = g(x, t) \qquad (4.8.19)$$

for all $n \geq n_0$, $x \geq 0$, and $t \geq 0$.

Let $g^{(0)}(x, t)$ be a function of $\langle x, t\rangle$ defined by $g^{(0)}(x, 0) = 1$ for all $x \geq 0$ and by $g^{(0)}(x, t) = 0$ for all $x \geq 0$ and $t \geq 1$. It is shown in Problem 20 that the function $g^{(0)}(x, t)$ is the identity with respect to the operation of age-dependent convolution. Successive age-dependent convolutions of the function $g(x, t)$ with itself are computed by setting $g^{(1)}(x, t) = g(x, t)$ and using the recursion relation

$$g^{(n)}(x, t) = \sum_{s=0}^{t} g^{(n-1)}(x, s) g(x+s, t-s), \qquad (4.8.20)$$

for $n \geq 2$, $x \geq 0$, and $t \geq 0$. A basic role in calculating the function $b(x, t)$ in (4.8.16), under the homogeneity assumption (4.8.19), is played by a function $f(x, t)$ defined by

$$f(x, t) = g^{(0)}(x, t) + g^{(1)}(x, t) + \ldots \qquad (4.8.21)$$

at all points $\langle x, t \rangle$ on a finite lattice. It turns out, see Problem 20, that the condition $g(x, 0) = 0$ for all $x \geq 0$ implies that the series in (4.8.21) contains only finitely many nonzero terms at each point $\langle x, t \rangle$. With $f(x, t)$ defined as in (4.8.21), it can be shown that for all points $\langle x, t \rangle$ on a finite lattice and $n_0 \geq 1$ the function $b(x, t)$ takes the form

$$b(x, t) = \sum_{v=1}^{n_0-1} h^{(v)}(x, t) + \sum_{s=0}^{t} h^{(n_0)}(x, s) f(x+s, t-s), \qquad (4.8.22)$$

provided the first sum on the right is interpreted as zero when $n_0 = 1$.

Even though Eq. (4.8.22) and the foregoing discussion demonstrate that the values of $b(x, t)$ on a finite lattice of $\langle x, t \rangle$-points may be calculated in finitely many steps, computer implementation may still not be practical unless an efficient method for calculating the values of the function $f(x, t)$ can be found. Littman, see Mode and Littman (1977), has shown that $f(x, t)$ can be computed in about the same number of steps as that required to compute $g^{(2)}(x, t)$. The computation of $b(x, t)$ thus becomes feasible for n_0 in the range $1 \leq n_0 \leq 5$, a range that covers many cases of interest.

Theorem 4.8.1 (Littman's Algorithm): Let the function $f(x, t)$ be defined on a finite lattice of $\langle x, t \rangle$-points as in (4.8.21). Then, $f(x, 0) = 1$ for all $x \geq 0$ and $f(x, t)$ may be calculated recursively from the equation

$$f(x, t) = \sum_{s=0}^{t-1} f(x, s) g(x+s, t-s), \qquad (4.8.23)$$

which is valid for all $x \geq 0$ and $t \geq 1$.

Proof. Equation (4.8.21) is analogous to the scalar equation $(1-x)^{-1} = 1 + x + x^2 + \ldots, |x| < 1$, for a geometric series. Clearly, $(1-x)^{-1}(1-x) = 1$. By analogy, it may be shown that

$$\sum_{s=0}^{t} f(x, s) [g^{(0)}(x+s, t-s) - g(x+s, t-s)] = g^{(0)}(x, t), \qquad (4.8.24)$$

using straight-forward algebra. But, Eq. (4.8.24) is equivalent to

$$f(x, t) = g^{(0)}(x, t) + \sum_{s=0}^{t} f(x, s) g(x+s, t-s). \qquad (4.8.25)$$

Equation (4.8.23) now follows by noting that $g^{(0)}(x, t) = 0$ for all $x \geq 0$ and $t \geq 1$ and $g(x, 0) = 0$ for all $x \geq 0$. ☐

Example 4.8.1. *Maternity Histories in a Nineteenth Century Belgian Commune – La Hulpe*

Among the essential steps in the development of stochastic models of human reproduction are their applications to the analysis and interpretation of real data. Whenever it is feasible, attempts to validate the model should also be included in these applications. Pickens (1980) has applied a version of the age-dependent potential birth process outlined above to the analysis of data on maternity histories in the nineteenth century Belgian commune, La Hulpe, for the period 1846–1881. It will be recalled that some of the estimates of the parameters in the lognormal distribution displayed in Table 4.6.1 were based on marriage data from the La Hulpe population. Analyzing maternity histories based on historical data requires careful attention to details, regarding the methods used in its collection, so that the possibility of biases may be minimized. Any discussion of the quality and possible defects in the La Hulpe data on maternity histories is beyond the scope of this example. Suffice it to say that the data are believed to be of good quality; the paper of Pickens and the references contained therein may be consulted for further details.

After eliminating cases with illegitimate children, maternity histories on 753 women were included in the analysis. As yet, useful parametric forms of the distribution functions $H(n-1, x, t)$, corresponding to the p.d.f.'s on the right in (4.8.6), have not been developed. Consequently, it was necessary to use a nonparametric procedure to obtain the desired estimates of the distribution functions $H(n-1, x, t)$. Following a widespread practice in demography, the nonparametric methods used in the estimation procedures were of the life table type. Due to limitations of sample size, it was possible to estimate $H(n-1, x, t)$ as a function of $n \geq 1$ and $t > 0$, on a time scale of one-tenth year, only by grouping the ages x when women attained parity $n-1$. Because women in La Hulpe tended to marry relatively late in life, the age groups chosen were [20, 24], [25, 29], [30, 34], [35, 39], and [40, 44]. The age group [15, 19] was not considered due to very few women marrying at these ages. For women attaining a given parity at ages 40 or more, it was sometimes necessary to group ages in the entire interval [40, 49] to obtain sample sizes sufficiently large for estimation purposes.

Two versions of the parity homogeneity assumption in (4.8.19) were used in the estimation procedures. In one version, called the parity homogenous case, it was assumed n_0 in (4.8.19) equalled one so that, following the first live birth, distributions for waiting times among subsequent live births were only age-dependent. In a second version, called the parity dependent case, it was assumed that n_0 equalled three.

By present-day Western European standards, infant mortality in nineteenth century La Hulpe was high; for, during the first year following birth, about 15% of the infants died. When mothers nurse infants and do not use contraception, as was the case in La Hulpe, deaths of infants tend to shorten waiting times to the next live birth, due to the cessation of nursing and the resumption of ovulation sooner than would have otherwise been the case. To get some measure of the change in pace of childbearing due to infant mortality, distribution functions of waiting times among live births were again estimated, after all births straddling an infant death were eliminated from the sample. Version one of the parity homogeniety assumption was used in this attempt to measure the impact of infant mortality on the pace of childbearing.

From the expository point of view, a disadvantage of nonparametric estimation procedures is that they lead to rather large numerical arrays whose information content cannot readily be communicated on a printed page of standard size. Barriers to communication thus arise. Computer generated graphs of nonparametric estimates of distribution functions are often useful in overcoming these barriers. Figure 2 in the paper of Pickens contains graphs of the nonparametric estimates of the distribution functions, $H(n-1, x, t)$, $n \geq 1$, for version one of the parity homogeniety assumptions. Briefly, the estimates represented by the graphs in this figure suggest that, among those women in La Hulpe who married for the first time during the age intervals [20, 24] and [25, 29], well over ninety percent experienced at least one live birth. But, for women marrying at ages in the intervals [30, 34] and [35, 39], this percentage dropped to about seventy-five; and, for women marrying at ages in the interval [40, 44], less than twenty-five percent experienced at least one live birth. Graphs of estimates of these distribution functions for $n \geq 2$ indicated, however, that among women experiencing at least one live birth, the probability of experiencing at least one more was uniformly higher across all age groups than for women of parity zero, suggesting heterogeneity in abilities to conceive.

Table 1 in the paper of Pickens contains estimates of the distribution functions $H(n-1, x, t)$ for $n \geq 2$, under version two of the parity homogeneity assumptions and for the case all intervals straddling infant deaths were eliminated from the sample. For each $n \geq 2$ and x belonging to one of the five age intervals, estimates of $H(n-1, x, t)$ were reported at $t = 1, 2, \ldots, 5$ expressed in years since attaining the previous parity. The estimates $\hat{H}(n-1, x, 5)$ contain useful information on whether women of age x when attaining parity $n-1$ will experience at least one more live birth in that these estimates are lower bounds on the conditional parity progression ratios expressed in (4.8.14).

Presented in Table 4.8.1 are the estimates $\hat{H}(n-1, x, 5)$ taken from Table 2 in Pickens' paper. Observe that in version two of the parity homogeneity assumptions, birth intervals for women of parity one or two were actually pooled in one class, due to limitations of sample size; while intervals for women of parity three or higher were pooled into another, see the third and fourth column of Table 4.8.1. The estimates presented in Table 4.8.1 tend to document, in a more precise digital form, statements to the effect that women of parity one or greater have larger parity progression ratios than those of parity zero. For example, among women in the age group [40, 44] attaining parity one or more, about 33–35% move on to a higher parity as compared to less than 25% for women in this age group marrying for the first time. The estimates in Table 4.8.1 also supply information on the magnitudes of decreases in parity progression ratios with increasing age. Judging whether differences in estimates in Table 4.8.1 are statistically significant is not straight-forward, because the usual assumptions of independence do not hold. Use of data from the same maternity histories, for example, in calculating the estimates in Table 4.8.1, makes them dependent within and across columns.

Pickens also estimated the density $b(x, t)$ on a lattice of $\langle x, t \rangle$-points, using formula (4.8.22) and Littman's algorithm. Having estimates of the density $b(x, t)$ on this lattice, it was possible to estimate the mean function $v(x, t)$ in (4.8.13) by computing cumulative sums. An estimate of $v(20, 30)$, for example, would correspond to the mean number of live births experienced by women marrying at age 20, surviving to age 50, and remaining married for 30 years. Under version one of the parity homogeneity assumption, $\hat{v}(20, 30) = 7.5$, under version two $\hat{v}(20, 30) = 7.7$, but, when intervals straddling infant deaths were elim-

Table 4.8.1. Lower bounds on parity progression ratios of order one or more for La Hulpe women

Age group	Intervals not straddling infant deaths	Parities one or two	Parities three or more
20–24	0.954	0.958	–[a]
25–29	0.914	0.935	0.913
30–34	0.872	0.922	0.878
35–39	0.820	0.815	0.827
40–44	0.330	–[a]	0.347

[a] Either no observations or insufficient dample size.

inated from the sample, $\hat{v}(20, 30) = 6.85$. This latter estimate suggests that the increase in fertility attributable to infant mortality was slight in the La Hulpe population, the difference $7.5 - 6.85 = 0.65$ being less than one live birth. At ages of marriage greater than 20, this difference would be expected to be even less.

Age at marriage for women in La Hulpe was relatively late as illustrated by the graph of the p.d.f. in Fig. 4.6.1. When this distribution for age at first marriage and the distribution functions estimated under the version one of the parity homogeneity assumptions were used to estimate $v(50)$, see (4.2.16), the value $\hat{v}(50) = 3.5$ was obtained. According to this estimate, on the average, married women in La Hulpe, surviving and remaining married to age 50, would experience about 3.5 live births. It is thus clear that, under the conditions of natural fertility experienced by the La Hulpe population, delaying marriage can be a decisive factor in decreasing fertility.

The mean $v(50)$ was also estimated when a marriage could be dissolved by the death of either the wife or husband, under the simplifying condition that the ages of brides and grooms were identical. When La Hulpe mortality was taken into account in this fashion, the estimate $\hat{v}(50) = 2.4$ was obtained, suggesting that the rather high level of adult mortality combined with a late age at marriage was a factor of some significance in depressing fertility.

In order to provide some insights into variability in family sizes, Pickens also displayed histograms of the p.d.f. $q_n(x, t)$ in (4.8.12) in his Fig. 5 at chosen values of x and t. A successful attempt to validate the model is also discussed in Sect. 6 of his paper. The results presented by Pickens illustrate that the model for the age-dependent potential birth process introduced in this section is a valid and useful tool for analyzing and interpreting data on maternity histories.

4.9 An Evolutionary Potential Birth Process

So far, in the development of stochastic models of the potential birth process, attention has been focused on a single birth cohort of females with a time invariant law of evolution. Rarely, however, are data on maternity histories collected for a single birth cohort of females and there is much evidence that laws governing the evolution of maternity histories change in time. In sample surveys taken over a period of several months, for example, data are usually collected on a sample of women with ages spanning the childbearing years, resulting in information on a mixture of complete and incomplete maternity histories. Data on complete maternity histories, i.e., on those women who cannot or will not have any more children, may possibly reflect changing patterns of childbearing for cohorts born at different epochs. But, data on incomplete maternity histories will also reflect the uncertainties of childbearing. For, some fertile women in the sample may be uncertain regarding their desires and future prospects for childbearing.

Whether women in the childbearing ages, especially in developed countries, have children, or additional children, depends not only on their abilities to control fertility but also on perceived abilities of a family to support them emotionally and economically. These perceived abilities may also change in time, due in part to changing perceptions as cohorts advance in age and in part to broad changes in economic and social conditions common to all cohorts living during a particular time period. It thus seems widely agreed that data on maternity histories should not be viewed as realization of some stochastic process with time invariant laws of evolution but as arising from some evolutionary or developmental stochastic process, whose laws of evolution may change with time. Indeed, these laws of evolution will be influenced by the time periods, epochs, in which the cohorts live.

In many developed countries, annual births, or perhaps births during a given period, are registered according to age of the mother at her last birthday. Estimates of the female population by age for each year, or period, are also frequently available. From these two pieces of information, age-specific as well as total fertility rates can be calculated for each period under study. Evidence for laws governing the evolution of maternity histories changing in time may also be obtained from the study of time series of such period and age-specific and total fertility rates. Examples of these time series abound in the demographic literature. A recent example has been supplied by the paper of Blackwood (1981), containing a yearly listing of age-specific and total fertility rates for Alaska Natives, during the period, 1959–1978. In 1959, the period total fertility rate of Alaska Natives was 7.225 births per woman; by 1978, this rate has declined to 3.731.

The development, computer implementation, and testing of an evolutionary potential birth process accommodating a set of biological, social, and economic factors, known or believed to affect human reproduction, is a difficult task, requiring the efforts of many investigators over an extended time period. Only a brief outline of such a process will be attempted in this section. More precisely, an attempt will be made to extend some of the results in Sects. 4.5, 4.3, 4.4, and 4.8 to an evolutionary potential birth process.

A useful way to approach the problem of formulating a potential birth process, with laws of evolution that may change in time, is to index each cohort by the epoch of birth. Accordingly, as in the study of mortality in Chap. 2, $t \in R$ will represent an epoch. Among those females in a cohort born at epoch $t \in R$ who marry and have children, let $T_0(t)$ be a random variable standing for age at marriage and let $T_n(t)$, $n \geq 1$, be random variables standing for the ages at which these females experience live births. In analogy with the formalism of Sect. 4.8, the random variables $Z_n(t) = T_n(t) - T_{n-1}(t)$, $n \geq 1$, will be waiting times among live births for a cohort of females born at epoch $t \in R$.

To provide for laws of evolution that may change with time, the joint distribution of the random variables

$$T_0(t) \leq T_1(t) \leq \ldots \leq T_n(t), \quad n \geq 1,$$

will be determined by age and period effects experienced by a cohort born at epoch $t \in R$. A convenient device for incorporating these effects mathematically is through risk or intensity functions. Consider, for example, age at marriage and let $\theta_f(t, x)$ be a continuous and nonnegative risk function for a female of age $x \in R^+$ to marry at epoch $t \in R$. Then, for a female cohort born at epoch $t \in R$, the distribution function of the random variable $T_0(t)$ is determined by

$$G_0(t, x) = P[T_0(t) \leq x] = 1 - \exp\left[-\int_0^x \theta_f(t+s, s)ds\right], \tag{4.9.1}$$

for $x > 0$. The probability that a female in this cohort eventually marries is

$$G_0(t, \infty) = 1 - \exp\left[-\int_0^\infty \theta_f(t+s, s)ds\right] \leq 1. \tag{4.9.2}$$

Age and period effects are incorporated into the model in the sense that the integrals in (4.9.1) and (4.9.2) are determined by all epochs $t+s$ and ages $s > 0$ that members of a cohort may assume following their birth at epoch $t \in R$.

It seems plausible that, in an evolutionary birth process, the distribution of waiting times to the n-th birth $Z_n(t)$, $n \geq 1$, could depend on n, an epoch or period effect, and the age of mothers when attaining parity $n-1$. Thus, one is led to consider continuous and nonnegative risk functions $\theta(n-1, t, x, y)$ at epoch $t \in R$ for women attaining parity $n-1$ at age x. For a cohort of females born at epoch $t \in R$, this four-variable risk function is to determine the conditional distribution function

$$H(n-1, t, x, y) = P[Z_n(t) \leq y \mid T_{n-1}(t) = x] \tag{4.9.3}$$

for all $n \geq 1$, $x > 0$, and $y > 0$. The equation connecting the risk function $\theta(n-1, t, x, y)$ with the distribution function (4.9.3) at the point $\langle n-1, t, x, y \rangle$ is assumed to have the form

$$H(n-1, t, x, y) = 1 - \exp\left[-\int_0^y \theta(n-1, t+x+s, x, s)ds\right]. \tag{4.9.4}$$

From equation (4.9.4), it follows that

$$p_n(t, x) = 1 - \exp\left[-\int_0^\infty \theta(n-1, t+x+s, x, s)ds\right] \le 1 \qquad (4.9.5)$$

is the conditional probability that a woman born at epoch t and attaining parity $n-1$ at age x will eventually experience the n-th live birth, given survival to the end of the childbearing ages.

The integrals in (4.9.1), (4.9.2), (4.9.4), and (4.9.5) are intended to convey the notion that probabilities of events in the life cycle of a cohort of females, depends not only on their epoch $t \in R$ of birth but also on conditions at each epoch following their birth. Probabilities of events in the life cycle of a cohort are, therefore, not forcordained but result from underlying laws of evolution determined by risk functions which change with time and depend on other variables. Finding appropriate mathematical forms for these risk functions and testing their validity is a research problem that will, no doubt, require extensive effort from many investigators before it can be brought to a satisfactory resolution.

Given the probability density functions

$$h(n-1, t, x, y) = \frac{\partial H(n-1, t, x, y)}{\partial y}, \qquad (4.9.6)$$

for $n \ge 1$, $t \in R$, $x > 0$, and $y > 0$, the joint conditional density of the random variables $T_1(t) \le T_2(t) \le \ldots \le T_n(t)$, given that $T_0(t) = x_0$ may be written down, under the assumption cohorts evole according to an age-dependent process of the form discussed in Sect. 4.8. Thus, for a cohort born at epoch $t \in R$, this joint p.d.f. is

$$\prod_{v=1}^{n} h(v-1, t, x_{v-1}, x_v - x_{v-1}), \qquad n \ge 1, \qquad (4.9.7)$$

where $x_0 \le x_1 \le x_2 \le \ldots \le x_n$, see (4.8.8). Hence, under the age-dependence assumption expressed in (4.9.7), for every $t \in R$, $T_1(t) \le T_2(t) \le \ldots$ is a conditional Markov process, given that $T_0(t) = x_0$.

Up until now, the phrase, laws governing the evolution of maternity histories, has been used in the sense of an undefined term. Mathematically, these laws refer to choices of the consistent finite dimensional distributions determining a stochastic process. The conditional Markov process just defined is just one choice among many, having some advantages with regard to computer implementation.

When the foregoing age-dependence assumption is in force, formulas analogous to the parity distribution in (4.8.12) and the mean function in (4.8.13) may be written down and, in principle, implemented on a computer for an evolutionary potential birth process. Of particular interest is the mean function in (4.8.13) and its density. Needed in the computation of this mean function and its density are the conditional distribution functions

$$H^{(n)}(t, x, y) = P[Z_1(t) + \ldots + Z_n(t) \le y \mid T_0(t) = x] \qquad (4.9.8)$$

and their densities

$$h^{(n)}(t, x, y) = \frac{\partial H^{(n)}(t, x, y)}{\partial y},$$
(4.9.9)

defined for $n \geq 1$, $t \in R$, $x > 0$, and $y > 0$. The sequence of p.d.f.'s in (4.9.9) are determined recursively by

$$h^{(1)}(t, x, y) = h(0, t, x, y),$$
(4.9.10)

and by

$$h^{(n)}(t, x, y) = \int_0^y h^{(n-1)}(t, x, s) h(n-1, t, x+s, y-s) ds$$
(4.9.11)

for $n \geq 2$.

Among women in a cohort born at epoch t who marry at age x and survive to age $x + y$, the function

$$v_c(t, x, y) = \sum_{n=1}^{\infty} H^{(n)}(t, x, y)$$
(4.9.12)

represents the mean number of live births experienced by these women after y time units of marriage. The density of the mean function in (4.9.12) is given by

$$b_c(t, x, y) = \frac{\partial v_c(t, x, y)}{\partial y} = \sum_{n=1}^{\infty} h^{(n)}(t, x, y).$$
(4.9.13)

As before, only finitely many terms in the series on the right in (4.9.12) and (4.9.13) will be nonzero. In practice, values of the density in (4.9.13) would be computed at selected values of t on a finite lattice of $\langle x, y \rangle$-points, using a version of formula (4.8.22) and Littman's algorithm.

Now consider women in a cohort born at epoch t who marry and survive to age y. Let $v_c(t, y)$ be the mean number of live births experienced by these women by age $y > 0$. Then,

$$v_c(t, y) = \int_0^y g_0(t, x) v_c(t, x, y-x) dx,$$
(4.9.14)

where $g_0(t, x)$ is the p.d.f. of the marriage distribution function in (4.9.1). The density of the mean function in (4.9.14) is determined according to the formula

$$b_c(t, y) = \frac{\partial v_c(t, y)}{\partial y} = \int_0^y g_0(t, x) b_c(t, x, y-x) dx,$$
(4.9.15)

defined for $t \in R$ and $y > 0$. The total fertility rate for a cohort born at epoch t is, by definition,

$$v_c(t, \infty) = \int_0^{\infty} b_c(t, y) dy.$$
(4.9.16)

Writing down the above formulas would seem unnecessary and redundant, because their analogues have appeared many times in the foregoing sections of this chapter. This apparent redundancy seems advisable, however, in order that cohort and period functions be clearly distinguished in the formalism of an evolutionary potential birth process. Estimation of the cohort fertility density in (4.9.15) would require observing a cohort of females over an extended time period of prohibitive length. Consequently, as indicated above, age-specific fertility rates, based on a mixture of birth cohorts, are frequently estimated for a given set of periods. One way of viewing such an exercise is that an attempt is being made to estimate a period fertility density $b_p(t, y)$ such that at epoch $t \in R$

$$R_p(t, x, h) = \frac{1}{h} \int_x^{x+h} b_p(t, y) dy \tag{4.9.17}$$

is the age-specific birth rate for women whose ages fall in the interval $[x, x+h)$, $h > 0$, at epoch t. The period total fertility rate of the population at epoch $t \in R$ is, by definition,

$$v_p(t, \infty) = \int_0^\infty b_p(t, y) dy. \tag{4.9.18}$$

It seems natural to view the period fertility density function $b_p(t, y)$ as being determined by cohort fertility density functions. If the members of a cohort are age x at epoch t, then they were born at epoch $t - x$. Therefore, the fundamental relationship connecting period and cohort fertility density functions is

$$b_c(t - y, y) = b_p(t, y), \tag{4.9.19}$$

where $t \in R$ and $y > 0$. It is of interest to observe that in view of (4.9.19), the period total fertility rate at epoch t now takes the form

$$v_p(t, \infty) = \int_0^\infty b_c(t - y, y) dy. \tag{4.9.20}$$

For mathematical convenience, all integrals in this section have been taken over the entire interval $(0, \infty)$ or one of its subintervals. These integrals are perfectly valid, when it is realized that the fertility densities in (4.9.20) vanish for y outside the childbearing ages.

Estimates $\hat{R}_p(t, x, h)$ of the period age-specific birth rates in (4.9.17) are usually calculated by forming $\hat{R}_p(t, x, h) = B(t, x, h)/K(t, x, h)$, the ratio of the number of births to women whose last birthday fell in the age interval $[x, x + h)$ to an estimate $K(t, x, h)$ of the population of females in the age interval $[x, x+h)$ at epoch $t \in R$, see (4.3.13). Sometimes h is chosen as $h = 1$ and x runs in yearly increments over the childbearing ages; but, usually, h is chosen as 5 and the childbearing ages are partitioned into five-year age intervals. Much research effort has been expended in fitting parametric forms for period fertility density functions to data. In this connection, the papers of Brass (1960), (1978), and Coale and Trussel (1974), (1978) may be consulted.

Recently, Hoem et al. (1981) have reviewed the literature on proposed parametric forms for a fertility density function and have compared the relative merits of several of these forms in computer experiments on Danish fertility for the period, 1962-1971. Throughout these experiments birth rates based on single-year age intervals were used. Goodness of fit, as determined by a nonlinear least squares procedure, was used as a criterion for differentiating among those parametric models studied.

Following widespread practice, Hoem and his coworkers did not always work directly with a parameterization of the period fertility density function $b_p(t, y)$ in (4.9.17). Rather, it was sometimes assumed that this density took the form

$$b_p(t, y) = v_p(t, \infty) f_p(t, y), \tag{4.9.21}$$

where $f_p(t, y)$ is a p.d.f. in the variable y for every $t \in R$. The period total fertility rate $v_p(t, \infty)$ as well as any parameters in the p.d.f. $f_p(t, y)$ were estimated by a nonlinear least squares procedure. At this juncture, it should be emphasized that the cohort and period fertility density functions $b_c(t, y)$ and $b_p(t, y)$, satisfying Eq. (4.9.19), may be thought of as existing in their own right and not necessarily determined by a set of assumptions underlying some model of the potential birth process.

Among the functions tested by Hoem and his coworkers, which included the Coale-Trussel function, two proposed by Brass, the Hadwiger function, as well as beta and gamma densities, a cubic spline fit empirical age-specific fertility rates better than all others. In terms of the notation of an evolutionary potential birth process, the period fertility density $b_p(t, y)$ could be represented as the cubic spline

$$b_p(t, y) = \sum_{k=0}^{3} \alpha_k y^k + \sum_{k=1}^{3} \beta_k (y - \xi_k)^3_+, \tag{4.9.22}$$

where $x_+ = x$ if $x > 0$ and $x_+ = 0$ if $x \leq 0$. The α, β, and ξ parameters, moreover, depend on t. Observe, that for each epoch $t \in R$ considered, there are ten parameters in (4.9.22) to estimate.

Although splines of the form appearing in (4.9.22) may fit period age-specific birth rates well, their value seems to be limited in that it seems very difficult to interpret the ten parameters in terms of economic, social, and biological factors underlying observed cohort and period fertility patterns. The principal value of splines in the analysis of fertility data appears to be their promise for faithfully condensing large quantities of data into a simple mathematical expression, depending on relatively few parameters.

It is of interest to observe that the choice of fertility function to be parameterized has, evidently, not been standardized. Keyfitz (1968), for example, in his Chap. 6, discussed the parameterization of a net maternity function, see Sect. 4.3. Hobcraft et al. (1982) have provided a review of the demographic literature on age, period, and cohort effects. A point of view emphasized in this paper for dealing with these effects was the log-linear models used in the analysis of contingency table data. Bishop et al. (1975) have given an extensive account of this methodology. To what extent the techniques of discrete multivariate analysis may be used to aid in fleshing out the skeletal formulation considered in this section is a question worthy of investigation.

Before proceding to a concrete example, it will be worthwhile to identify

the gross and net maternity functions, defined in Sect. 4.3, within the context of
an evolutionary potential birth process. In the present context, $b_c(t, y)$, the
density in (4.9.15), is the cohort gross maternity function for a cohort born at
epoch $t \in R$. Similarly, $b_p(t, y)$, defined by Eq. (4.9.19), is the period gross
maternity function at epoch $t \in R$.

Let $S_f(t, y)$ be a cohort survival function for a cohort born at epoch t. Then,
the net maternity function for this cohort is

$$\varphi_c(t, y) = S_c(t, y) b_c(t, y), \quad y > 0. \tag{4.9.23}$$

Let $S_p(t, y)$ be the period survival function at epoch t. By analogy, the period
net maternity function at epoch t is defined by

$$\varphi_p(t, y) = S_p(t, y) b_p(t, y), \quad y > 0. \tag{4.9.24}$$

As will be illustrated in a subsequent chapter, a period net maternity
function is sometimes used to provide a current estimate of a future rate of
population growth. Just as a speedometer in an automobile provides an indication of the current speed but supplies no information regarding future
speeds of the vehicle, so estimates based on a period net maternity function
may not provide a reliable indication of future population growth. When laws
underlying the evolution of a population change with time, it is easy to
understand why the metaphor is true.

4.10 The Evolution of Period Fertility in Sweden – 1780 to 1975

Valuable insights into the empirical nature of evolutionary birth processes may be
obtained by studying time series of period fertility. Accordingly, the purpose of
this section is to present, in the form of computer generated graphs, certain
aspects of the evolution of period fertility in Sweden during the time span,
1780 to 1975, a span covering 195 years. Two sources of data were used as
input to the computer generated graphs. One source was the numbers of births
in Sweden by age group of mother along with estimates of the size of the
female population in each age group as reported by Keyfitz and Flieger (1968)
for five year periods centered at the years 1780, 1785, ..., 1965. The corresponding data for the years 1970 and 1975 were taken from a Historical
Supplement to the United Nations Demographic Yearbook, 1979, which is
available in most libraries.

In both sources of data, births were attributed to mothers in the nine age
groups, 10–14, 15–19, ..., 50–54. As explained in Sect. 4.9, age specific fertility
rates were estimated by calculating the ratios of the numbers of births to the
size of the female population in each of the nine age groups. These ratios were
then multiplied by five in order to estimate the contribution to the period total
fertility rate for each age group. In other words, the resulting numbers were

intepreted as estimates of

$$hR_p(t, x, h) = \int_x^{x+h} b_p(t, y)dy, \tag{4.10.1}$$

for $h = 5$, $x = 10, 15, \ldots, 50$, and $t = 1780, 1785, \ldots, 1975$. Experience with the Swedish historical data suggests that the set of ages y for which the period fertility density $b_p(t, y)$ is nonzero is almost all of the open interval (10, 55).

Ideally, it would be nice to have a three-dimensional plot of the period fertility density $b_p(t, y)$; but, given the grouped data available, such a plot would be difficult to obtain. Insights into evolutionary changes in period age patterns of childbearing may be obtained, however, by plotting estimates of the integral on the right in (4.10.1) as a function of the epoch t for each fixed age x. Presented in Fig. 4.10.1 are such plots of the contributions of the various age groups to the period total fertility rates corresponding to the epochs 1780, 1985, ..., 1975. Ages of mothers less than 20 and greater than 39 have been summed into two groups in order to make the number of simultaneous plots manageable. For each age group, points at each epoch have been connected by straight lines to aid the eye in detecting time trends.

An inspection of the graphs in Fig. 4.10.1 shows that period age patterns of childbearing for Swedish women remained relatively unchanged during the century 1780 to 1880. Throughout this century, the three age groups contributing most to the period total fertility rates were 25–29, 30–34, and 35–39. The contribution of the age group, 30–34, in this century dominated all others; and,

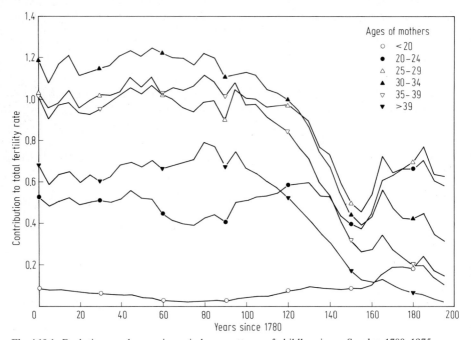

Fig. 4.10.1. Evolutionary changes in period age patterns of childbearing – Sweden 1780–1975

with few exceptions, the contribution of the age group, 35–39, dominated that of 25–29. In terms of their contributions to the total period fertility rates, the ordering of the remaining age groups was: >39, 20–24, and <20, see Fig. 4.10.1. In summary, the period age patterns of childbearing for Swedish women during the century, 1780 to 1980, seems to be those of a population marrying at a relatively late age and exercising little control over their natural fertility.

A further inspection of the graphs in Fig. 4.10.1 during the time span, 1880 to 1935, shows that there was a steady decrease in the contributions to the period total fertility rates for all age groups except those for the groups, 20–24 and <20. These decreases for almost all age groups graphically illustrates the overall decline in Swedish fertility that occurred during the time span, 1880 to 1935. During the time span, 1935 to 1975, however, the graphs exhibit relatively large fluctuations, except that for the age group, >39, which shows an almost steady decline. These fluctuations clearly illustrate that the rate of flow of births into the Swedish population during the time span, 1935 to 1975, was far from constant. Also illustrated in Fig. 4.10.1 are changes in period age patterns in childbearing for Swedish women. In the period centered at 1960, for example, the contribution to the total period fertility rate of the age group, 25–29, dominated all others and was followed in order by the age groups, 20–24, 30–34, 35–39, <20, and >39.

It is apparent from the graphs in Fig. 4.10.1 that fertility in Sweden did not decline to levels that might resemble realizations of a stochastic process governed by stationary laws of evolution. Rather, the graphs in Fig. 4.10.1 after 1935 seem to resemble realizations of a process whose laws of evolution change with time. In practical terms, the development of capabilities for understanding

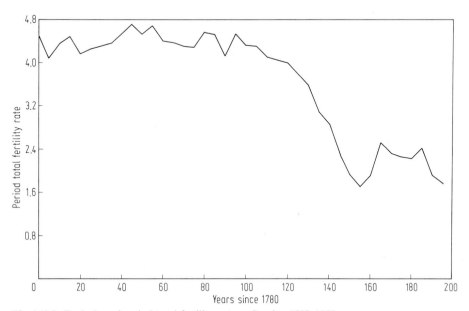

Fig. 4.10.2. Evolution of period total fertility rates – Sweden 1780–1975

and forecasting fertility in modern societies, such as that of Sweden, is one of the central problems of mathematical demography.

Another perspective on the evolution of fertility may be obtained by graphing the period total fertility rates, the sum of the quantities graphed in Fig. 4.10.1 for each age group of mothers. Figure 4.10.2 contains the graph of the period total fertility rates for Sweden during the time span, 1780 to 1975. Around 1780, this rate for Sweden was 4.53. At no time during the century, 1780 to 1880, did the rate fall below 4.00. By 1880, it stood at 4.32; and, by 1900, it had declined to 3.98. During the first three decades of the twentieth century, there was a steady decline in the period total fertility rate, with a low of 1.70 being reached around 1935, a time of world wide economic depression. In the next decade, 1935 to 1945, there was a steep rise in the rate; and, a relative peek of 2.52 was reached around 1945, the end of World War II. After 1945 there was a slight decline in the rate and then another relative peek of 2.41 was reached around 1965. During the next decade, 1965 to 1975, the rate again declined; in 1975 it stood at 1.78, a value slightly higher than the low of 1.70 recorded around 1935. From an inspection of the graph in Fig. 4.10.2, it is apparent that the problem of forecasting fertility during the century, 1780 to 1880, differed markedly from that of forecasting fertility during the forty year time span, 1935 to 1975, a span of time characterized by swings in fertility associated with changing social and economic conditions in the developed world.

Another perspective on the evolution of fertility in Sweden may be obtained by plotting cross-sections of the graphs in Fig. 4.10.1 at each epoch. The cross-sections corresponding to the epochs 1800, 1900, and 1960 are presented in Fig. 4.10.3. In each of these graphs, the contributions to the period total fertility rate for each of the nine age groups, 10–14, 15–19, ..., 50–54, are plotted against the midpoints of these intervals. The points were then connected by straight lines to produce period age patterns of childbearing. It is apparent that the period age patterns of childbearing for the epochs, 1800 and 1900, differed relatively little; while that for the more recent epoch, 1960, was

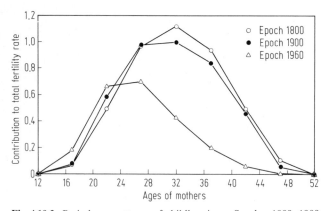

Fig. 4.10.3. Period age patterns of childbearing – Sweden 1800, 1900, and 1960

distinctly different from the other two. On the average, Swedish infants born around 1960 had younger mothers than those born around 1800 and 1900. The period total fertility rates associated with the epochs 1800, 1900, and 1960 were 4.16, 3.98, and 2.23, respectively.

4.11 Further Reading

Tabulations of period age specific fertility rates by five year age groupings of mothers are easy to find. Tabulations of age specific fertility rates for birth cohorts are, however, relatively rare. In this connection, an interesting set of cohort fertility tables were published in 1976 by the U.S. National Center for Health Statistics, DHEW Publication No. (HRA) 76–1152. Contained in this document are extensive cohort fertility tables, classified by color and ages of mothers in single years, covering the age span 14–49. Births registered in the United States during the time span, 1917–1973, were used in the tabulations. Consequently, all cohorts born between 1903 and 1924, inclusive, passed through the childbearing ages during this time span so that complete listings of age specific fertility rates for these cohorts are available. Only a partial listing of age specific fertility rates was possible, however, for cohorts born during the time spans 1867 to 1902 and 1925 to 1959, the other two sets of birth cohorts contributing births to the U.S. population during the time span, 1917–1973. Methods used in calculating age specific fertility rates were discussed in detail in an appendix included in the documents. Farid (1973) has also discussed methods for constructing cohort fertility tables.

Mixtures of waiting time distributions with covariates is a research topic that has recently received considerable attention, see, for example, Heckman and Singer (1982). Applications of this class of models within the context of the present chapter would include distributions of waiting times among live births in a cohort of females. Included in the cases studied by Heckman and Singer was a waiting time distribution of the form

$$P[T \leq t | \mathbf{X}] = \int_{\Lambda} f(\lambda) F(\lambda, \alpha, \boldsymbol{\beta}; t) d\lambda, \tag{4.11.1}$$

where

$$F(\lambda, \alpha, \boldsymbol{\beta}; t) = 1 - \exp[-t^\alpha \exp((\boldsymbol{\beta}, \mathbf{X}) + \lambda)]$$

and $t > 0$. In the above, Λ is a finite interval, α is a positive parameter, and $(\boldsymbol{\beta}, \mathbf{X}) = \beta_1 X_1 + \ldots + \beta_k X_k$. The X's are covariates and the β's are sometimes referred to as structural parameters.

Suppose a sample of n individuals are observed and with each individual there is associated an observed duration t_i and a vector X_i of covariates. Given the sample, $\langle t_i, \mathbf{X}_i \rangle$, $1 \leq i \leq n$, of independent observations, the estimation problem is that of estimating the unknown p.d.f. $f(\lambda)$, the parameter α, and the vector $\boldsymbol{\beta}$ of structural parameters. Heckman and Singer used a maximum likelihood procedure for estimating all unknowns.

Based on Monte Carlo experiments designed to test the performance of the estimation procedure, these authors arrived at the following general conclusions. If a parametric form of the mixing p.d.f. $f(\lambda)$ was assumed, then the estimates of α and the structural parameters depended in a crucial way on the choice of $f(\lambda)$. For any given choice of $f(\lambda)$, α and the structural parameters tended to be estimated accurately. If, on the other hand, the p.d.f. $f(\lambda)$, α, and the structural parameters were estimated simultaneously, the estimators of α and the structural parameters behaved well but the estimator of the mixing distribution behaved poorly, even in large samples. These results suggest that, when dealing with models of the type displayed in (4.11.1), an investigator should consider statistical procedures for finding the most plausible choice of mixing distribution among a collection of possibilities suggested by substantive considerations regarding heterogeneity in a population.

Problems and Miscellaneous Complements

1. The random function $K(t)$ is finite valued for every $t>0$, if, and only if, the condition

$$\sum_{n=0}^{\infty} P[K(t)=n]=1,$$

is satisfied for all $t>0$. Because the maximum number of offspring per female in human populations never exceeds some finite integer n_0, $H^{(n)}(t)=0$ for all $n\geq n_0$ and $t>0$. Use (4.2.7) and this observation to show that the above condition is always satisfied in human populations.

2. Let $p_n(t)$, $n\geq 0$, be the probabilities defined in (4.2.12) and given in (4.2.13). By using an argument similar to that used in the derivation of (4.3.4), show that

$$P[N(t)=n]=\int_0^t p_n(x)g(x)dx+S(t)p_n(t)$$

for $n=0, 1, 2, \ldots$ and $t>0$.

3. A well-known property of the convolution integral in (4.2.5) is that

$$\lim_{t\uparrow\infty} H^{(2)}(t)=\lim_{t\uparrow\infty} H(0, t)H(1, t)=p_1 p_2,$$

see (4.2.1). Equation (4.4.2) then follows from (4.2.6) by an induction argument.

4. Suppose, that in a desired family composition of $d=m+f$ of m males and f females, the first $d+x-1$ births consists of $m-1$ males and $f+x$ females. The probability of this event is

$$\binom{d+x-1}{m-1} p^{m-1}q^{f+x}.$$

Then, the desired number of males is attained in a family size of $d+x$ offspring if the next child is a male. Multiplying the above expression by p, the probability an offspring is male, yields the first term on the right in (4.4.7). An analogous argument is used to derive the second term.

5. (a) The convolution of two exponential densities $f_i(t)=\lambda_i \exp[-\lambda_i t]$, $t>0$, is

$$f^{(2)}(t)=\int_0^t f_1(t-s)f_2(s)ds$$
$$=\lambda_1\lambda_2(\psi(1,2)\exp[-\lambda_1 t]+\psi(2,2)\exp[-\lambda_2 t]),$$

where $\psi(1,2)=(\lambda_2-\lambda_1)^{-1}$, $\psi(2,2)=(\lambda_1-\lambda_2)^{-1}$, and $\lambda_1\neq\lambda_2$.

(b) Let $\beta(n)=\lambda_1\lambda_2\dots\lambda_n$ and by the induction hypothesis suppose

$$f^{(n-1)}(t)=\beta(n-1)\sum_{k=1}^{n-1}\psi(k,n-1)\exp[-\lambda_k t],$$

where $n>1$ and the ψ's are constants. Then, the validity of (4.5.5) and the recursion relations (4.5.6) follow from

$$f^{(n)}(t)=\int_0^t f^{(n-1)}(t-s)f_n(s)ds, \qquad t>0,$$

when all the λ's are unequal.

6. (a) Recall that the Laplace transform of an exponentially distributed random variable Y with parameter $\lambda>0$ is

$$E[\exp[-sY]]=\lambda(\lambda+s)^{-1}, \qquad s\geq0.$$

Therefore, if $Y_1, Y_2, \dots, Y_n$, $n\geq1$, are independent random variables with a common distribution equal to that of Y, the Laplace transform of the sum $Y_1+Y_2+\dots+Y_n$ is $\lambda^n(\lambda+s)^{-n}$, $s\geq0$. Verify that this function is also the Laplace transform of the gamma density,

$$g_n(y)=\frac{\lambda^n}{\Gamma(n)}y^{n-1}\exp[-\lambda y], \qquad y>0.$$

Hence, because the Laplace transform completely determines the density, the sum has a gamma distribution.

(b) Let the distribution function $\Gamma_n(\lambda,t)$ be defined by

$$\Gamma_n(\lambda,t)=\int_0^t g_n(y)dy, \qquad t>0.$$

Use integration by parts to show that

$$\Gamma_n(\lambda,t)=\Gamma_{n-1}(\lambda,t)-\exp[-\lambda t]\frac{(\lambda t)^{n-1}}{(n-1)!}, \qquad n\geq2.$$

(c) By iterating the result in (b), show that

$$\Gamma_n(\lambda,t)=1-\exp[-\lambda t]\sum_{v=0}^{n-1}\frac{(\lambda t)^v}{v!}.$$

for any positive integer $n \geq 1$. This result is of both theoretical and computational interest. It may, for example, be used to make the formulas (4.5.10) more explicit.

7. Consider a double exponential distribution function of the form

$$G(x) = \exp[-\exp(\alpha - \beta x)],$$

for $x > 0$ and $G(x) = 0$ for $x \leq 0$. The parameters α and β are assumed to be positive.

(a) Show that the p.d.f. of $G(x)$ is

$$g(x) = \beta \exp[\alpha - \beta x - \exp(\alpha - \beta x)]$$

for $x > 0$ and $g(x) = 0$ for $x \leq 0$.

(b) Show that the risk function associated with the distribution function $G(x)$ is

$$\theta(x) = \frac{\beta \exp[\alpha - \beta x - \exp(\alpha - \beta x)]}{1 - \exp[-\exp(\alpha - \beta x)]}$$

for $x > 0$. Investigate whether $\theta(x)$ approaches a limit as $x \uparrow \infty$. Find the limit if it exists.

(c) Points of inflexion of $G(x)$ are those values of x such that $G''(x) = 0$. Show that $G''(x) = 0$ for $x > 0$ implies that $x = \alpha/\beta$. Observe that α/β is also the mode of the density in part (a).

(d) Let $0 < q < 1$. A number $x_q > 0$ such that $G(x_q) = q$ is called the q-th quantile of the distribution. Show that

$$x_q = \frac{\alpha - \log_e(-\log_e q)}{\beta}.$$

8. The following is a brief outline of the ideas proposed by Braun and Hoem (1979) in their analysis of maternity history data for Danish women.

(a) Suppose that the risk function for women who are age x_v when reaching parity $v \geq 0$ has the form

$$\theta(y) = \gamma \beta_v (\beta_v y)^{\alpha - 1} \exp[-\beta_v y],$$

for $0 < y < \eta - x_v$ and $\theta(y) = 0$ for $y \notin (0, \eta - x_v)$, where α and γ are positive parameters and η is the upper limit of the childbearing ages, say age 45.

(b) Suppose further that the scale parameter β_v depends not only on the age x_v at which parity v is attained but also on the length of the previous birth interval. Define a function $\lambda(x)$ by $\lambda(x) = 0$ for $x \leq 14$, $\lambda(x) = 1$ for $14 < x \leq 25$,

$$\lambda(x) = \left(1 - \left(\frac{x - 25}{20}\right)^\tau\right)^2,$$

for $25 < x \leq 45$, and $\lambda(x) = 0$ for $x > 45$. Let a_v be a covariate measuring the length of the previous birth interval for women who have experienced at least one live birth. Then, the form of the scale parameter proposed by Braun and Hoem was

$$\beta_0 = \delta \lambda(x_0)$$

and

$$\beta_v = \delta \lambda(x_v) a_v^{-\rho}$$

for $v \geq 1$, where, evidently, δ and ρ were positive parameters.

(c) Observe that the five parameters, α, γ, δ, ρ, and τ, determine the system. Using a method of maximum likelihood, Braun and Hoem estimated these parameters according to a threeway classification, consisting of parities, age cohorts of women, and lengths of cohabitational spells.

9. The following observations pertain to the distribution determined by the double exponential risk function in (4.6.1).

(a) Let $\eta(x)$ be the risk function in (4.6.1). Show that

$$\frac{d\eta(x)}{dx} = \beta\gamma\delta \exp[-\delta x]\eta(x)$$

for $x > 0$. Using the above formula, conclude that $\eta(x)$ is strictly monotone increasing for $x > 0$.

(b) Let $S(x)$ be the survival function defined in (4.6.3). Show that

$$\frac{dS(x)}{dx} = -\eta(x)S(x)$$

for $x > 0$, where $S(0) = 1$. When the integral of a risk function cannot be expressed in terms of elementary functions, it is sometimes useful to compute $S(x)$ by solving the above differential equation numerically.

10. Among other things Coale and McNeil (1972) showed that the distribution based on the double exponential distribution in (4.6.1) was related to a distribution with the double exponential density,

$$f(y) = \frac{\lambda}{\Gamma\left(\frac{\alpha}{\lambda}\right)} \exp[-\alpha y - \exp(-\lambda y)],$$

defined for all $y \in R$. The parameters α and λ are assumed to be positive.

(a) Suppose a random variable T has a standard gamma p.d.f.

$$h(t) = \frac{1}{\Gamma(\theta)} t^{\theta - 1} \exp[-t],$$

for $t > 0$, where $\theta = \alpha/\lambda$. Show that the random variable $Y = -\lambda^{-1} \log_e T$ has the p.d.f. $f(y)$ as defined above.

(b) In Monte Carlo simulations, it is sometimes useful to simulate realizations of a random variable by transforming realizations of a random variable with a known distribution such as the standard gamma. Age at marriage is always nonnegative and difficulties may arise if negative values were simulated. If Y defined in part (a) were used to simulate ages at marriage in a cohort, would negative values arise with positive probability?

11. *Properties of Lognormal Distribution*
(a) Many statistical libraries have computer programs for generating realizations of a standard normal random variable Z with mean zero and variance one. Show that the random variable $T = \alpha + \exp[\mu + \sigma Z]$ has the p.d.f. in (4.6.7) with $p = 1$. This result is useful for generating realization of a lognormal random variable in Monte Carlo simulations.

(b) The moment generating function of a standard normal random variable Z is $E[\exp(tZ)] = \exp[t^2/2]$ for $t \in R$. With reference to the random variable in part (a), show that

$$E[(T - \alpha)^n] = \exp\left[n\mu + \frac{n^2}{2}\sigma^2\right]$$

for $n \geq 0$.

(c) Use the formula in part (b) to show that

$$E[T] = \alpha + \exp\left[\mu + \frac{\sigma^2}{2}\right],$$

and

$$\mathrm{Var}[T] = \exp[2\mu + \sigma^2](\exp[\sigma^2] - 1).$$

(d) Show that the mode of the p.d.f. in (4.6.7) is $\alpha + \exp[\mu - \sigma^2]$.

12. *Properties of Bivariate Lognormal Distribution*
(a) Let $Q(x_1, x_2)$ be the quadratic form in the exponent of the bivariate lognormal denisty, see (4.6. 10) and (4.6.11). Show that $(1 - \rho^2)Q(x_1, x_2)$ equals

$$\left(\frac{x_1 - \mu_1}{\sigma_1}\right)^2 - 2\rho\left(\frac{x_1 - \mu_1}{\sigma_1}\right)\left(\frac{x_2 - \mu_2}{\sigma_2}\right) + \left(\frac{x_2 - \mu_2}{\sigma_2}\right)^2.$$

(b) If a vector random vector $\mathbf{X}$ follows a bivariate normal distribution with mean vector $\boldsymbol{\mu}$ and covariance matrix $\boldsymbol{\Sigma}$, then the moment generating function of $\mathbf{X}$ is

$$E[\exp[\mathbf{V}'\mathbf{X}]] = \exp[\mathbf{V}'\boldsymbol{\mu} + \tfrac{1}{2}\mathbf{V}'\boldsymbol{\Sigma}\mathbf{V}],$$

for all vectors $\mathbf{V}' = \langle v_1, v_2 \rangle$ in the $v_1 v_2$-plane. Suppose T_1 and T_2 follow a bivariate lognormal distribution with location vector $\boldsymbol{\alpha}' = \langle \alpha_1, \alpha_2 \rangle$ and parameters $\boldsymbol{\mu}$ and $\boldsymbol{\Sigma}$. Use the formula for the moment generating function of the bivariate normal distribution to show that

$$E[(T_1 - \alpha_1)^{n_1}(T_2 - \alpha_2)^{n_2}] = \exp[\mathbf{n}'\boldsymbol{\mu} + \tfrac{1}{2}\mathbf{n}'\boldsymbol{\Sigma}\mathbf{n}],$$

where $\mathbf{n}' = \langle n_1, n_2 \rangle$ is any vector of nonnegative integers.

(c) Use the result in part (b) to show that

$$\mathrm{Cov}(T_1, T_2) = E[(T_1 - E[T_1])(T_2 - E[T_2])],$$

the covariance of T_1 and T_2, equals

$$\exp[\mu_1 + \mu_2 + \tfrac{1}{2}(\sigma_1^2 + \sigma_2^2)](\exp[\rho\sigma_1\sigma_2] - 1).$$

(d) Derive the formula for

$$\eta = \text{cov}(T_1, T_2)/(\text{Var}[T_1]\,\text{Var}[T_2])^{1/2},$$

the correlation between T_1 and T_2, displayed in (4.6.15).

13. *Conditional Distributions of the Bivariate Lognormal*
As the following example shows, the bivariate lognormal distribution inherits many nice properties from the bivariate normal.

(a) Suppose the random variables X_1 and X_2 follow a bivariate normal distribution with mean vector μ and covariance matrix Σ. By completing the square in the quadratic form in Problem 12 (a), show that

$$(1-\rho^2)Q(x_1, x_2) = (1-\rho^2)\left(\frac{x_1-\mu_1}{\sigma_1}\right)^2 + \left(\frac{x_2 - h(x_1)}{\sigma_2}\right)^2,$$

where

$$h(x_1) = \mu_2 + \frac{\rho\sigma_2}{\sigma_1}(x_1 - \mu_1).$$

(b) From the result in part (a), deduce that the conditional distribution of X_2, given that $X_1 = x_1$, is normal with mean $h(x_1)$ and variance $\sigma_2^2(1-\rho^2)$.

(c) Suppose the random variables T_1 and T_2 follow a bivariate lognormal distribution as described in Problem 12 (b). Use Eqs. (4.6.11) and (4.6.12) along with the result in part (b) to conclude that $g(t_2|t_1)$, the conditional p.d.f. of T_2 given that $T_1 = t_1$, is

$$\frac{1}{\sqrt{2\pi}\,\sigma_2(1-\rho^2)^{1/2}(t_2-\alpha_2)}\exp\left[-\frac{1}{2}\left(\frac{\log_e(t_2-\alpha_2)-h(\log_e(t_1-\alpha_1))}{\sigma_2(1-\rho^2)^{1/2}}\right)^2\right],$$

for $t_1 > \alpha_1$ and $t_2 > \alpha_2$. Hence, the conditional p.d.f. of T_2, given that $T_1 = t_1$, is lognormal with parameters $h(\log_e(t_1-\alpha_1))$ and $\sigma_2(1-\rho^2)^{1/2}$ and location parameter α_2.

(d) The conditional expectation, $E[T_2|T_1=t_1]$, considered as a function of t_1, is sometimes called the regression of T_2 on t_1. From Problem 11 (c) and part (c), it follows that

$$E[T_2|T_1=t_1] = \alpha_2 + \exp\left[h(\log_e(t_1-\alpha_1)) + \frac{\sigma_2^2(1-\rho^2)}{2}\right],$$

for $t_1 > \alpha_1$. Show that

$$E[T_2|T_1=t_1] = \alpha_2 + (t_1-\alpha_1)^\beta \exp\left[\mu_2 - \beta\mu_1 + \frac{\sigma_2^2(1-\rho^2)}{2}\right],$$

where $\beta = (\rho\sigma_2)/\sigma_1$. It is of interest to note that the regression function on the right is nonlinear in t_1.

14. *A Stopping Time for the Potential Birth Process*
In societies where divorce is rare, the death of the husband, as well as death of the wife, is an event that may stop the potential birth process introduced in Sect. 4.2. The results in Problem 13 may be applied to finding a survival function for such a stopping time among brides of age t_1.

(a) Suppose a husband was married at age t_2. If $S^{(m)}(t)$ is the survival function for males in a cohort, then

$$S^{(m)}(t|t_2) = S^{(m)}(t+t_2)/S^{(m)}(t_2), \quad t \geq 0,$$

is the conditional survival function for males marrying at age t_2. One way of considering survivorship among husbands of females marrying at age t_1 is to inspect $S^{(m)}(t|t_2)$ for plausible values of t_2, the ages of grooms among brides of age t_1.

(b) The conditional density $g(t_2|t_1)$ in 13 (c) may be used to take into account the variation in the ages of grooms among brides of age t_1. Let $X^{(m)}$ be a random variable representing the number of time units a husband survives following marriage. Then, the conditional survival function of $X^{(m)}$, for brides of age $T_1 = t_1$, is

$$P[X^{(m)} > t | T_1 = t_1] = \int_{\alpha_2}^{\infty} S^{(m)}(t|t_2) g(t_2|t_1) dt_2,$$

where $t \geq 0$. Among women marrying at age t_1, the random variable $X^{(m)}$ could be used as a stopping time for the potential birth process. By using the techniques outlined in Sect. 4.3, specific formulas could be derived.

15. (a) The formula

$$\int_0^{\infty} x^{\alpha-1} e^{-\beta x} dx = \frac{\Gamma(\alpha)}{\beta^{\alpha}},$$

which is valid for $\alpha > 0$ and $\beta > 0$, is fundamental when dealing with integrals involving the gamma p.d.f. Use this formula to derive the p.d.f. displayed in (4.7.7).

(b) Recall the following basic result from mathematical statistics. Suppose a random variable X has the p.d.f. $f(x)$ whose support is the set S, usually an interval of real numbers. Then $f(x) > 0$ if $x \in S$ and $f(x) = 0$ if $x \notin S$. Let $Y = \varphi(X)$ be a one-to-one transformation from the set S to the set T with the inverse $x = \varphi^{-1}(y)$, $y \in T$. Furthermore, suppose the derivative,

$$\frac{dx}{dy} = \frac{d\varphi^{-1}(y)}{dy},$$

exists and is continuous for all $y \in T$. Then, the p.d.f. of the random variable Y is

$$g(y) = f(\varphi^{-1}(y)) \left| \frac{d\varphi^{-1}(y)}{dy} \right| \quad \text{if } y \in T$$

and $g(y) = 0$ if $y \notin T$. Use this result to derive the p.d.f. in (4.7.9).

16. (a) In deriving formula (4.7.19), expressions for the conditional expectations, $E[Z_v^2|A_2, \Lambda = \lambda]$ and $E[Z_v^2|A_2]$, were found. Show that

$$E[Z_v^2|A_2, \Lambda = \lambda] = \frac{(\beta + \delta(2))^{\alpha} \theta_v(\theta_v + 1)}{\beta^{\alpha} \beta_v^2 \lambda^2} \exp[-\delta(2)\lambda].$$

By applying the formula in Problem 15(a), show that

$$E[\Lambda^{-2}\exp[-\delta(2)\Lambda]]=\frac{\beta^{\alpha}}{(\alpha-1)(\alpha-2)(\beta+\delta(2))^{\alpha-2}},$$

provided that $\alpha>2$. From this expression and the above conditional expectation conclude that

$$E[Z_v^2|A_2]=\frac{(\beta+\delta(2))^2\theta_v(\theta_v+1)}{\beta_v^2(\alpha-1)(\alpha-2)},\qquad\alpha>2.$$

This latter formula together with (4.7.18) yields the formula in (4.7.19) for the conditional variance of Z_v, given the event A_2.

(b) A formula for the conditional expectation, $E[Z_1Z_2|A_2]$, played a basic role in the derivation of the formula in (4.7.20) for the conditional covariance of the random variables Z_1 and Z_2, given the event A_2. Use the techniques outlined in part (a) to show that

$$E[Z_1Z_2|A_2]=\frac{(\beta+\delta(2))^2\theta_1\theta_2}{\beta_1\beta_2(\alpha-1)(\alpha-2)},\qquad\alpha>2.$$

17. (a) Substitute (4.7.4) and (4.7.6), for $v=1,2$, into (4.7.3) and show that the joint p.d.f. of the random variables Z_1 and Z_2 may be represented in the form

$$h(t_1,t_2)=\frac{\beta^{\alpha}\Gamma(\alpha(2))(\beta(2))^{\alpha}(\beta_1t_1)^{\theta_1}(\beta_2t_2)^{\theta_2}}{(\beta(2))^{\alpha}\Gamma(\alpha)\Gamma(\theta_1)\Gamma(\theta_2)(\beta(2,t_1,t_2))^{\alpha(2)}t_1t_2}$$

for $t_1>0$ and $t_2>0$. Here $\alpha(2)=\alpha+\theta_1+\theta_2$, $\beta(2)=\beta+\delta(2)$, $\delta(2)=\delta_1+\delta_2$, and $\beta(2,t_1,t_2)=\beta(2)+\beta_1t_1+\beta_2t_2$

(b) The transformation

$$w_1=\frac{\beta_1t_1}{\beta(2,t_1,t_2)}\quad\text{and}\quad w_2=\frac{\beta_2t_2}{\beta(2,t_1,t_2)}$$

is a one-to-one mapping of the set $R_2^+=[\langle t_1,t_2\rangle:t_1>0,t_2>0]$ onto the set $S=[\langle w_1,w_2\rangle:0<w_1+w_2<1]$. Show that the inverse transformation is

$$t_1=\frac{\beta(2)w_1}{1-w_1-w_2}\quad\text{and}\quad t_2=\frac{\beta(2)w_2}{1-w_1-w_2}.$$

This transformation is of interest, because it provides a means of transforming the p.d.f. in part (a) into that for a Dirichlet distribution, see Sect. 7.7 in Wilkes (1962).

(c) Define a pair of random variables $\langle W_1,W_2\rangle$ by

$$W_1=\frac{\beta_1Z_1}{\beta(2,Z_1,Z_2)}\quad\text{and}\quad W_2=\frac{\beta_2Z_2}{\beta(2,Z_1,Z_2)}.$$

Then, as is well-known from mathematical statistics, the joint p.d.f. of the random variables W_1 and W_2 is

$$g(w_1,w_2)=h(t_1,t_2)|J(w_1,w_2)|,$$

for $\langle w_1, w_2\rangle \in S$, where $J(w_1, w_2)$ is the Jacobian of the transformation in part (b). The Jacobian of this transformation is the determinant of the two-by-two matrix

$$\left(\frac{\partial t_i}{\partial w_j}\right), \quad i, j = 1, 2.$$

Show that

$$J(w_1, w_2) = (\beta(2))^2 (1 - w_1 - w_2)^{-3}.$$

(d) Use the results in part (c) to show that $g(w_1, w_2)$, the joint p.d.f. of the random variables W_1 and W_2, is

$$\frac{\beta^\alpha \Gamma(\alpha(2))}{(\beta(2))^\alpha \Gamma(\alpha)\Gamma(\theta_1)\Gamma(\theta_2)} w_1^{\theta_1 - 1} w_2^{\theta_2 - 1} (1 - w_1 - w_2)^{\alpha - 1},$$

where $\langle w_1, w_2\rangle \in S$. This result is of interest, because it shows that the distribution function of the random variables Z_1 and Z_2 may be expressed in terms of incomplete Dirichlet integrals. The extension of these results to the joint distribution function of the random variables $Z_1, Z_2, ..., Z_n$, $n > 2$, is tedious but straight-forward.

18. *A Negative Binomial Distribution*
(a) With respet to the gamma mixture of gamma distributions considered in Sect. 4.7, suppose that $\theta_v = 1$, $\beta_v = 1$, and $\delta_v = 0$ for all $v \geq 1$. Then, given that $\Lambda = \lambda$, the random variables $Z_1, Z_2, ...$ are conditionally independent with a common exponential distribution whose scale parameter is λ. From Problem 6 (c), it follows that the conditional distribution function of the random variable $Z_1 + ... + Z_n$, given that $\Lambda = \lambda$, is

$$H^{(n)}(\lambda; t) = \Gamma_n(\lambda; t).$$

Let $K(t)$ be a random function, representing family size, number of live births, as a function $t > 0$ of duration of marriage. Use the above result to show that $K(t)$ is a Poisson process with intensity parameter $\lambda > 0$. Thus,

$$P[K(t) = n \mid \Lambda = \lambda] = \frac{\exp[-\lambda t](\lambda t)^n}{n!}$$

for $n = 0, 1, 2, ...$, a Poisson p.d.f. with parameter λt.
(b) Define a sequence of constants by

$$\alpha^{(n)} = \frac{\Gamma(\alpha + n)}{\Gamma(\alpha)}$$

for $n = 0, 1, 2, ...$. Under the assumption the random variable Λ follows a gamma distribution with index parameter α and scale parameter β, use the result in part (a) to show that

$$P[K(t) = n] = \frac{\alpha^{(n)}}{n!} [p(t)]^\alpha [q(t)]^n, \quad n = 0, 1, 2, ...,$$

where

$$p(t) = \frac{\beta}{\beta + t} \quad \text{and} \quad q(t) = \frac{t}{\beta + t}.$$

But, this is the p.d.f. of a negative binomial distribution with parameters α and $p(t)$.

19. Used in the derivation of the joint conditional p.d.f. in (4.8.8) of the random variables $T_1 \leq T_2 \leq \ldots \leq T_n$, given that $T_0 = x$, is an extention of the theory of Jacobian, briefly discussed in Problem 17 (c), to densities of an arbitrary number of variables $n \geq 2$.

(a) The transformation connecting the density in (4.8.7) with that in (4.8.8) is $t_v = x_v - x_{v-1}$, $1 \leq v \leq n$. Show that for the case $n = 2$, the 2×2 Jacobian matrix is

$$\left(\frac{\partial t_i}{\partial x_j} \right) = \begin{bmatrix} 1 & 0 \\ -1 & 1 \end{bmatrix}, \quad 1 \leq i, j \leq 2.$$

Hence, for $n = 2$, the Jacobian of the transformation is one.

(b) Conclude that the p.d.f. in (4.8.8) is valid for any $n > 2$ by generalizing the result in (a).

20. *Properties of Age-Dependent Convolutions*
(a) To fill in some of the details regarding Eqs. (4.8.20) and (4.8.21), it will suffice to consider a set of functions $g(x, t)$ defined on a finite lattice of $\langle x, t \rangle$-points with nonnegative coordinates. A function $g^{(0)}(x, t)$, defined on this lattice, is said to be an identity with respect to the operation of age-dependent convolution if

$$\sum_{s=0}^{t} g^{(0)}(x, s) g(x + s, t - s) = \sum_{s=0}^{t} g(x, s) g^{(0)}(x + s, t - s) = g(x, t)$$

at all points $\langle x, t \rangle$ and every function $g(x, t)$ in the set. Show that if $g^{(0)}(x, t)$ is defined for all $x \geq 0$ by $g^{(0)}(x, 0) = 1$ and $g^{(0)}(x, t) = 0$ if $t \geq 1$, then it is an identity.

(b) In all applications considered in Sect. 4.8, all functions $g(x, t)$, going into the age-dependent convolutions in (4.8.20), have the property $g(x, 0) = 0$ for all $x \geq 0$. Suppose further that for all $x \geq 0$, $g(x, t) \neq 0$ if $t \geq 1$. Use induction and (4.8.20) to conclude that, for every $n \geq 1$ and $x \geq 0$, $g^{(n)}(x, t) = 0$ if $0 \leq t < n$ and $g^{(n)}(x, t) \neq 0$ if $t \geq n$.

(c) Let $f(x, t)$ be the sum of the "geometric" series in (4.8.21). Use the result in part (b) to conclude that

$$f(x, t) = g^{(0)}(x, t) + g^{(1)}(x, t) + \ldots + g^{(t)}(x, t)$$

for all $x \geq 0$ and $t \geq 0$.

References

Aitchison, J. and Brown, J.A.C. (1963): *The Lognormal Distribution.* Cambridge University Press, Cambridge

Anderson, T.W. (1958): *An Introduction to Multivariate Statistical Analysis.* John Wiley and Sons, New York

Bishop, Y.M.M., Fienberg, S.E. and Holland, P.W. (1975): *Discrete Multivariate Analysis – Theory and Practice.* M.I.T. Press, Cambridge, Mass.

Blackwood, L. (1981): Alaska Native Fertility Trends, 1950–1978. *Demography* 18:173–178

Brass, W. (1958): The Distribution of Births in Human Populations. *Population Studies* 12:51–72

Brass, W. (1960): The Graduation of Fertility Distributions by Polynomial Functions. *Population Studies* 14:148–162

Brass, W. (1978): Population Projections for Planning and Policy. Papers of the East-West Population Institute, #55. Honolulu, Hawaii

Braun, H.I. (1980): Regression – Like Analysis of Birth Interval Sequences. *Demography* 17:207–221

Braun, H.I. and Hoem J.M. (1979): Modelling Cohabitation Birth Intervals in the Current Danish Population. *Working Paper No. 24*, Laboratory of Actuarial Mathematics, University of Copenhagen

Coale, A.J. (1971): Age Patterns of Marriage. *Population Studies* 25:193–214

Coale, A.J. (1972): *The Growth and Structure of Human Populations.* Princeton University Press, Princeton, New Jersey

Coale, A.J. and McNeil, D.R. (1972): The Distribution by Age of the Frequency of First Marriage in a Female Cohort. *Journal of the American Statistical Association* 67:743–749

Coale, A.J. and Trussell, T.J. (1974): Model Fertility Schedules – Variation in the Age-Structure of Childbearing in Human Populations. *Population Index* 40:185–258

Coale, A.J. and Trussell, T.J. (1978): Technical Note – Finding the Two Parameters That Specify a Model Schedule of Marital Fertility. *Population Index* 44:203–213

Curtin, L.R., Frockt, I.J. and Koch, G.G. (1979): A Note on the Analysis of Parity Progression Ratios. *Demography* 16:481–484

Dandekar, K. (1963): Analysis of Birth Intervals of a Set of Indian Women. *Eugenics Quarterly* 10:73–78

de Finetti, B. (1973): La prévision: ses lois logiques, ses sources subjectives. Annales de l'Institut Henri Poincaré 7:1–68

Farid, S.M. (1973): On the Pattern of Cohort Fertility. *Population Studies* 27:159–168

Feller, W. (1966): *An Introduction to Probability Theory and Its Applications,* Vol. II, John Wiley and Sons, New York

Feller, W. (1968): *An Introduction to Probability Theory and Its Applications.* Vol. I, Third Edition. John Wiley and Sons, New York

Galambos, J. (1978): *The Asymptotic Theory of Extreme Order Statistics.* John Wiley and Sons, New York

Galton, F. (1879): The Geometric Mean in Vital and Social Statistics. *Proceedings of the Royal Society, London* 29:365–367

George, A. and Mathai, A.M. (1975): A Generalized Distribution for Inter-Live Birth Interval. *Sankhyā* 37B:332–342

Heckman, J.J. and Singer, B. (1982): Population Hetzerogeneity in Demographic Models, pages 567 to 595, in *Multidimensional Mathematical Demography,* edited by K.C. Land and A. Rogers. Academic Press, New York and London

Hewitt, E. and Savage, L.J. (1956): Symmetric Measures on Cortesian Products. *Transactions of the American Mathematical Society* 80:470–501

Hobcraft, J., Menken, J. and Preston, S. (1982): Age, Period, and Cohort Effects in Demography – A Review. *Population Index* 48:4–43

Hobcraft, J. and Rodriguez, G. (1980): Methodological Issues in Life Table Analysis of Birth Histories. Seminar on *The Analysis of Maternity Histories,* London School of Hygiene and Tropical Medicine

Hoem, J.M., Madsen, D., Nielsen, J.L., Ohlsen, E.M., Hansen, H.O. and Rennermalm, B. (1981): Experiments in Modelling Recent Danish Fertility Curves. *Demography* 18:231–244

Kennedy, W.J. and Gentle, J.E. (1980): *Statistical Computing*. Marcel Dekker, Inc., New York and Basel

Keyfitz, N. (1968): *Introduction to the Mathematics of Population*. Addison-Wesley, Reading, Mass.

Keyfitz, N. and Flieger, W. (1968): *World Population*. The University of Chicago Press, Chicago, Ill.

Lee, C.F. and Lin, K.H. (1976): Parity Patterns of Birth Interval Distributions. *Demography* 13:45–64

Leridon, H. (1977): *Human Fertility – The Basic Components*. The University of Chicago Press, Chicago and London

Littman, G.S. and Mode, C.J. (1977): A Non-Markovian Model for the Taichung Medical IUD Experiment. *Mathematical Biosciences* 34:279–302

Lotka, A.J. (1939): A Contribution to the Theory of Self-Renewing Aggregates With Special Reference to Industrial Replacement. *Ann. Math. Stat.* 10:1–25

Marquardt, D.W. (1963): An Algorithm for Least Squares Estimation of Nonlinear Parameters. *Journal of the Society for Industrial and Applied Mathematics* 11:431–441

McAlister, D. (1879): The Law of the Geometric Mean. *Proceedings* of the Royal Society, London. 29:367–375

McClelland, G.H. (1979): Determining the Impact of Sex Preferences on Fertility: A Consideration of Parity Progression Ratio, Dominance, and Stopping Rule Measures. *Demography* 16:377–388

Mode, C.J. (1967): A Renewal Density Theorem in the Multi-Dimensional Case. *Journal of Applied Probability* 4:62–76

Mode, C.J. (1975): Perspectives in Stochastic Models of Human Reproduction – A Review and Analysis. *Theoretical Population Biology* 8:247–291

Mood, A.M. and Graybill, F.A. (1963): *Introduction to the Theory of Statistics*. McGraw-Hill, New York

Pathak, K.B. (1971): A Stochastic Model for the Study of Open Birth Interval – A Cohort Approach. *Sankhyā* 33 B:305–314

Pickens, G.T. (1980): A Stochastic Model of Natural Fertility – Theory and Analysis of Some Historical Population Data. *Mathematical Biosciences* 48:129–151

Poole, W.K. (1973): Fertility Measures Based on Birth Interval Data. *Theoretical Population Biology* 4:357–387

Potter, R.G. (1963): Birth Intervals – Structure and Change. *Population Studies* 17:155–166

Pretorius, S.J. (1930): Skew Bivariate Frequency Surfaces Examined in the Light of Numerical Illustrations. *Biometrika* 22:109–223

Pullum, T.W. (1979): Adjusting Stated Fertility Preferences for the Effect of Actual Family Size With Application to World Fertility Survey Data. Paper presented at Annual Meeting of PAA, Philadelphia, Pa.

Révész, P. (1968): *The Laws of Large Numbers*. Academic Press, New York and London

Ross, J.A. and Madhavan, S. (1981): A Gompertz Model for Birth Interval Analysis. *Population Studies* 35:439–454

Ruzicka, L.T. (1976): Age at Marriage and Timing of the First Birth. *Population Studies* 30:527–538

Sheps, M.C. (1963): Effects on Family Size and Sex Ratio of Preferences Regarding Sex of Children. *Population Studies* 17:66–72

Sheps, M.C. (1965): An Analysis of Reproductive Patterns in an American Isolate. *Population Studies* 19:65–80

Sheps, M.C. and Menken, J.A. (1973): *Mathematical Models of Conception and Birth*. The University of Chicago Press

Singh, S.N. (1964): On the Time of First Birth. *Sankhyā* 26 B:95–102

Singh, S.N. (1968): A Chance Mechanism of Variation in the Number of Births per Couple. *Journal of American Statistical Association* 63:209–213

Singh, S.N., Bhattacharya, B.N. and Yadava, R.C. (1974): A Parity Dependent Model for Number of Births and Its Applications. *Sankhyā* 36 B:93–102

Singh, S.N., Bhattacharya, B.N. and Yadava, R.C. (1979): An Adjustment of a Selection Bias in Post-Partum Amenorrhea From Follow-Up Studies. *Journal of American Statistical Association* 74:916–920

Singh, S.N., Chakrabarty, K.C. and Barman, D. (1981): On Estimating the Risk of Conception From Observations on First Contraceptive Delays. *Journal of Theoretical Biology* 90:1–7

Singh, S.N., Chakrabarty, K.C. and Singh, V.K. (1976): A Modification of a Continuous Time Model for First Conception. *Demography* 13:37–44

Suchindran, C.M. and Lachanbruch, P.A. (1974): Estimates of Parameters in Probability Model for First Live Birth Interval. *Journal American Statistical Association* 69:507–513

Suchindran, C.M. and Linger, J.W. (1977): On Comparison of Birth Interval Distributions. *Journal of Biosocial Science* 9:25–31

Svinivasan, K. (1968): A Set of Analytical Models for the Study of Open Birth Intervals. *Demography* 5:34–44

Tuan, C.H. (1958): Reproductive Histories of Chinese Women in Rural Taiwan. *Population Studies* 12:40–50

Venkatacharya, K. (1972): Some Problems in the Use of Open Birth Intervals as Indicators of Fertility Change. *Population Studies* 26:495–505

Watkins, S.C. and McCarthy, J. (1980): The Female Life Cycle in a Belgian Commune – La Hulpe, 1847–1866. *Journal of Family History* 5:167–179

Wilks, S.S. (1962): *Mathematical Statistics*. John Wiley and Sons, New York

Wolfers, D. (1968): Determinants of Birth Intervals and Their Means. *Population Studies* 22:253–262

Chapter 5. A Computer Software Design Implementing Models of Maternity Histories

5.1 Introduction

Continued in this chapter is the study of models of maternity histories. After providing a brief introduction to the theory of semi-Markov processes in discrete time with stationary transition densities, an extended model of maternity histories, based on this theory, is developed. This model is an extension of those in chapter four to the extent that birth intervals, following marriage, are decomposed into certain subsegments of maternity histories. These subsegments include: waiting times to pregnancies; the possibilities that a pregnancy ends in a live birth, induced abortion, or a spontaneous abortion; and temporary sterile periods following each type of pregnancy outcome.

Taken together, the model just described and those in chapter four may be viewed as a set of mathematical tools awaiting the writing of software so that they may be implemented on a computer. But, the writing of software is a labor-intensive and time-consuming task, making it necessary to give careful consideration as to which models are actually chosen for computer implementation. Described in this chapter is a hierarchical software design based on renewal-type models that involve only ordinary convolutions. The rationale for choosing this design is discussed along with twenty-six illustrative computer runs. Attention is also given to the description of models for each subsegment of maternity histories considered, which are used to provide input to the hierarchical system. Finally, some notes on a structure for marking fertility projections, taking into account both period and cohort effects, are offered.

5.2 Semi-Markov Processes in Discrete Time with Stationary Transition Probabilities

Some concepts, taken from the theory of semi-Markov processes, will be used subsequently in the construction and computer implementation of stochastic models of human reproduction in which different types of pregnancy outcomes, as well as use of contraception, are taken into account. A brief sketch of the required theory will, therefore, be set down in this section. Continuous time formulations have great aesthetic appeal; but, in order to bring the mathematics more closely into line with arithmetic done by computers, a discrete time formulation will be considered. In addition to making the mathematics and computer operations match more closely, discrete time formulations have an appealing simplicity.

A basic step in formulating a model based on concepts involving a semi-Markov process is to define a finite set $\mathscr{I}$ of states. The elements of $\mathscr{I}$ may be defined in many ways and should be thought of as "states" individuals may move among over a period of time. Concrete examples of the set $\mathscr{I}$ will be given in subsequent developments. At an initial time, an individual enters some state $i_0 \in \mathscr{I}$; after a random sojourn time in i_0, he moves to state $i_1 \in \mathscr{I}$; and so the process continues. A record of states visited by an individual, along with the sojourn times in each state, is his sample path.

As a further step toward developing concepts, it will be helpful to set down some notation for sample paths. Let i_ν, $\nu \geq 0$, be states of $\mathscr{I}$ for some sample path; and, let y_ν be the time taken to go from $i_{\nu-1}$ to i_ν, $\nu \geq 1$. Observe that y_ν is the sojourn time in state $i_{\nu-1}$. The pair $\langle i_\nu, y_\nu \rangle$, describing the ν-th step among the states of $\mathscr{I}$, $\nu \geq 1$, will be denoted by ω_ν; the initial state will be denoted by the singleton $\omega_0 = \langle i_0 \rangle$. With these definitions of symbols, a sample path may be represented as the sequence $\omega = \langle \omega_0, \omega_1, \omega_2, \ldots \rangle$. In what follows, it will be useful to consider truncated sequences of the form $\langle \omega_0, \omega_1, \ldots, \omega_n \rangle$, which will be denoted by $\omega^{(n)}$, $n \geq 0$.

A notation for sample paths has been set down, because the finite dimensional distributions determining a semi-Markov process are defined directly on the sample paths. To define these distributions, it will be helpful to introduce pairs of random variables $\langle X_n, Y_n \rangle$, describing the process at the n-th step, $n \geq 1$. The random variable X_n indicates the state entered at the n-th step, the random variable Y_n is the sojourn time in state X_{n-1}, $n \geq 1$, and X_0 will indicate the initial state. For $n \geq 1$, a particular realization of the random variables X_0, $\langle X_1, Y_1 \rangle$, ..., $\langle X_n, Y_n \rangle$ will be denoted by the truncated sample path $\omega^{(n)}$.

A fundamental object underlying a semi-Markov process in discrete time is the one-step transition matrix of density functions $\mathbf{a}(t) = (a_{ij}(t))$, where i and j are states in $\mathscr{I}$, $t = 0, 1, 2, \ldots$, and $a_{ij}(t)$ is a function with values such that $0 \leq a_{ij}(t) \leq 1$ for all i, j, and $t \geq 0$. Given the one-step transition matrix $\mathbf{a}(t)$, the basic assumption characterizing the finite dimensional distributions determining a semi-Markov process with stationary transition probabilities may be stated as follows. Let $\omega^{(n-1)}$, $n \geq 1$, represent a realization of the random variables X_0, $\langle X_1, Y_1 \rangle$, ..., $\langle X_{n-1}, Y_{n-1} \rangle$. Then, if $X_{n-1} = i$,

$$P[X_n = j, \ Y_n = t | \omega^{(n-1)}] = a_{ij}(t) \tag{5.2.1}$$

for all states i and j in $\mathscr{I}$, $n \geq 1$, and $t \geq 0$. When $n = 1$, the conditional probability on the left in (5.2.1) is to be interpreted as $P[X_1 = j, \ Y_1 \leq t | X_0 = i]$. It is essential to observe that the righthand side of (5.2.1) depends on i, the last state visited, but it does not depend on n. Nondependence on n makes the transition probabilities of the process stationary. From (5.2.1) it follows that the conditional joint p.d.f. of the random variables $\langle X_1, Y_1 \rangle$, ..., $\langle X_n, Y_n \rangle$, given that $X_0 = i_0$, is

$$P[X_\nu = i_\nu, \, Y_\nu = y_\nu \, 1 \leq \nu \leq n \,|\, X_0 = i_0] = \sum_{\nu=1}^{n} a_{i_{\nu-1} i_\nu}(y_\nu), \qquad (5.2.2)$$

for all states $i_0, i_1, \ldots, i_n$ in $\mathscr{I}$ and nonnegative integers $y_1, \ldots, y_n$, $n \geq 1$.

Before giving further interpretations of basic Eqs. (5.2.1) and (5.2.2), it will be helpful to classify the states of $\mathscr{I}$. For all pairs of states i and j in $\mathscr{I}$ and $t \geq 0$ define a function $A_{ij}(t)$ by

$$A_{ij}(t) = \sum_{s=0}^{t} a_{ij}(s); \qquad (5.2.3)$$

and let

$$A_i(t) = \sum_j A_{ij}(t), \qquad j \in \mathscr{I}. \qquad (5.2.4)$$

Equations (5.2.1) and (5.2.3) imply that $A_i(t)$ is the distribution function of the sojourn time in state i. State i will be called transient, if an individual moves out of i eventually with probability one. Thus, state $i \in \mathscr{I}$ is transient if, and only if,

$$\lim_{t \uparrow \infty} A_i(t) = 1. \qquad (5.2.5)$$

A state i will be called absorbing, if transitions out of the state are impossible. Mathematically, an absorbing state i may be characterized by the condition

$$A_i(t) = 0 \qquad (5.2.6)$$

for all $t \geq 0$.

If state i is entered at $t = 0$, then $1 - A_i(t)$ is the probability an individual is still in i at time $t > 0$. Condition (5.2.6) implies that $1 - A_i(t) = 1$ for all $t \geq 0$. Hence, if state $i \in \mathscr{I}$ is absorbing, then after entering i an individual remains there with probability one. When the state space $\mathscr{I}$ contains absorbing states, it will be useful to partition $\mathscr{I}$ into two disjoint subsets $\mathscr{I}_1$ and $\mathscr{I}_2$, where $\mathscr{I}_1$ and $\mathscr{I}_2$ are the sets of absorbing and transient states, respectively.

For the sake of completeness, the transient states will be classified further. Two transient states i and j are said to communicate, if each can be reached from the other with positive probability in finitely many steps. In almost all the examples of semi-Markov processes considered in this chapter, the transient states in $\mathscr{I}_2$ will be a set of communicating states.

The finite dimensional distributions of the sequences of random variables, $\langle X_\nu \rangle$, $\nu \geq 0$, and $\langle Y_\nu \rangle$, $\nu \geq 1$, are completely determined by the fundamental assumption expressed in Eq. (5.2.1). Thus, Eq. (5.2.1) implies that the sequence of random variables $\langle X_\nu \rangle$, $\nu \geq 0$, is a Markov chain with a finite matrix (p_{ij}) of stationary transition probabilities. If $i \neq j$, then

$$p_{ij} = \lim_{t \uparrow \infty} A_{ij}(t); \qquad (5.2.7)$$

but, if $i=j$, then

$$p_{ii} = \lim_{t \uparrow \infty} (1 - A_i(t)). \qquad (5.2.8)$$

Hence, if i is a transient state, then $p_{ii}=0$; whereas, if i is absorbing, then $p_{ii}=1$.

Another implication of (5.2.1) is that the sequence of random variables $\langle Y_v \rangle$, $v \geq 1$, representing sojourn times in states, are independent. It is interesting to observe that condition (5.2.6) implies that sojourn times in absorbing states are infinite valued with probability one. Problem 1 may be consulted for further technical details.

From the practical point of view, as soon as the state space $\mathcal{I}$ has been defined, the next step in formulating a semi-Markov process in discrete time is to specify the finite matrix $\mathbf{a}(t)=(a_{ij}(t))$ of density functions. It is, therefore, necessary to characterize absorbing and transient states in terms of the functions in the matrix $\mathbf{a}(t)$. Suppose, for example, that state i is absorbing. Then, for every $j \in \mathcal{I}$ and $t \geq 0$, the condition, $a_{ij}(t)=0$, must be satisfied. It is easy to see that this condition implies (5.2.6). Next suppose state i is to be transient. Suppose further that it takes at least one time unit to observe a transition from one state to another. Then, for every transient state $i \in \mathcal{I}$ and state $j \in \mathcal{I}$, $a_{ij}(0)=0$ and $a_{ij}(t) \geq 0$ for all $t \geq 1$. Transitions of a transient state i into itself, without a visit to another transient state, will be ruled out. Therefore, for every transient state $i \in \mathcal{I}$, $a_{ii}(t)=0$ for all $t \geq 0$. The final step in assuring state i is a transient state is to choose the i-th row of the matrix $\mathbf{a}(t)$ of density functions such that condition (5.2.5) is satisfied.

Because the steps of the process are not synchronized with the discrete time scale $t=0, 1, 2, \ldots$, it is of interest to consider a random function $Z(t)$, indicating the state in $\mathcal{I}$ an individual occupies at time $t>0$. Some authors refer to the random function $Z(t)$, $t=0, 1, 2, \ldots$, as a semi-Markov process. Suppose state i_0 is entered at $t=0$ by an individual. If $t<Y_1$, the time of the first step, then the individual is still in state i_0 at time $t>0$ and $Z(t)=i_0$. The time the second step occurs is $Y_1 + Y_2$. Thus, if $Y_1 \leq t < Y_1 + Y_2$, then $Z(t)=i_1$. By continuing this line of reasoning, it can be seen that after the first step, $Z(t)=j$, if, and only if,

$$X_n = j \quad \text{and} \quad Y_1 + \ldots + Y_n \leq t < Y_1 + \ldots + Y_{n+1} \qquad (5.2.9)$$

for some $n \geq 1$. The conditional probabilities,

$$P_{ij}(t) = P[Z(t)=j \mid X_0 = i], \qquad (5.2.10)$$

defined for all pairs of states i and j in $\mathcal{I}$ and $t>0$, will be referred to as the current state probabilities for a semi-Markov process.

Having defined a set of states $\mathcal{I}$ and determined numerical specifications of the density functions in the transition matrix $\mathbf{a}(t)$, it is often of interest to calculate the values of the current state probabilities in (5.2.10), when implementing a semi-Markov process on a computer. As will be illustrated below, the calculation of the current state probabilities involves finding numerical solutions to systems of renewal type equations; equations that are, in fact, discrete analogues of renewal type integral equations. Two cases need to be considered in deriving these equations. One case arises when both i and j belong to the set $\mathscr{I}_2$ of transient states. Another case arises when i belongs to $\mathscr{I}_2$ and j belongs to $\mathscr{I}_1$, the set of absorbing states.

Suppose both i and j are transient states in $\mathscr{I}_2$. If an individual enters state i at $t=0$, then $1-A_i(t)$ is the probability he is still in state i at time $t>0$. The other possibility is that the individual moves to another transient state $v\in\mathscr{I}_2$ after s time units with probability $a_{iv}(s)$, $0\leq s\leq t$. Because all transition probabilities are assumed to be stationary, the process begins anew in state v and $P_{vj}(t-s)$ is the conditional probability of being in state j at time t, if state v is entered at time s. The conditional probability of starting out in state i at $t=0$, making a transition to state v after s time units, and then being in state j at time $t>0$ is $a_{iv}(s)P_{vj}(t-s)$. By observing that the two possibilities are disjoint events and summing over s and v, it follows that the current state probabilities, $P_{ij}(t)$, satisfy the system of renewal equations

$$P_{ij}(t)=\delta_{ij}(1-A_i(t))+\sum_v \sum_{s=0}^{t} a_{iv}(s)P_{vj}(t-s), \tag{5.2.11}$$

where $v\in\mathscr{I}_2$, i and j are transient states in $\mathscr{I}_2$, and δ_{ij} is the Kronecker delta. That is, $\delta_{ii}=1$ and $\delta_{ij}=0$ if $i\neq j$. Ideas of the type sketched in the derivation of Eqs. (5.2.11) are frequently referred to as a renewal argument.

Next consider the case i is a transient state in $\mathscr{I}_2$ and j is an absorbing state in $\mathscr{I}_1$. If state i is entered at $t=0$, then $A_{ij}(t)$ is the probability an individual moves to state j on the first step at or before time $t>0$. The other possibility is that, on the first step, an individual moves to another transient state $v\in\mathscr{I}_2$ after s time units, $0\leq s\leq t$. Given that state i is entered at $t=0$, another renewal argument yields the system of renewal type equations

$$P_{ij}(t)=A_{ij}(t)+\sum_v \sum_{s=0}^{t} a_{iv}(s)P_{vj}(t-s), \tag{5.2.12}$$

where $v\in\mathscr{I}_2$, $i\in\mathscr{I}_2$, and $j\in\mathscr{I}_1$.

From a practical point of view, renewal type equations, as exemplified by (5.2.11) and (5.2.12), will be useful only if efficient numerical algorithms for solving them can be found and implemented on a computer. Problem 2 is devoted to a discussion of such algorithms.

It should also be pointed out that absorbing states have uses other than stopping a stochastic process as discussed above. Sometimes, as a computational device, it is useful to make a transient state absorbing when calculating the distribution of times of first entrance into the state. An example will be provided in the next section.

Equations (5.2.11) and (5.2.12) suggest that the mathematical structure outlined in this section is a blend of the theory of Markov chains and renewal theory. In fact, some authors refer to this structure as Markov renewal theory. Çinlar (1969, 1975) has provided general discussions of the type of stochastic process considered here in continuous time.

Historically, the term, semi-Markov process, arose when investigators generalized the theory of continuous time Markov processes with a discrete state space and stationary transition probabilities. Doob (1953), for example, in his Chap. III, developed the theory of this class of stochastic processes and showed that sojourn times in states would necessarily be exponentially distributed. Semi-Markov processes are generalizations of the class of processes considered by Doob in that sojourn times in states may have arbitrary distributions. For the discrete time formulation of semi-Markov processes considered here, the geometric distribution on the positive integers, $1, 2, \ldots$, is the analogue of the exponential distribution.

5.3 A Decomposition of Birth Intervals

Ignored in the discussions of distributions of waiting times among live births in Chap. IV were factors in the life cycle of the human female that can and do affect these distributions. One of these factors are pregnancies with outcomes other than a live birth. In this section, pregnancy outcomes will be classified according to whether a pregnancy ended in a live birth, an induced abortion, or other types of pregnancy outcomes. Induced abortions will be included in the model, because, in many populations, an acceptable method of terminating an unwanted pregnancy is an induced abortion. Spontaneous abortions will, usually, be the predominant type in the "other" classification. To a lesser extent, stillbirths will also be included in this classification.

When more than one type of pregnancy outcome is recognized, a woman may occupy at least five states during a time interval between two live births. Among these states is the fecundable state, a state in which a woman is fertile and capable of becoming pregnant. To simplify the initial discussion, it will be supposed that a woman may exit from the fecundable state only by entering the pregnant state. Exits from the pregnant state will, in turn, occur when a pregnancy ends in either a live birth, an induced abortion, or another type of outcome. Upon exiting from the pregnant state, a temporary physiologically sterile state is entered immediately; the sojourn time in this sterile state depends on the type of pregnancy outcome. To further simplify the discussion, it will be assumed that the only possible exit from each of these temporary sterile states will be the fecundable state.

Viewing human reproduction as moving among a set states according to some semi-Markov process dates back at least to the early sixties, see, for example, the well-known paper of Perrin and Sheps (1964). One approach to formulating human reproduction as a semi-Markov process is to view the set of states mentioned above as one class of communicating states through which women cycle during their reproductive careers. If it is also assumed that the transition densities, governing movement among the states, do not change with age and parity, then limit theorems from classical renewal theory may be used to make statements about the long-term behavior of the process with a minimal number of calculations. The assumption that the densities governing transitions among states do not change with age and parity is, unfortunately, an over-simplification.

When it is possible to do a large number of computations at a relatively low cost, models based on more realistic assumptions may be implemented on a computer. Models based on the assumption that transition densities change with parity but not with age are, for example, easily implemented on a computer. As will be demonstrated subsequently, when the physiologically sterile state following a live birth is treated as absorbing, distributions of waiting times among live births may be easily calculated under this assumption. It is of interest to observe that when the vast majority of women marry in a relatively short age interval, there is a rather close correspondence between age and parity.

Whether a state is treated as transient or absorbing will affect the way it is labeled in the set $\mathscr{I}$ of states. Table 5.3.1 contains the definitions and ordering of states that will be used in this section. For women of parity $v \geq 0$, it will be assumed that the one-step transition matrix of densities, governing waiting times to live birth $v+1$, has the form

$$\mathbf{a}(v, t) = \begin{bmatrix} 0 & 0 & 0 & 0 & 0 \\ 0 & 0 & a_{23}(v, t) & 0 & 0 \\ a_{31}(v, t) & 0 & 0 & a_{34}(v, t) & a_{35}(v, t) \\ 0 & a_{42}(v, t) & 0 & 0 & 0 \\ 0 & a_{52}(v, t) & 0 & 0 & 0 \end{bmatrix}, \tag{5.3.1}$$

for $t \geq 0$, where $\mathbf{a}(v, 0) = \mathbf{0}$, a 5×5 zero matrix.

To further organize ideas, each row of the matrix function $\mathbf{a}(v, t)$ in (5.3.1) will be discussed in turn. Row one, containing all zeros, merely reflects that state E_1 is being treated as absorbing; while, row two expresses the condition that the only exit possible from the fecundable state E_2 is the pregnant state E_3. The function $a_{23}(v, t)$, $t \geq 1$, is the p.d.f. of the waiting times to pregnancy after entering the fecundable state E_2 at $t=0$. Among other things, this p.d.f. may reflect the use of contraception to space live births and prevent unwanted pregnancies. Row three of the matrix function in (5.3.1) expresses the condition that the three possible outcomes of a pregnancy are live birth, induced abortion, or some other type of outcome. That is, a pregnancy ends when one of the states E_1, E_4, or E_5 is entered, with the timing of the entrance being governed by the densities $a_{3j}(v, t)$, $j=1$, 4, or 5. Finally, rows four and five of the matrix $\mathbf{a}(v, t)$ express the condition that the only possible exit from

Table 5.3.1. Definitions of states for a semi-Markovian model governing waiting times among life births

E_1	Physiologically sterile period following life birth
E_2	Fecundable – fertile and capable of becoming pregnant
E_3	Pregnant
E_4	Physiologically sterile period following induced abortion
E_5	Physiologically sterile period following other types of pregnancy outcomes

states E_4 and E_5 is the fecundable state E_2. The densities, $a_{j2}(v, t)$, $j=4, 5$, govern the lengths of stay in the states E_4 and E_5, physiological sterile periods following an induced abortion or another type of pregnancy outcome.

Having defined the elements in the matrix $\mathbf{a}(v, t)$ of functions in (5.3.1), the next problem to consider is that of computing the distributions of waiting times among live births. Suppose a woman of parity $v \geq 0$ enters the fecundable state E_2 at $t=0$. Given this event, let $h_{21}(v, t)$ be the conditional probability that a live birth $v+1$ occurs at time $t>0$. Various possibilities need to be taken into account when deriving formulas for calculating the p.d.f. $h_{21}(v, t)$. One possibility is that the first pregnancy after attaining parity v ends in a live birth. Let $f_{21}(v, t)$ be the conditional probability that this event occurs at time $t>0$. Then,

$$f_{21}(v, t) = \sum_{s=0}^{t} a_{23}(v, s) a_{31}(v, t-s). \tag{5.3.2}$$

Alternatively, this first pregnancy may end when either state E_4 or E_5 is entered at $t>0$. Let $f_{2j}(v, t)$ be the conditional probability that state E_j, $j=4$ or 5, is entered at $t>0$. Then,

$$f_{2j}(v, t) = \sum_{s=0}^{t} a_{23}(v, s) a_{3j}(v, t-s). \tag{5.3.3}$$

After entering state E_j, $j=4, 5$, a live birth may occur at a subsequent time $t>0$. Suppose state E_j is entered at $t=0$; and, let $h_{j1}(v, t)$ be the conditional probability that a live birth occurs at time $t>0$, given this event. Then, by conditioning on the outcome of the first pregnancy and using a renewal argument, it follows that

$$h_{21}(v, t) = f_{21}(v, t) + \sum_{j=4}^{5} \sum_{s=0}^{t} f_{2j}(v, s) h_{j1}(v, t-s). \tag{5.3.4}$$

From (5.3.4) it can be seen that the density $h_{21}(v, t)$ may be calculated only if the densities $h_{j1}(v, t)$, $j=4, 5$, are known.

To calculate the densities $h_{j1}(v, t)$ for $j=4$ or 5, a 2×2 system of renewal type equations will be derived. Two possibilities must be considered in deriving this system of equations. One possibility is that after entering either state E_4 or E_5, a live birth occurs before either of these states is re-entered. Let $f_{i1}(v, t)$ be the conditional probability of this event, given that state E_i, $i=4, 5$, is entered at $t=0$. Then,

$$f_{i1}(v, t) = \sum_{s=0}^{t} a_{i2}(v, s) f_{21}(v, t-s), \tag{5.3.5}$$

for $i=4, 5$. The other possibility is that either state E_4 or E_5 may be reentered after leaving state E_i before the next live birth occurs. Suppose state E_i, $i=4, 5$, is entered at $t=0$. Given this event, let $f_{ij}(v, t)$ be the conditional probability that state E_j is entered for the first time at $t>0$ after leaving E_i without a visit to state E_1. Then, for $i, j=4, 5$,

$$f_{ij}(v, t) = \sum_{s=0}^{t} a_{i2}(v, s) f_{2j}(v, t-s). \tag{5.3.6}$$

Another conditioning and renewal argument yields the 2×2 system of renewal equations

$$h_{i1}(v, t) = f_{i1}(v, t) + \sum_{j=4}^{5} \sum_{s=0}^{t} f_{ij}(v, s) h_{j1}(v, t - s), \tag{5.3.7}$$

for determining the densities $h_{i1}(v, t)$, $i = 4, 5$.

To see that Eqs. (5.3.7) may be solved efficiently, it will be convenient to cast them in vector-matrix notation. Define two 2×1 column vectors of functions by $\mathbf{f}(v, t) = (f_{i1}(v, t))$ and $\mathbf{h}(v, t) = (h_{i1}(v, t))$ for $i = 4, 5$. Let $\mathbf{b}(v, t) = (f_{ij}(v, t))$, $i, j = 4, 5$, be a 2×2 matrix of functions. Then, Eqs. (5.3.7) may be represented in the succinct form

$$\mathbf{h}(v, t) = \mathbf{f}(v, t) + \sum_{s=0}^{t} \mathbf{b}(v, s) \mathbf{h}(v, t - s), \tag{5.3.8}$$

for $v \geq 0$ and $t \geq 0$. For every $v \geq 0$, Eqs. (5.3.8) may be solved by an application of the matrix version of Littman's algorithm, see Problem 2 of this chapter for a discussion of the details. After solving Eqs. (5.3.8), the functions $h_{i1}(v, t)$, $i = 4, 5$, may be substituted into the righthand side of Eq. (5.3.4) to determine the desired function $h_{21}(v, t)$. The densities $h_{41}(v, t)$ and $h_{51}(v, t)$ can be of interest in their own right; but, if only the function $h_{21}(v, t)$ is desired, then there is a more efficient algorithm. Problem 3 may be consulted for details.

The densities in the one-step transition matrix $\mathbf{a}(v, t)$ defined in (5.3.1) are not sufficient to determine the distributions of waiting times among live births following marriage within the framework to be developed in this section. To complete the specification of the model, it will be necessary to introduce three additional components. One of these components is the distribution for the length of stay in the state E_1. For women of parity $v \geq 1$, let $a_{12}(v, t)$ be the p.d.f. of this distribution. A second component is a function $A_s(t)$, representing the fraction of fertile women in a cohort as a function t of age. A third component consists of parity progression ratios interpreted in terms of family planning intentions; consult the discussion in Sect. 4.4 for further details.

As a next step in the development of the model, formulas for the density functions of waiting times among live births will be derived, under the assumption that the evolution of maternity histories are governed only by the densities in the matrix $\mathbf{a}(v, t)$ defined in (5.3.1) and the p.d.f.'s $a_{12}(v, t)$ for the lengths of stay in the temporarily sterile state E_1. Let $g(v, t)$ be the p.d.f. of the waiting times to live birth $v + 1$ for women of parity $v \geq 0$. Then,

$$g(0, t) = h_{21}(0, t) \tag{5.3.9}$$

for all $t \geq 0$. After the first live birth, there will always be a sojourn time in the state E_1 before the state E_2 is reentered. Therefore,

$$g(v, t) = \sum_{s=0}^{t} a_{12}(v, s) h_{21}(v, t - s), \tag{5.3.10}$$

for all $v \geq 1$ and $t \geq 0$.

Distributions of waiting times to the n-th, $n \geq 1$, live birth following marriage will also be needed, when the assumptions of the above paragraph are in force. For marriages occurring at $t=0$, let $g^{(n)}(t)$ be the p.d.f. of the waiting times to the n-th live birth following marriage. Then,

$$g^{(1)}(t) = g(0, t); \tag{5.3.11}$$

and, for $n \geq 2$ and $t \geq 0$, $g^{(n)}(t)$ may be determined recursively according to the formula

$$g^{(n)}(t) = \sum_{s=0}^{t} g^{(n-1)}(s) g(n-1, t-s). \tag{5.3.12}$$

Imagine maternity histories in a cohort evolving according to the sequence of p.d.f.'s, $g(v, t)$, $v \geq 0$, until the birth process is stopped by either the completion of desired family size or the onset of permanent sterility. If K is interpreted as a random variable, representing desired numbers of offspring in a cohort of women, and birth intentions are represented by parity progression ratios, p_n, $n \geq 1$, then

$$P[K \geq n] = p_1 p_2 \cdots p_n = \alpha(n) \tag{5.3.13}$$

is the probability that at least $n \geq 1$ live births are desired. To simplify the formulas and reduce the number of convolutions that must be calculated in the computer implementation of the model, it will also be assumed that the length of a pregnancy ending in a live birth is some constant γ.

For women in a cohort marrying at age x, let $h^{(n)}(x, t)$ be the p.d.f. of the waiting times to the n-th live birth, $n \geq 1$, following marriage at $t=0$. Then, the n-th live birth may arrive at t time units following marriage, if and only if, at least n live births are desired, with probability $\alpha(n)$, and a woman is still fertile at age $x+t-\gamma$, with probability $A_s(x+t-\gamma)$. Therefore, under independence assumptions,

$$h^{(n)}(x, t) = \alpha(n) A_s(x+t-\gamma) g^{(n)}(t), \tag{5.3.14}$$

for all $x \geq 0$ and $t \geq 0$.

Define a function $d(t)$ by

$$d(t) = \sum_{n=1}^{\infty} \alpha(n) g^{(n)}(t); \tag{5.3.15}$$

and, let $b(x, t)$ be the density for the discrete analogue of the mean $v(x, t)$ defined in (4.2.14), see also (4.8.16). Then, for the discrete type of model under consideration,

$$b(x, t) = \sum_{n=1}^{\infty} h^{(n)}(x, t) = A_s(x+t-\gamma) d(t). \tag{5.3.16}$$

Consequently, the gross maternity function for the cohort has the form

$$b(t) = \sum_{x=0}^{t} g_0(x) b(x, t-x) = A_s(t-\gamma) \sum_{x=0}^{t} g_0(x) d(t-x), \tag{5.3.17}$$

where $g_0(x)$ is the p.d.f. for age at first marriage in a cohort. Compare formula (5.3.17) with (4.2.16) and (4.3.5). Observe, that in (5.3.17) an age-dependent convolution has been reduced to an ordinary convolution, greatly reducing the number of arrays that need to be considered in computer implementation. Finally, if $S(t)$ is the cohort survival function, then

$$\varphi(t)=S(t)b(t), \qquad t\geq 0, \tag{5.3.18}$$

is the cohort net maternity function.

If the function $c(t)$ is defined by

$$c(t)=\sum_{x=0}^{t} g_0(x)d(t-x), \tag{5.3.19}$$

then the cohort net maternity function takes the form

$$\varphi(t)=S(t)A_s(t-\gamma)c(t), \tag{5.3.20}$$

for all ages $t\geq 0$.

Formula (5.3.20) for the cohort net maternity function may also be interpreted as follows. Suppose a potential birth process has the birth rate density function $c(t)$ and may be stopped by either death of a female or the onset of permanent sterility. Let X_1 be a random variable representing age at death in a cohort of females and let T_s be an independent random variable representing age at onset of permanent sterility. If the random variable X is a stopping time of the process, then

$$P[X>t]=P[X_1>t]P[T_s>t-\gamma]=S(t)A_s(t-\gamma). \tag{5.3.21}$$

Continuous time analogues of formula (5.3.20) may thus be viewed as a generalization of formula (4.3.7).

Formulas (5.3.17) and (5.3.20) are useful from the point of view of computer implementation, because one computer program will encompass many cases. Cases that may be considered in the computer implementation of these formulas will be discussed subsequently.

5.4 On Choosing Component Functions of the Model

Before the model of human reproduction developed in Sect. 5.3 can be implemented on a computer, it will be necessary to devise methods for specifying all component functions of the model in numerical form. As will become apparent in what follows, searching for ways of specifying these functions numerically raises a host of interesting questions. The purpose of this section is to discuss some of the simplest choices of the functions in question. Initially, the discussion will be organized according to the rows of the matrix-valued function $\mathbf{a}(v, t)$ in (5.3.1).

Considered first is the function $a_{23}(v, t)$, a p.d.f. governing passage times from the fecundable state E_2 to the pregnant state E_3, in the second row the matrix in (5.3.1). Just as in Sect. 1.1, passage from state E_2 to state E_3 will be interpreted as a recognizable pregnancy experienced by a married woman continuously exposed to the risk of pregnancy. Consequently, one possible choice of $a_{23}(v, t)$ would be the p.d.f. for the geometric distribution given in (1.2.2), where θ, the fecundability parameter, could depend on v, the parity of a woman. An alternative choice of the p.d.f. $a_{23}(v, t)$ could be that for the beta-geometric distribution defined by (1.5.5), (1.5.6), and (1.5.7), with the parameters α and β depending on v. If coital patterns were of essential interest, $a_{23}(v, t)$ could depend on patterns as described in Chap. 1. Subsequently, models in which $a_{23}(v, t)$ depends on patterns of contraceptive use will be discussed.

Row three of the matrix-valued function in (5.3.1) is concerned with the three types of pregnancy outcomes under consideration. One approach to choosing numerical specifications for the functions in this row is to assume that they are of the form

$$a_{3j}(v, t) = \pi_{3j}(v) b_{3j}(v, t), \tag{5.4.1}$$

for all $v \geq 0$, $t \geq 0$, and $j = 1, 4$, and 5. In (5.4.1), $\pi_{3j}(v)$ are nonnegative constants such that

$$\pi_{31}(v) + \pi_{34}(v) + \pi_{35}(v) = 1 \tag{5.4.2}$$

for all $v \geq 0$; and, $b_{3j}(v, t)$ is a proper p.d.f. for each v and j. The parameter $\pi_{3j}(v)$ is the conditional probability a pregnancy ends in the j-th outcome for a woman of parity v. Given that a woman of parity v has a pregnancy outcome of type j, $b_{3j}(v, t)$ is the conditional p.d.f. governing the timing of this outcome.

Various approaches may be used in specifying the densities in (5.4.1) numerically. The most desirable approach is to get estimates of the densities in (5.4.1) from data that have been carefully gathered and analyzed. Leridon (1977), in his Table 4.1, has reported on a longitudinal study of pregnancies conducted by French and Bierman (1962) in Hawaii, during the time span, 1953 to 1956. According to the life table procedure used by Leridon, the fractions of pregnancies ending in either a spontaneous abortion or stillbirth was 0.2373. The complementary fraction, $1 - 0.2373 = 0.7627$, of pregnancies ended in a live birth. Whenever pregnancies are allowed to follow their natural course for women of parity v, these fractions could be taken as estimates of the conditional probabilities $\pi_{31}(v)$ and $\pi_{35}(v)$. In these circumstances, the conditional probability that a pregnancy ends in an induced abortion would, of course, be zero.

When induced abortions are used to terminate unwanted pregnancies, direct estimates of the conditional probabilities in Eq. (5.4.2) are not always possible, due to a lack of information on spontaneous abortions and stillbirths. In some circumstances, however, both induced abortions and live births are registered so that the ratio of induced abortions to live births may be estimated. Let r be an estimate of this ratio. Then, r may be viewed as an estimate of the ratio of conditional probabilities, $\pi_{34}(v)/\pi_{31}(v)$, for some parity v. If it is assumed that $\pi_{35}(v)$, the conditional probability that a pregnancy ends in either a spontaneous abortion or stillbirth, is known, then the relation, $\pi_{34}(v) = r\pi_{31}(v)$, may be substituted into (5.4.2). Solving the resulting system of equations for $\pi_{34}(v)$ and $\pi_{31}(v)$ yields the formulas

$$\pi_{31}(v) = \frac{1 - \pi_{35}(v)}{1 + r},$$

$$\pi_{34}(v) = \frac{r(1 - \pi_{35}(v))}{1 + r}. \tag{5.4.3}$$

If the ratio r were estimated by parity, then the estimates in (5.4.3) could be further refined. Equations (5.4.3) seem to have been used extensively; Inoue (1977) is among the investigators who have used these equations. A possible choice for the value of the conditional probability on the right in (5.4.3) would be the French-Bierman estimate.

When designing computer simulation experiments, it would be useful to have a capability for holding the densities $b_{3j}(v, t)$ $j = 1, 4, 5$, constant while varying the parameters $\pi_{3j}(v)$ in order that parity dependence may be accommodated in a parsimonious way. One way of achieving this capability is to estimate these densities from archival data and assume they do not depend on the π-probabilities. If parity is ignored, then estimates of the densities $b_{31}(t)$ and $b_{35}(t)$ may be obtained by normalizing the last two columns of Table 4.1 in Leridon (1977), i.e., the entries in each column were divided by their sum. An estimate of the density $b_{34}(t)$, describing the timing of induced abortions, may be obtained from some Swedish data as reported by Tietze (1981). The estimates of the densities in question are reported in Table 5.4.1.

As can be seen from the second column of Table 5.4.1, the vast majority of live births occur in the 9th and 10-th months following conception. In computer implementation of models of the type described in Sect. 5.3, requiring the computation of a large number of convolutions, many computations can be eliminated, if it can be assumed that a p.d.f. is concentrated at some point. From the evidence presented in the second column of Table 5.4.1, it can be seen that little information would be lost if the p.d.f. $b_{31}(t)$ were concentrated

Table 5.4.1. Estimates of conditional densities governing timing of pregnancy outcomes

Month	Live birth[a] b_{31}	Induced abortion[b] b_{24}	Other[a] b_{35}
1	0	0	0.4555
2	0	0.371	0.2625
3	0	0.574	0.1568
4	0	0.055	0.0447
5	0.0004	0	0.0278
6	0.0016	0	0.0105
7	0.0094	0	0.0097
8	0.0266	0	0.0093
9	0.3889	0	0.0105
10	0.5731	0	0.0127

[a] Based on data of French and Bierman (1962).
[b] Based Swedish data as reported by Tietze (1981).

Table 5.4.2. Hypothetical density functions governing timing in sterile states following induced abortions and other types of pregnancy outcomes

Month	Induced abortion a_{42}	Other a_{32}
1	0.25	0
2	0.50	0.10
3	0.25	0.25
4	0	0.30
5	0	0.25
6	0	0.10

at, say, 9 or 9.5 months. Unlike the estimate of $b_{31}(t)$, that for the p.d.f. $b_{35}(t)$, governing the timing of spontaneous abortions and stillbirths, is concentrated in the first four months following conception, see column four in Table 5.4.1. As one would expect from sound medical practice, the estimate of the p.d.f. $b_{34}(t)$ for the timing of induced abortions in column three of Table 5.4.1. is concentrated in the first four months following conception.

Rows four and five of the matrix-valued function in (5.3.1) are concerned with the p.d.f.'s of the sojourn times in temporary sterile states following induced abortions and other types of pregnancy outcomes. Little recorded empirical information on these densities seems to be available. Because little can be gained from introducing additional assumed forms of these densities into the literature, the hypothetical densities suggested by Perrin and Sheps (1965) will be used in all computer simulations based on the model set forth in Sect. 5.4. Moreover, it will be assumed that the densities in question are not parity dependent. For ease of reference, the hypothetical densities suggested by Perrin and Sheps are listed in Table 5.4.2.

At this point, it seems appropriate to pause and discuss the validity of some of the assumptions underlying the model outlined in Sect. 5.3. As the model now stands, it is parity dependent but within each parity the laws governing the evolution of maternity histories are homogeneous in time. Leridon (1977) presents evidence that the level of intrauterine mortality, spontaneous abortions, depends not only on the order of the pregnancy, see his Table 4.9, but also on the age of a woman when the conception occurred, see his Table 4.12. Thus, the model set forth in Sect. 5.3 is realistic to the extent that parity reflects pregnancy order; it is unrealistic to the extent that aging is taken into account only through parity. It should be observed that the dependence of intrauterine mortality on parity can easily be accommodated in model of Sect. 5.3 by adjusting the π-probabilities in Eq. (5.4.2). Other deficiencies of the model will be discussed subsequently.

One of the most sensitive components of the model of Sect. 5.3, affecting the rate of flow of births into a noncontracepting population, is the p.d.f. $a_{12}(v, t)$, governing the timing of returns to the fecundable state E_2 from the postpartum sterile state E_1. For, when mothers nurse infants, the time spent in the state E_1 often exceeds the time spent in the pregnant state E_3. Consequently, in noncontracepting populations, long sojourn times in state E_1 tend to reduce the average number of live births experienced by women during their reproductive careers. Considerable attention has been given to the postpartum sterile period in the literature, see, for example, Chap. 5 in Leridon (1977). Many forms of the p.d.f. $a_{12}(v, t)$ could be considered; but, only variants of the negative binomial distribution will be discussed in this section.

A random variable X, taking values in the nonnegative integers, is said to follow a standard negative binomial distribution with parameters $\rho > 0$ and θ, where $0 < \theta < 1$, if its p.d.f. has the form

$$P[X = x] = f(x) = \frac{\rho^{(x)} \theta^\rho (1 - \theta)^x}{x!}, \tag{5.4.4}$$

for $x = 0, 1, 2, \ldots$. In (5.4.4), the function $\rho^{(x)}$ is defined by

$$\rho^{(x)} = \frac{\Gamma(\rho + x)}{\Gamma(\rho)}, \tag{5.4.5}$$

for $x \geq 0$.

Instant returns to the fecundable state E_2 from the state E_1 are impossible. Therefore, let $\tau \geq 1$ be the first month following a live birth that a woman is again at risk of pregnancy. Then, a random variable T, representing return times to the fecundable state E_2 from the state E_1, will be said to follow a negative binomial distribution with location parameter τ, if it may be represented in the form $T = \tau + X$, where X is a random variable having the p.d.f. $f(x)$ in (5.4.4). Obviously, the p.d.f. of the random variable T is

$$P[T = t] = a_{12}(v, t) = f(t - \tau), \tag{5.4.6}$$

for $t = \tau, \tau + 1, \ldots$. If necessary, the parameters ρ, θ, and τ could depend on v; but, more importantly, they could also depend on the level of nursing. It can be shown that the mean and variance of T are given by

$$E[T] = \tau + \frac{\rho(1 - \theta)}{\theta},$$

and   (5.4.7)

$$\mathrm{Var}[T] = \frac{\rho(1 - \theta)}{\theta^2}.$$

When designing computer simulation runs, an investigator is frequently compelled to use scanty information reported in the literature. One might, for example, have an estimate $\hat{\tau}$ of the location parameter τ and an estimate $\bar{t}$ of the mean $E[T]$. Setting $\bar{t}$ equal to the expression for $E[T]$ in (5.4.7), yields the equation

$$\bar{t} = \hat{\tau} + \frac{\hat{\rho}(1 - \hat{\theta})}{\hat{\theta}}, \tag{5.4.8}$$

where $\hat{\rho}$ and $\hat{\theta}$ are estimates of the parameters ρ and θ. But, in order to determine $\hat{\rho}$ and $\hat{\theta}$ uniquely, an additional equation in $\hat{\rho}$ and $\hat{\theta}$ is needed.

Provided that an estimate $\hat{\sigma}^2$ of Var$[T]$ is available, a second equation,

$$\hat{\sigma}^2 = \frac{\hat{\rho}(1-\hat{\theta})}{\theta^2},\tag{5.4.9}$$

may be derived from (5.4.7). Together, Eqs. (5.4.8) and (5.4.9) determine $\hat{\rho}$ and $\hat{\theta}$ uniquely for given values of $\bar{t}$ and $\bar{\tau}$. In doing sensitively analyses, an investigator, would, of course, be free to consider arbitrary values of Var$[T]$ in order to calculate trial values of the parameters ρ and θ as determined by Eqs. (5.4.8) and (5.4.9).

An alternative approach to determining $\hat{\rho}$ and $\hat{\theta}$ could be based on some feeling for the mode of the p.d.f. of the random variable T. It is shown in Problem 4, where some properties of the negative binomial distribution are discussed, that the double mode of the random variable T is

$$t_0 = \tau + \frac{\rho(1-\theta)-1}{\theta},\tag{5.4.10}$$

provided that the expression on the right is a positive integer. Thus, by assigning a value $\bar{t}_0$ to t_0, Eqs. (5.4.8) and (5.4.10) may be used to determine $\hat{\rho}$ and $\hat{\theta}$ for the given values $\bar{t}$ and $\hat{\tau}$. In fact, the solution is

$$\hat{\rho} = \frac{(\bar{t}-\hat{\tau})\hat{\theta}}{1-\hat{\theta}}$$

and   (5.4.11)

$$\theta = \frac{1}{\bar{t}-\hat{t}_0},$$

provided that $\bar{t}-\hat{t}_0 > 1$. Actually, Eqs. (5.4.11) would yield useful results even if t_0 in (5.4.10) were not a positive integer.

As an example of the way the foregoing theory could be applied to obtain rough estimates of the parameters ρ, τ, and θ, the information given in Table 5.4 of Leridon (1977) could be considered. Contained in this table compiled by Leridon is a listing of means and medians for durations of amenorrhea, following childbirth, expressed in days and classified by durations of breastfeeding. At this point, it should be mentioned that the resumption of menstruation following childbirth is not a sure indicator of a return to the fecundable state E_2; because, the first menstrual period may or may not be ovulatory.

According to Leridon's table, among French women who practiced no breastfeeding, the mean duration of amenorrhea was reported as 58 days, or approximately two months. From such information a likely guess for the τ corresponding to these women would be $\hat{\tau}=1$, one month. For women who breastfed their infants, one might expect τ to have a higher value, say, $\tau=2$ or perhaps $\tau=3$. Given estimates $\hat{\tau}$ of τ and estimates $\bar{t}$ of $E[T]$ for various levels of breastfeeding, either one of the methods outlined above could be used to obtain trial estimates of the parameters ρ and θ for each level of breastfeeding considered. It should be emphasized that the above methods should be used only when datailed information is not readily accessible.

One special case deserves mention; when $\rho=2$ and $\tau=2$, the negative binomial distribution is sometimes referred to as the Pascal distribution. Observe that when $\hat{\rho}=2$, $\hat{\tau}=2$, and $\bar{t}$ is given, then $\hat{\theta}$ may be determined uniquely from Eq. (5.4.8); in fact, $\hat{\theta}=2/\bar{t}$. Due to the importance of the postpartum sterile period, more elaborate models for the p.d.f. $a_{1\,2}(v,t)$ will be discussed subsequently.

The last component function of the model developed in Sect. 5.3 to be discussed is $A_s(t)$, the fraction of women in a birth cohort who are still fertile at age t. Chosen as a model for this function was the four parameter family,

$$A_s(t)=\alpha\,\exp\left[-\left(\frac{x-\beta}{\gamma}\right)^{\delta}\right],\quad t\geq\beta,\tag{5.4.12}$$

based on the well-known Weibull distribution. In (5.4.12), α is the fraction of females in a cohort who are fertile, i.e., capable of bearing children once they reach the childbearing ages; the parameters β, γ, and δ are the positive location, scale, and shape parameters, respectively, of the Weibull distribution.

Finding data which may be used, for estimating the parameters in the Weibull distribution, or in any other parametric model for $A_s(t)$, is difficult, because in contracepting populations an observer cannot ascertain with certainty the age at which a fertile woman becomes permanently sterile. One approach to finding estimates of the parameters in (5.4.12) is to use data on the last confinement for married women in noncontracepting populations, such as the Hutterites, who survive to age 50 and beyond. In Hutterite societies, married women continue to bear children for as long as they are able, so that age at last confinement would be a pretty good indicator of the age of onset of permanent sterility. Eaton and Meyer (1953) and especially Hostetler (1974) have given detailed accounts of the social biology and nature of Hutterite society.

The estimates of the parameters β, γ, and δ, presented in Table 5.4.3, were obtained by applying an APL nonlinear least squares procedure, described in Sect. 4.6 to the adjusted data on the age of Hutterite wives at last confinement as reported in Table 1 of the paper, Tietze (1957). In implementing the least squares procedure, the t-values were chosen as the midpoints of the four-year age intervals used by Tietze. The mean square deviation of the model from the

Table 5.4.3. Nonlinear least squares estimates of parameters in Weibull distribution based on age of Hutterite wives at last confinement

Parameter[a]			Mean square deviation
β	γ	δ	
6.06×10^{-7}	41.7555	12.973	4.5×10^{-4}

[a] The parameter α would have to be estimated from other data sources. In most societies, the vast majority of women are capable of bearing children. Some value of α over 0.9 would appear to be a reasonable guess for α.

actual values was 4.5×10^{-4}, indicating that the function in (5.4.12) fit the observations fairly well.

The Weibull model of permanent sterility presented in (5.4.12) was, evidently, first used in this connection in simulations reported by Busby and Mode (1983). An alternative choice for the function $A_s(t)$ could be the exponential described by Pittenger (1973), a model that was roughly based on several sets of historical data. Inoue (1977) may also be consulted for numerical examples of the function $A_s(t)$. Choosing forms of the function $A_s(t)$ remains problematic; Chap. 6 in Leridon (1977) may be consulted for further details.

5.5 An Overview of MATHIST – A Computer Simulation System

The extent to which a computer simulation system is useful depends not only on its mathematical design, which is a reflection of reality, but also on whether it is feasible to use the system repeatedly. Feasibility of repeated use will, in turn, depend on ease of implementation and computational efficiency. In human terms, ease of implementation revolves around the problem of interfacing human beings and machines, a problem that will be touched upon subsequently. Computational efficiency, on the other hand, depends largely on the mathematical design underlying the system and the class of mathematical operations that require the most computations.

A class of mathematical operations common to all the stochastic models of maternity histories discussed so far is that of convolutions. Two types of convolutions, ordinary and age-dependent, have been considered. Ordinary convolutions are by far the easiest to implement on a computer, because only one-dimensional arrays need to be considered. To implement age-dependent convolutions, however, computations involving two-dimensional arrays must be carried out. It is also well known that Fast Fourier Transform (FFT) algorithms may be used to compute ordinary convolutions very efficiently. Therefore, high computational efficiency may be obtained if attention is restricted to those models involving ordinary convolutions. Because the models of maternity histories developed in Sects. 4.2, 4.3, 4.4, and 5.3 require only the computation of ordinary convolutions, they will underlie the mathematical design of the computer simulation system to be described in this section. For ease of reference, the acronym, MATHIST, indicating that maternity histories in a cohort are the objects of study, will be used as a name for this system.

Choosing models of maternity histories for computer implementation that require only the computation of ordinary convolutions is, in fact, a trade-off between realism and computational efficiency. Age-dependence in maternity histories is, for example, a natural phenomenon to consider, when developing stochastic models of human reproduction. One would think that models of maternity histories accommodating both age and parity dependence should be more realistic than those accommodating only parity dependence. But, part of the price for constructing a model to include both age and parity dependence is the computation of many age-dependent convolutions; consult Sect. 4.8 for details.

When basing MATHIST on mathematical models requiring only the computation of ordinary convolutions, one would hope that the system would be sufficiently robust so that gains made in computational efficiency will offset losses in realism due to simplified assumptions. Judgments regarding the degree to which MATHIST is robust can best be made by making many simulation runs and comparing the results against reliable observations. Examples of computer runs, using MATHIST, will be given in subsequent sections. Although FFT algorithms were used in the computer implementation of MATHIST, any discussion of the mathematics underlying these algorithms is beyond the scope of this book. Interested readers may consult the book, Brigham (1974), for details.

One way of viewing MATHIST is that of a hierarchical software system superimposed on a class of mathematical models. This hierarchical system was motivated by the types of mathematical models that result from varying the interpretation of components making up Eq. (5.3.14). If, for example, the sterility function $A_s(t)$ is taken identically equal to one and the densities $g(v, t)$, $v \geq 0$, as discrete analogues of the densities corresponding to the distribution functions on the right in (4.2.2), then $h^{(n)}(x, t)$ in (5.3.14) is mathematically equivalent to $h^{(n)}(t)$, a discrete analogue of the density corresponding to the distribution function $H^{(n)}(t)$ in (4.2.6). Consequently, the potential birth process discussed in Sects. 4.2, 4.3, and 4.4 is formally contained within the structure developed in Sect. 5.3, provided the component functions are suitably interpreted. ·

The interpretation of the functions going into $g^{(n)}(t)$, the n-fold convolution in (5.3.14), thus provides a basis for the first breakdown in the hierarchical structure of MATHIST. A computer run will be said to belong to class one if the densities $g(v, t)$, $v \geq 0$, going into the n-fold convolution $g^{(n)}(t)$, reflect only birth interval information as described in Eq. (4.2.2), i.e., densities for waiting times among live births. A computer run will be said to belong class two, if the densities $g(v, t)$, $v \geq 0$, are computed according to the model in Sect. 5.3, where a live birth interval was decomposed into some of the events ocurring in the reproductive cycle of the human female. For computer runs in class two, the algorithm implemented in MATHIST for computing the densities $h_{21}(v, t)$, going into Eqs. (5.3.9) and (5.3.10), was that described in Problem 3. Whatever be the interpretation of the densities $g(v, t)$, $v \geq 0$, after they are entered into MATHIST, a single subroutine, based on FFT algorithms, is used to compute the n-fold convolutions $g^{(n)}(t)$, $n \geq 1$, in (5.3.12).

A further breakdown in the classification of computer runs within classes one and two may be based on interpretations and assignments of values to parity progression ratios, as they enter into the system through the products, $\alpha(n) = p_1 p_2 \ldots p_n$, in (5.3.13) and (5.3.14). Within class one, a computer run will be said to be of type one, if the parity progression ratios p_v, $v > 1$, are interpreted as those that could be observed *a posteriori* in a sample of completed maternity histories. Mathematically, p_v in this case is defined as in Eq. (4.2.1). Among other things, these parity progression ratios reflect the onset of permanent sterility, particularly in populations practicing no birth control. Thus, the sterility function, $A_s(t)$, would be taken as the constant, 1.

Estimates of *a posteriori* parity progression ratios, such as those appearing in Table 2.2 of Leridon (1977), could, in some cases, be used as input into MATHIST. Conditional densities for waiting times among live births, given that they occur, could be chosen as the double exponential distribution, discussed in Example 4.5, and used to generate numerical input into MATHIST, provided that suitable estimates of the parameters were available. Given such birth interval information and a density for age at first marriage in a cohort, MATHIST could be used to simulate a cohort gross maternity function as defined in Sect. 4.3 and displayed more explicity in the discretized formula, (5.3.17), implemented in MATHIST.

A computer run within class one will be said to be of type two, if parity progression ratios are given *a priori* and reflect family planning intentions as described in Sect. 4.4. Realized or *a posteriori* parity progression ratios will in general differ from *a priori* parity progression ratios in a cohort, because, among other things, the birth process may be stopped by the onset of permanent sterility before intended family size is attained. For runs of type two within class one, the sterility function $A_s(t)$ would not be taken as the constant one. Incidentally, these latter runs are based on a mathematical model that has not been discussed explicitly; but, it is implicit in the argumentation used in deriving Eq. (5.3.14).

Within class two, a computer run will be said to be of type one, if family planning intentions are accommodated through waiting times to conception, which, in turn, reflect the use of contraception to limit and space births. When an unwanted pregnancy occurs, induced abortion may be used to terminate it. In these runs, it will be assumed that, for those females surviving the childbearing ages, the potential birth process ends with the onset of permanent sterility. Parity progression ratios are thus realized *a posteriori*. When making a type one run within class two, the products, $\alpha(n)$, in (5.3.14) would be put equal to one.

A computer run within class two will be said to be of type two, if family planning intentions are reflected, in part, through *a priori* parity progression ratios. Spacings of live births, could be reflected through waiting times to conception. If d is the desired number of births, then $p_v = 1$ for all v such that $1 \leq v \leq d$; if $v > d$, then $0 < p_v < 1$. The situation discussed in Sect. 4.4, in which sex composition of the children was a deciding factor in determining the number of live births desired, provides a concrete method for calculating parity progression ratios *a priori*. The use or nonuse of induced abortion to terminate unwanted pregnancies could be reflected through the π-probabilities on (5.4.1). For desired births, induced abortions would not be used to terminate pregnancies so that $\pi_{34}(v) = 0$ and $\pi_{31}(v) + \pi_{35}(v) = 1$ for all v such that $1 \leq v \leq d$. For $v > d$, unwanted pregnancies could be terminated with an induced abortion with positive probability. In these cases, $\pi_{34}(v) > 0$ and $\pi_{31}(v) + \pi_{34}(v) + \pi_{35}(v) = 1$.

Within either class one or two, many computer runs may be made by selecting the components of the models of maternity histories in different combinations. Some of the components needed for computer runs in class two, however, may not vary greatly among populations and will, therefore, be treated as constants in MATHIST. Two of these components are the conditional densities for the timing of induced abortions and other types of pregnancy outcomes. These densities could vary with parity in a cohort of women; but, definite information on such variation seems to be lacking. Consequently, in MATHIST, the numerical forms of the conditional densities $b_{34}(v, t)$ and $b_{35}(v, t)$ on the right in (5.4.1) were taken as those in columns four and five of Table 5.4.1 for all parities $v \geq 0$. Definite information also seems to be lacking on the density functions governing sojourn times in states of temporary sterility following induced abortions and other types of pregnancy outcomes. For lack of better information, in MATHIST, the numerical forms chosen for the densities $a_{42}(v, t)$ and $a_{52}(v, t)$ were those in columns two and three of Table 5.4.2 for all parities $v \geq 0$.

Whether a computer simulation system will be regarded as easy to implement, user-friendly, will depend, to a large extent, on the computing environment in which it is used. Many computing facilities have progressed to the point where interactive video terminals directly connected to a computer, or perhaps connected by telephone, are available to a user. Although the use of MATHIST is not limited to computing facilities with video terminals, for the sake of concreteness, the discussion that follows will center on the monitor, T.V. screen, of a video terminal. Relative to older style computing facilities, requiring punched cards and batch processing, those with interactive video terminals would be regarded as more user-friendly and a significant step forward in coming to grips with the problem of interfacing human beings and machines.

Computer runs in either class one or two require that numerical forms of the components of the models not being held constant be entered into MATHIST. Finding a solution to the problem of making MATHIST easy to use has been approached through the design of a main menu, containing instructions as to what numerical information is to be entered into MATHIST. Presented in Table 5.5.1 is the main menu for implementing MATHIST.

Suppose a user has the main menu before him on a monitor and wishes to make a computer run for the first time. To guide such a user in executing the main menu the option, 12; Enter All, which automatically guides a user through all the numerical inputs needed to run MATHIST, has been provided. If a user enters the number 12 into his terminal, then the instructions given in the upper panel of Table 5.5.2 will appear on the monitor.

These instructions call for entering the run constants into MATHIST. The first run constant is the length of the gestation period in months, indicating that the sojourn time in the pregnant state, is always treated as a constant in MATHIST. A commonly chosen value for this constant would be nine months. The other run constants for a cohort of women are earliest age at first marriage, greatest fertile age, percent ever married, and maximum allowable births. MATHIST will accommodate up to twenty live birth intervals in a maternity history. The run constants are to be entered according to the order in which their descriptions appear on the monitor. A typical set of run constants could be entered as the numbers: 9 15 50 95 15. Due to the nature of FORTRAN, the programming language in which the programs of MATHIST have been written, all run constants must be positive integers.

After the run constants are entered at the terminal, the instructions in the middle panel of Table 5.5.2 will appear on the monitor, asking the user whether he wishes to make a computer run in either class one or two. Choosing the class of computer run is a branch point in MATHIST, which determines the type of birth interval information that is to be entered into the simulation system. Apart from birth interval information, all other inputs into MATHIST require the same operating procedures for both class one and class two runs. The instructions for entering birth interval information into MATHIST for runs in classes one and two will be discussed in detail subsequently. Therefore, attention will first be focused on procedures for entering that numerical information into MATHIST common to both classes of runs.

As indicated in the main menu displayed in Table 5.5.1, instructions for entering the sterility function, the density for age at first marriage, birth interval information, parity progression ratios, and the cohort survival function for females will appear on the monitor. A user has three options, displayed in the lower panel of Table 5.5.2, for entering numerical information on each component making up MATHIST.

If, for example, a user has a numerical specification of some component that has not been previously stored in the computer, then this information may be entered at the terminal by typing the number 1 followed by entering the numbers in question. If a disk file of a numerical specification of some component has been made and stored on the computer on a previous occasion, then a user may type the number 2 followed by the typed name of the disk file to enter it into MATHIST. Special programs for implementing some parametric formula may also have been written and stored in the computer on a previous occasion. To implement these programs, values of the parameters must be assigned. Numerical information generated by such programs could be entered into MATHIST by typing the number 3 followed by the name of the formula and values of the parameters.

Considered next is the information on birth intervals required to differentiate computer runs in class one from those in class two. Table 5.5.3 contains the headings for the menus differentiating the information required for runs in these two classes. The heading in the upper panel of Table 5.5.3 is that for computer runs in class one; while the headings in the lower panel are those for runs in class two.

Suppose the heading for runs in class one has appeared on the monitor and a user has entered 20 as the run constant, indicating the maximum number of birth intervals allowed is 20. If a user then typed the integers 1 through 20, the program in MATHIST would require him to enter twenty numerical forms of the conditional densities for waiting times among live births, given that they occur. Entering twenty distinct densities is technically feasible in MATHIST, because the FFT algorithms compute convolutions very efficiently. Rarely, however, would a user have this kind of detailed information available. More commonly, a user may have a conditional density for the waiting time to the first birth following marriage and a second density governing waiting times among subsequent births, given that they occur. In this case, a user would type the numbers, 1 2. MATHIST would then ask him to enter the two densities in question. If the number, 15, were the run constant previously entered for the maximum number of birth intervals, then MATHIST would calculate all convolutions needed for maternity histories with fifteen birth intervals.

Considered next is the lower panel in Table 5.5.3, containing the headings for entering birth interval information for computer runs in class two. In its present state of development, MATHIST requires only one density for the temporary sterile period following a live birth. In terms of the model developed in Sect. 5.3, it is assumed that densities $a_{12}(v, t)$ are the same for all parities $v \geq 0$. The second heading in the menu calls for entering the densities $a_{23}(v, t)$ for the waiting times to conception, following entrance into the fecundable state. These densities may be distinct up to the maximum number of live birth intervals; or, just as in the densities for runs within class one, they may be distinct only for some number of live birth intervals less than the maximum. The third heading in the menu for entering birth interval information requires the numerical specification of pregnancy outcome probabilities. MATHIST is sufficiently flexible so that only spontaneous pregnancy wastage is allowed for intended live births; furthermore, induced abortion may be used to terminate unwanted pregnancies. The general rules for entering pregnancy outcome probabilities into MATHIST are the same as those for the conditional densities of waiting times to conception.

Instructions in Table 5.5.3 for entering birth interval information into MATHIST seem clear, in that either conditional densities or pregnancy outcome probabilities are required. Ambiguities may exist, however, regarding the mathematical terms in the main menu as presented in Table 5.5.1. Table 5.5.4 is devoted to assigned strict mathematical meanings to some of the terms in the main menu.

After all the required information has been entered into MATHIST, the main menu in Table 5.5.1 will again appear on the monitor. If a user wishes to execute MATHIST, the number, 99, is entered at his terminal. Alternatively, after entering all the required information into MATHIST, a user may have second thoughts and wish to change some entries before executing the program. This he may do by typing the number in the main menu corresponding to the input that he wishes to change, entering the desired changes, and then typing the number, 99, to execute the program. Should a user wish to repeat a previous run after changing the input for some selected components of the system, MATHIST is so structured that such repeat runs may be made in either class one or class two. Output from MATHIST will be discussed in subsequent sections.

To complete this overview of MATHIST, it will be necessary to discuss the options in the second column of the main menu presented in Table 5.5.1. These options call utility programs which process information into the form required by the mathematical formulas programmed in MATHIST. Suppose, for example, survival information in an historical life table, given at five year age intervals, is to be used in MATHIST. Option 11 in the main menu has a utility program, whereby such information may be interpolated down to the age units required for running MATHIST. Moreover, the interpolated values are monotone decreasing, a necessary mathematical property for values of a survival function. There are also cases in which a user may wish to compute *a posteriori* parity progression ratios from the distribution of final parity in a cohort of women. Option 9 in the main menu has a utility program for converting such a distribution into parity progression ratios.

It should also be mentioned that the classification scheme, presented in this section, for entering information into MATHIST, should be viewed as a set of

Table 5.5.1. The main menu for MATHIST

1:	Enter run constants		
2:	Enter sterility function	3:	Process sterility info.
4:	Enter AFM[a] density	5:	Process AFM info.
6:	Enter birth inter. info.	7:	Process birth interval info.
8:	Enter parity progression ratios	9:	Process parity progress info.
10:	Enter survival function	11:	Process survival info.
12:	Enter all	99:	Begin execution

[a] Age at first marriage in a cohort

Table 5.5.2. Sub-menus for MATHIST

Run constants

Enter in order
1: Length of gestation period (months)
2: Earliest age at first marriage (years)
3: Greatest fertile age (years)
4: Percent ever married (integer, not fraction)
5: Maximum allowable birth intervals

Choose class of run
1: Class one (only birth interval info.)
2: Class two (decomposed birth intervals)

Choose method of inputing information
1: Terminal   2: Disk file
3: Parametric formula

Table 5.5.3. Birth interval information menus for computer runs in classes one and two

Class one

Enter numbers of birth intervals where distinct densities of times between life births are in effect. (Time from marriage to first birth is interval 1.)

Class two

Input density for post partum sterile period. (Common to all parities.)

Enter numbers of birth intervals where distinct densities of waiting times to conception are in effect. (Time from marriage to first birth is interval 1.)

Enter numbers of birth intervals where distinct pregnancy outcome probabilities are in effect.

Table 5.5.4. Mathematical meanings for some terms used in main menu for MATHIST

Term	Mathematical meaning
Sterility function	$A_s(x)$ – The fraction of women in a cohort still fertile at age x
AFM density	$g_0(x)$ – The conditional P.D.F. for age at marriage in a cohort, given that marriage occurs
Parity progression ratios	p_v – The *a priori* or *a posteriori* probability that a woman of parity $v-1$ attains parity v, $v \geq 1$
Survival function	$S(x)$ – The fraction of females in a cohort who are still alive at age x

guidelines. These guidelines do not, in fact, exhaust all the possibilities for obtaining computer output which is mathematically correct. Users who become familiar with both the mathematics and the software underlying MATHIST will, no doubt, develop other schemes for entering information, which yield output that is mathematically correct.

5.6 Applications of MATHIST – Two Simulation Runs in Class One

Reported in this section is some computer output from MATHIST for two simulation runs in class one. As described below, all information, regarding birth intervals, was taken from archival sources and used as computer input to these runs. Some of this information was in a tabular form not suitable for use as computer input. Also illustrated is the use of some utility programs for processing tabular information into a form that could be used as input to MATHIST.

Eaton and Mayer (1953), in their Table 10, have reported the number of children ever born to married ethnic Hutterite women forty-five years of age and over in 1950. From their table, it was possible to estimate the *a posteriori* parity progression ratios for the Hutterite population presented in Table 5.6.1. Because the Hutterites have experienced some of the highest fertility levels ever recorded for a human population, these estimates of parity progression ratios are of intrinsic interest.

Table 5.6.1. Estimates of Hutterite parity progression ratios

Parity		Parity		Parity		Parity	
v	p_v	v	p_v	v	p_v	v	p_v
1.	0.9706	5.	0.9674	9.	0.8692	13.	0.6700
2.	0.9878	6.	0.9529	10.	0.7952	14.	0.6724
3.	0.9725	7.	0.9292	11.	0.7562	15.	0.5138
4.	0.9684	8.	0.9012	12.	0.6612	16.	0.2111

Watkins and McCarthy (1980), in their Table 3, have presented the cumulative probability of the first legitimate birth, by duration of marriage expressed in months, for the La Hulpe population, a 19-th century Belgian commune. The probabilities presented in their table were based on data grouped by time intervals of unequal length. In order to transform this tabular information into a form that could be used as input to MATHIST, these cummulative probabilities were interpolated down to single months, using a utility program in which a monotone interpolation procedure was implemented. Since women in the La Hulpe population did not use contraception in the modern sense of the term, this interpolated empirical distribution function could be used to describe waiting times to the first births, given that they occur, in a cohort following natural fertility.

Tietze (1957), in his Table 4, presented a percentage distribution for intervals among confinements for Hutterite women. The data were recorded in months and grouped into time intervals of unequal length. After converting percentages to decimals, the resulting numbers were interpolated down to single months, using a utility program in MATHIST in which a spline interpolation procedure was implemented. After normalization, the resulting empirical density could be used to describe waiting times among live births following the first, given that they occur, in a cohort reproducing according to a natural fertility regime.

Given the numerical information described above, two computer runs within class one were possible. In a type one run, the *a posteriori* parity progression ratios for the Hutterites, presented in Table 5.6.1, were used as computer input. Because the onset of permanent sterility is reflected in these ratios, the sterility function $A_s(t)$ was taken as $A_s(t)=1$ for all $t \geq 0$. In a run of type two, the potential birth process was stopped by applying a sterility function $A_s(t)$ of the Weibull type presented in (5.4.12). The parameter estimates used in this computer run were those in Table 5.4.3, estimated from ages of last confinement for the Hutterite population.

A common lognormal distribution for age of first marriage was used in both computer runs. Example 4.6.2 may again be consulted for a discussion of this distribution. The estimates of the three parameters in the lognormal distribution were based on data for the Hutterite population as presented in Table 11 of Eaton and Mayer (1953). These marriage data were not for cohorts. Rather, the data consisted of fractions of currently married women in five-year age groups, ranging from age 15 to 49 during the periods 1926–1930, 1936–1940, and 1946–1950. Although these data do not correspond to cohorts, they are at least indicative of a population in which marriage is early and nearly universal but divorce is rare.

Midpoints in the five-year age intervals, as reported by Eaton and Mayer, were used in a nonlinear least squares procedure, which yielded the parameter estimates $\hat{\alpha}=17.92$, $\hat{\mu}=1.2901$, and $\hat{\sigma}=1.040$, with a mean square deviation of 7.8377×10^{-4}. The least squares procedure used in obtaining these estimates was that discussed in Example 4.6.3. It is of interest to observe that, for a lognormal distribution with these parameter values, the median age at first marriage, $p=0.5$, is $x_p = 17.92 + \exp[1.2901] = 21.553$, indicating that marriage was early and at a rather rapid pace. The run constants used in both runs were: 9, 15, 50, 100, and 16.

Included in the computer output from MATHIST are simulated cohort age-specific fertility rates as determined by a numerical evaluation of the integral in formula (4.3.5) for $h = 5$. It is, therefore, of interest to compare simulated rates with period age-specific fertility rates estimated from data on the Hutterite population. Comparing simulated with period rates presents at least two sets of problems. A first set of problems revolves around the fact that period age-specific fertility rates are usually estimated according to five-year age groupings of women in the childbearing ages. It is thus not entirely clear how period age-specific fertility rates, estimated from grouped data, should be related to a theoretical cohort birth rate function calculated on a finer time scale. The problem of determining this relationship will be touched upon briefly in a subsequent section. Related problems have already been discussed in Sect. 4.9. A second set of problems revolves around the observation that, even in populations with reproductive regimes that are presumably stable, such as the Hutterites, age-specific period fertility rates can fluctuate among periods. Despite these sets of problems, a comparison of period Hutterite and simulated age-specific fertility rates was undertaken.

To get some idea about the magnitude of the fluctuations in fertility rates among periods, age-specific fertility rates for the Hutterites, during the periods 1926–1930, 1934–1940, and 1946–1950, were borrowed from Table 11 in Eaton and Mayer (1953) and presented in the second, third, and fourth columns of Table 5.6.2. Simulated age-specific rates from the type one run are presented in the fifth column of Table 5.6.2; whereas the corresponding rates from the run of type two are presented in the sixth column of the table.

An inspection of Table 5.6.2 leads to the conclusion that simulated rates in the fifth and sixth columns for the age groups [25, 30), [30, 35), and [35, 40) fall, for the most part, within the observed ranges of the rates for the Hutterites in the second, third, and fourth columns. When compared to the observed Hutterite rates, the simulated rates for the age groups [15, 20), [20, 25), [40, 45), and [45, 50), however, appear problematic. What factors underlie these discrepancies is not entirely clear. One possibility is that there are errors, due to grouping ages, in calculating the simulated age-specific fertility rates from the values of the birth rate function presented on a yearly basis in the computer output. For example, if the values of the birth rate function for the ages 15, 16, and 17 were used, the rate 15.3 would result for a computer run of type one in the fifth column. This value is within the range observed in the Hutterites for the age group [15, 20). Similarly, if the values of the birth rate function for the ages 21, 22, and 23 were averaged, the rate would be 258, a value that is again in the range observed in the Hutterites for the age group [20, 25). Further comparisons of this type are given in the footnotes of Table 5.6.2.

Regarding the appropriateness of the sterility function used in the run of type two in the sixth column of Table 5.6.2, the low simulated rates for the age groups [40, 45) and [45, 50) suggest that the sterility function of the Weibull type, as determined by the parameter estimates in Table 5.4.3, may decrease

Table 5.6.2. Hutterite and simulated age-specific fertility rates – births per thousand females

Age group	Hutterite period rates			Computer runs	
	1926–1930	1936–1940	1946–1950	Type one	Type two
[15, 20)	16.9	13.1	12.0	41.4[a]	41.4
[20, 25)	268.0	259.1	231.0	320.0[b]	323.8
[25, 30)	417.4	465.6	382.7	429.6	447.2
[30, 35)	397.1	461.9	391.1	400.4	449.6
[35, 40)	355.0	430.6	344.6	316.2	386.0
[40, 45)	238.9	202.9	208.3	195.2[c]	160.2
[45, 50)	23.5	47.6	42.1	90.2[d]	8.0
TFR[e]	8.58	9.40	8.06	8.97	9.08

[a] Average of ages 15, 16, and 17 is 15.3.
[b] Average of ages 21, 22, and 23 is 258.0.
[c] Average of ages 40, 41, and 42 is 219.7.
[d] Average of ages 48 and 49 is 66.0.
[e] Total fertility rate per female.

too rapidly during the ages [45, 50). Another factor complicating comparisons among simulated and Hutterite fertility rates was the probability density function for the waiting times to the first birth. Recall that in the simulations, this p.d.f. was estimated from La Hulpe rather than Hutterite data.

A total fertility rate in a cohort is an important indicator of the rate of flow of births into a population. If simulated total fertility rates are reliable, then simulations may not be too misleading, even though fertility is not properly distributed among age groups. For the three periods in Table 5.6.2, 1926–1930, 1936–1940, and 1946–1950, Hutterite total fertility rates, births per female surviving to age 50, were 8.58, 9.40, and 8.06, respectively. Clearly the simulated TFR's in columns five and six of Table 5.6.2 fall within the tabulated range of the observed values. MATHIST thus seems to do a reasonable job of simulating Hutterite fertility.

5.7 A Factorial Experiment Based on Class Two Runs in MATHIST

Factorial experiments, designed to study the effects of combinations of factors, each at several levels, on some measurable phenomenon, are frequently carried out in science and technology. During the past fifty years, a field of statistics, called experimental design, has evolved to deal with the twin problems of designing factorial experiments and analyzing the data that they yield. Among the many applications of MATHIST are those computer simulation experiments, with a factorial structure, designed to study the impact of combinations of a set of factors on the flow of births into a population. Unlike microsimulation models, which utilize Monte Carlo methods to generate computer output needing statistical analyses, MATHIST, a macrosimulation system, estimates summary statistical entities, such as distributions, means, variances, and age-specific birth rates, which beg for interpretation but require no further statistical analyses. An example of such a factorial experiment will be presented in this section.

5.7.1 An Overview of Computer Input

Four factors affecting the flow of births into a population, called A, B, C, and D, were chosen as variables in MATHIST. Factor A was the distribution of ages at marriage in a female cohort; factor B was the distribution for the durations of postpartum sterile periods; factor C was pregnancy outcome probabilities; and, factor D was the distribution of waiting times to conception. Levels of each factor will be denoted by a subscript on the letter corresponding to that factor.

Used as a model for age at first marriage in a cohort of females was the lognormal distribution described in Example 4.6.2. Two levels of age at marriage were determined by assigning different values to the three parameters α, μ, and σ. Level A_1 was an early age at marriage determined by the parameter values $\hat{\alpha} = 12.982$, $\hat{\mu} = 2.040$, and $\hat{\sigma} = 0.398$ for U.S. white females listed in Table 4.6.1. Level A_2 was a later age at marriage determined by the parameter values $\hat{\alpha} = 15.971$, $\hat{\mu} = 2.487$, and $\hat{\sigma} = 0.472$ for La Hulpe females listed in Table 4.6.1. At this point, it would be useful to again view the graphs in Fig. 4.6.1 of the lognormal density for these two sets of parameter values. Among U.S. white females that married with ages in the interval 40–44 in 1973, the median age at marriage, as determined by the lognormal distribution, was 20.673 years; for the nineteenth century women of La Hulpe, the corresponding median was 27.996 years.

Among the factors affecting the distribution of durations of postpartum sterile periods is the length of time mothers nurse their infants. A nursing infant tends to prolong lactation; lactation, in turn, tends to prolong the period of anovulation following childbirth. As long as a married woman is anovulatory, her risk of becoming pregnant is zero. It is thus clear that the lengths of time mothers choose to nurse their infants can affect the flow of births into a population, especially in noncontracepting populations. Popkin et al. (1982) have provided extensive tables for durations of breastfeeding in low-income countries.

Two distribution functions, governing lengths of postpartum sterile periods, were chosen as levels of factor B. Level B_1 was a distribution determined by short nursing and short anovulatory times; level B_2 was a distribution determined by long nursing and long anovulatory times. Presented in Table 5.7.1 are selected values of these two distribution functions. Observe that in level B_1, 99.3 % of the women resume ovulation six months following childbirth; in level B_2, the corresponding percentage is 10.3. The two distribution functions presented in Table 5.7.1 were determined by a model for postpartum sterile periods, which will be discussed in a subsequent section.

Table 5.7.1. Selected values of two distribution functions governing lengths of postpartum sterile periods

Months after childbirth	Level B_1	Level B_2
6	0.993	0.103
12	1.000	0.295
18	1.000	0.482
24	1.000	0.692
30	1.000	0.884
36	1.000	0.978

Let $p_1 = 0.7627$ and $p_2 = 0.2373$, respectively, be the probabilities that a pregnancy ends in a live birth or another type of pregnancy outcome (mostly spontaneous abortions) as estimated from the observations of French and Bierman (1962). Two levels of pregnancy outcome probabilities, based on p_1 and p_2, were used as computer input. In level C_1, it was assumed that induced abortions were not used to terminate pregnancies. Therefore, in level C_1, the π-probabilities on the right in (5.4.1) were assigned the values $\pi_{31}(v) = p_1$, $\pi_{34}(v) = 0$, and $\pi_{35}(v) = p_2$ for all parities $v \geq 0$.

Level C_2 was characterized by two assumptions. A first assumption was that induced abortions were used to terminate pregnancies only for women of parity two or greater, $v \geq 2$. The probability of such a termination was assigned the value $\pi_{34}(v) = 0.4$, $v \geq 2$, a value in the range observed for some European countries, see Tietze (1981). A second assumption was that $\pi_{31}(v) = p_1$, $\pi_{34}(v) = 0$, and $\pi_{35}(v) = p_2$ for $v = 0, 1$; but, with induced abortions in force for $v \geq 2$, these probabilities were assigned the values $\pi_{31}(v) = p_1(1 - \pi_{34}(v))$, $\pi_{34}(v) = 0.4$, and $\pi_{35}(v) = p_2(1 - \pi_{34}(v))$. Observe that in level C_2, the ratio of live births to other types of pregnancy outcomes is the same for all parities $v \geq 2$, i.e., p_1/p_2, the ratio determined from the data of French and Bierman (1962). Summarized in Table 5.7.2 are the pregnancy outcome probabilities for levels C_1 and C_2.

Chosen as the distribution for waiting times to conception was the Beta-Geometric with positive parameters α and β, see Sect. 1.4. Instead of working

Table 5.7.2. Two levels of pregnancy outcome probabilities

Level	Type of outcome		
	live birth	induced abortion	other
C_1	0.7627[a]	0.0000	0.2373
C_2	0.4576[b]	0.4000	0.1424

[a] Based on French and Bierman (1962).
[b] Hypothetical values.

with the parameterization of the Beta distribution described in Sect. 1.4, it will be convenient to work with the reparameterization determined by the transformation, $\gamma = \alpha + \beta$ and $\delta = \alpha/\gamma$. In terms of this parameterization, it follows from (1.5.2) and (1.5.3) that the mean and variance of a random variable Θ following a Beta distribution are given by

$$E[\Theta] = \delta$$

and   (5.7.1)

$$\text{Var}[\Theta] = \frac{\delta(1-\delta)}{1+\gamma}.$$

It is interesting to observe that the larger the value of the parameter γ, the smaller is the variation of the random variable Θ about its mean.

Sometimes it is informative to consider the coefficient of variation defined by $\text{C.V.} = (\text{Var}[\Theta])^{1/2}/E[\Theta]$. The coefficient of variation for a random variable Θ following a Beta distribution is

$$\text{C.V.} = \left(\frac{1-\delta}{\delta(1+\gamma)}\right)^{1/2},$$ (5.7.2)

for $\gamma > 0$ and $0 < \delta < 1$.

Three levels of waiting times to conception, as determined by three assignments of values to the parameters in the pair $\langle \gamma, \delta \rangle$, were considered in the computer experiments under discussion. Corresponding to each pair $\langle \gamma, \delta \rangle$ of values are the values $\alpha = \gamma\delta$ and $\beta = \gamma(1-\delta)$ for the parameters in the original parameterization of the Beta distribution. Given values of the parameters α and β, the mean and variance of a random variable T, following a Beta-Geometric distribution, may be computed, provided they exist, by using formulas (1.5.10) and (1.5.11). Furthermore, $h(t)$, the p.d.f. of T, may be computed, using the recursive system defined by (1.5.5) and (1.5.7).

Level D_1 was determined by the parameter values $\gamma = 30$ and $\delta = 0.25$, the mean fecundability in a cohort. The C.V. of the random variable Θ, determined by these parameter values, is about 0.31, indicating a relatively low level of variation in fecundability among members of a cohort. The mode of a Beta p.d.f. with these parameter values is about 0.23; and, the corresponding values of the parameters α and β are $\alpha = 7.5$ and $\beta = 22.5$. Whenever $\alpha > 2$, both the mean and variance of T, the random variable representing waiting times to conception, are finite, see (1.5.10) and (1.5.11). In level D_1, the mean and variance of T, expressed in months, are $E[T] = 4.462$ and $\text{Var}[T] = 21.060$. When level D_1 was in force, it was assumed that the distribution of waiting times to conception was the same for all parities.

Level D_2 was determined by a mean fecundability of $E[\Theta] = \delta = 0.25$, the same as in level D_1, but the parameter γ was reduced to the value $\gamma = 4$, thereby increasing the variance of Θ. The C.V. of the random variable Θ in this case was about 0.77, indicating that the level of variation in fecundability was nearly 2.5 times greater than in level D_1. The values of the parameters α and β, corresponding to the values $\gamma = 4$ and $\delta = 0.25$, are $\alpha = 1$ and $\beta = 3$.

When $\alpha = 1$, the p.d.f. of the Beta distribution has the form

$$f(\theta) = \beta(1 - \theta)^{\beta - 1}, \tag{5.7.3}$$

for $\beta > 0$ and $0 < \theta < 1$. If $\beta > 1$, then the p.d.f. in (5.7.3) has a finite mode at $\theta = 0$. A reconsideration of the argument used in the derivation of formulas (1.5.10) and (1.5.11) leads to the conclusion that, in level D_2, both the mean and variance of the waiting times to conception T are infinite. Infinite means and variances are indicative of distributions on the positive integers – 1, 2, ... – with long right-hand tails. From the biological point of view, long right-hand tails result from a substantial fraction of women in a cohort having low fecundabilities. Throughout all computer runs using level D_2, it was assumed that the distribution of waiting times to conception was the same for all parities.

Let η, $0 < \eta < 1$, be the mode of the Beta density. Then, if the parameters γ and η are fixed in advance, the mean δ of the Beta distribution may be determined by expressing the formula for the mode, in Sect. 1.5.4, in terms of the parameters γ and δ. Level D_3 was determined by setting $\gamma = 30$ and $\eta = 0.01$. The value of δ determined by these values was about $\delta = 0.0427$ with a standard deviation of 0.0363. Not only is mean fecundability low in level D_3 but the variation about the mean is relatively small.

The rationale for this choice of parameter values in level D_3 was as follows. It seems plausible that if contraception were practiced efficiently in a cohort with variable fecundability, than a Beta distribution with parameter values $\gamma = 30$, $\eta = 0.01$, and $\delta = 0.0427$ might describe variability in fecundability under contraception. Corresponding to the parameter values for γ and δ are the values $\alpha = 1.28$ and $\beta = 28.72$. Because α satisfies the condition, $1 < \alpha < 2$, the mean waiting time to conception, $E[T] = 103.57$ months, is finite; but, the variance of the Beta-Geometric distribution, $\text{Var}[T]$, is infinite in level D_3. Just as in level D_2, the Beta-Geometric distribution for the parameter values, $\alpha = 1.28$ and $\beta = 28.72$, has a long right-hand tail. For those computer runs in which level D_3 was used, it was assumed that level D_1 of fecundability was in force until two live births occurred. After two live births, level D_3 of fecundability applied, indicating that couples were using contraception to prevent unwanted pregnancies.

Summarized in Table 5.7.3 are the three pairs $\langle \gamma, \delta \rangle$ of parameter values, along with C.V., the coefficient of variation, characterizing the three levels of fecundability D_1, D_2 and D_3. Also listed in Table 5.7.3 are the means and variances of waiting times to conception, for each level of fecundability, as determined by the Beta-Geometric distribution. The parameter value assignments in level D_1 are realistic to the extent that δ, the mean value of fecundability, is within the range reported by Leridon (1977) in his Table 3.2; but, the variance is slightly smaller than any reported by Leridon. On the other hand, the parameter value assignments in levels D_2 and D_3 are largely hypothetical in that they are not based on the analysis of data.

Problem 5 is devoted to a discussion of the properties of the Beta-Geometric distribution used in the foregoing discussion. Table 5.7.4 contains abbreviated descriptions of the four factors under study, along with the levels of each

Table 5.7.3. Three levels of fecundability and waiting times to conception

Fecundability				Conception[a]	
Level	γ	δ	C.V.	Mean	Variance
D_1	30	0.25	0.31	4.462[b]	21.060
D_2	4	0.25	0.77	Infinite	Infinite
D_3	30	0.0427	0.85	103.57[b]	Infinite

[a] Waiting times to conception.
[b] Months.

Table 5.7.4. A brief description of factors used in computer experiment

Factor A – Age at marriage in a cohort of females
Levels
A_1 – Early
A_2 – Late

Factor B – Duration of postpartum sterile period
Levels
B_1 – Short
B_1 – Long

Factor C – Pregnancy outcome probabilities
Levels
C_1 – No induced abortion
C_2 – Induced abortion

Factor D – Waiting times to conception
Levels
D_1 – High fecundability, small variance
D_2 – High fecundability, large variance
D_3 – Low fecundability, small variance

factor just described. Throughout all computer runs, the run constants – 9, 15, 50, 95, and 15 – for MATHIST were used, see Sect. 5.5. Also used throughout all computer runs was the Weibull sterility function displayed in (5.4.12) with the parameter values of Table 5.4.3. All parity progression ratios were also set equal to one. Thus, according to the definitions given in Sect. 5.5, all computer runs in class two were of type one.

When the levels of the four factors just described are combined in all possible ways, twenty-four, $2^3 \times 3 = 24$, computer runs may be considered. Some of these runs may be representative of modern populations, some may roughly correspond to historical populations, while others may represent purely hypothetical populations. Even though some of the computer runs are purely hypothetical, valuable insights, as to how the four factors under study interact to affect the flow of births into a population, may be gained by studying computer output from all twenty-four runs. For, roughly speaking, historical data on maternity histories may be viewed as a realization of some part of a hypersurface. Only by studying collections of points on this hypersurface can we hope to begin to understand its nature more completely.

5.7.2 Phenomenological and Population Policy Implications of Simulated Cohort Total Fertility Rates and Their Variances

MATHIST has a capability for listing a rather large variety of computer output; but, only some of this output will be discussed in this subsection. Included in this output is the mean and variance of the potential birth process $K(t)$ as a function of age t, expressed in years, in a cohort of females. When fifty is the greatest age of childbearing, then the numbers, $E[K(50)]$ and $\text{Var}[K(50)]$, represent the mean and variance of the number of live births experienced by women reaching age fifty. The number $E[K(50)]$ is calculated by a numerical evaluation of the integral in (4.3.11); while, $\text{Var}[K(50)]$ is calculated from a numerical evaluation of the distribution of $K(50)$ as expressed in Eq. (4.2.13). Actually, $E[K(50)]$ may also be calculated from this distribution. Another name for the mean, $E[K(50)]$, is the cohort total fertility rate.

Listed in Table 5.7.5 are the mean and variance of the random variable $K(50)$, corresponding to each of the twenty-four computer runs. Each computer run is designated by four letters with subscripts, indicating the level of each factor. A computer run with the designation, $A_1 B_1 C_1 D_1$, indicates, for example, that the computer input for all factors was at the first level. To facilitate making comparisons among the entries in the table, the arrangement of the twenty-four runs has been systematized. Level A_1, representing early marriage, has been held constant in the upper three panels of the table; whereas, level A_2, representing late marriage, has been held constant in the lower three panels. Within each panel of the table, the four possible combinations of levels of the factors B and C has been listed at a given level of the factor D.

A question of phenomenological interest may be stated as follows. What was the impact of factors B and C on the values of $E[K(50)]$ at an early age of marriage (A_1) and a high level of fecundability (D_1)? To answer this question, attention would be focused on the first, upper-most, panel in Table 5.7.5. Listed in this panel are the mean values, 10.33, 7.58, 6.58, and 5.44, corresponding, respectively, to the computer runs $A_1 B_1 C_1 D_1$, $A_1 B_1 C_2 D_1$, $A_1 B_2 C_1 D_1$, and $A_1 B_2 C_2 D_1$. The difference, $\mu(A_1 B_1 C_1 D_1) - \mu(A_1 B_1 C_2 D_1) = 10.33 - 7.58 = 2.75$, is the mean decrease in live births attributable to induced abortions (C_2) in a female cohort that marries early (A_1), experiences short anovulatory times following live births (B_1), and has a high level of fecundability (D_1). The difference, $\mu(A_1 B_2 C_1 D_1) - \mu(A_1 B_2 C_2 D_1) = 6.58 - 5.44 = 1.14$, is the mean decrease in live births attributable to induced abortions (C_2) when long anovulatory times (B_2) are in force at the first levels of the factors A an D. From these two comparisons it is clear that, against a background of levels A_1 and D_1 for the factors A and D, the impact of induced abortions (C_2) on the cohort total fertility rate, $E[K(50)]$, depends on the level of the factor B.

A question, having population policy implications, is whether induced abortions or long anovulatory periods following childbirth are most effective in reducing the mean number of live births experienced in a female cohort surviving to age 50. The difference, $\mu(A_1 B_1 C_1 D_1) - \mu(A_1 B_1 C_2 D_1) = 2.75$, suggests that long anovulatory periods (B_2) are more effective in reducing fertility than induced abortions (C_2), particularly in cohorts marrying early (A_1) and having high fecundability (D_1).

By continuing as in the preceding two paragraphs, other phenomenological and policy implications of the means in Table 5.7.5 could be discussed. No attempt will be made, however, to exhaustively discuss the implications of the numerous comparisons that could be made among the means and variances listed in Table 5.7.5. Rather, the discussion will be limited to several general statements concerning the phenomenological and policy implications.

5.7.2.1 Phenomenological Implications

1. With the exception of computer runs $A_1 B_2 C_2 D_2$ and $A_2 B_2 C_1 D_1$, which produced the same mean, 4.17, but different variances, 4.39 versus 6.36, all other runs produced different means and variances. Therefore, all factors listed in Table 5.7.4 had an effect on the mean number, as well as the variation in, the number of births experienced in a cohort of women surviving to age 50.

2. With few exceptions, when the means and variances in Table 5.7.5, corresponding to two different levels of a given factor, are compared at constant levels of the other three factors, the magnitude of the difference depends on the combination of levels for the other three factors. Therefore, there is interaction among the four factors, regarding their impact on the mean and variance of the number of births experienced in a cohort of women. For the most part, the magnitudes of these differences are greatest among the computer runs in the upper three panels of Table 5.7.5, where early age at marriage (A_1) was in force.

3. High variation in fecundability among women in a cohort may have a significant effect on the flow of births into a population. Compare, for example, the means in the first panel with those in the second, representing, respectively, low (D_1) and high (D_2) variability in fecundability about the same mean.

4. In general, the larger the mean, the greater the variance in the number of live births experienced in a cohort of women surviving to age fifty. At every combination of levels of the factors A and D, the combination $B_1 C_1$, representing short anovulatory periods and no induced abortions, produced the largest variances, see variances in first column of Table 5.7.5. Similarly, from the fourth column in Table 5.7.5, it may be seen that the combination $B_2 C_2$, representing long anovulatory periods and induced abortions, produced the smallest variance for every combination of levels of the factors A and D.

5. Displayed in panels three and six of Table 5.7.5 are the means for computer runs at level D_3 of low fecundability due to contraception. Recall that, in these runs, it was supposed that only two live births was desired and contraception was used only after the desired family size had been attained. Despite the use of contraception to prevent unwanted pregnancies after attaining desired family size, women, on the average, still experienced more than two live births.

Table 5.7.5. Means and variances of numbers of live births for women surviving to age fifty for twenty-four computer runs

$A_1 B_1 C_1 D_1$ [a]	$A_1 B_1 C_2 D_1$	$A_1 B_2 C_1 D_1$	$A_1 B_2 C_2 D_1$
10.33 [b]	7.58	6.58	5.44
15.39 [c]	9.53	6.56	4.63
$A_1 B_1 C_1 D_2$	$A_1 B_1 C_2 D_2$	$A_1 B_2 C_1 D_2$	$A_1 B_2 C_2 D_2$
7.14	5.26	5.09	4.17
14.73	8.07	6.53	4.39
$A_1 B_1 C_1 D_3$	$A_1 B_1 C_2 D_3$	$A_1 B_2 C_1 D_3$	$A_1 B_2 C_2 D_3$
4.34	3.41	3.73	3.09
4.50	2.60	2.80	1.82
$A_2 B_1 C_1 D_1$	$A_2 B_1 C_2 D_1$	$A_2 B_2 C_1 D_1$	$A_2 B_2 C_2 D_1$
6.52	4.91	4.17	3.56
15.93	8.95	6.36	4.45
$A_2 B_1 C_1 D_2$	$A_2 B_1 C_2 D_2$	$A_2 B_2 C_1 D_2$	$A_2 B_2 C_2 D_2$
4.62	3.56	3.30	2.82
11.34	6.38	5.27	3.71
$A_2 B_1 C_1 D_3$	$A_2 B_1 C_2 D_3$	$A_2 B_2 C_1 D_3$	$A_2 B_2 C_2 D_3$
3.17	2.63	2.70	2.36
3.92	2.51	2.63	1.92

[a] Computer run.
[b] Mean.
[c] Variance.

5.7.2.2 Implications for Population Policy

1. A comparison of the means in the upper three panels in Table 5.7.5 with those in the lower three suggests that the implementation of a population policy, advocating late marriage, would tend to reduce the flow of births into a population. The magnitude of this reduction in flow would depend on the levels of the factors B, C, and D.

2. A comparison of means in the first, second, and third columns of Table 5.7.5 at fecundability levels D_1 and D_2, representing no contraception, suggests that the implementation of a population policy advocating long periods of nursing to increase anovulatory periods, following childbirth, would be more effective in decreasing the flow of births into a population than a policy advocating short periods of nursing and induced abortions as a means of birth control.

3. A corollary of statement 2 is that in noncontracepting populations, practicing long periods of breastfeeding, the implementation of a policy advocating early weaning of infants could lead to an increase in the flow of births into a population, even if induced abortions occurred at level C_2.

4. Suppose members of a population desire two live births. Then computer run $A_2 B_2 C_2 D_3$, with a mean of 2.36, suggests that the implementation of a population policy, advocating late marriage (A_2), long periods of nursing (B_2) and induced abortion (C_2) as a back-up measure when contraception (D_3) fails, would be most effective in attaining an average of two live births. In populations in which two live births were desired and contraception was practiced more effectively than level D_3, it seems plausible that the same degree of birth control could be attained by delaying childbearing to the late twenties, even if marriage occurred early and infants were weaned at an early age.

5.7.3 Comparisons of Simulated Cohort and Period Age-Specific Fertility Rates

Archival age-specific fertility rates, for a given period, are usually reported by five-year age groupings for mothers of infants born during that period. As such, these rates reflect partial information on many birth cohorts, with maternity histories that may have been affected by factors not taken into account in the numerical specifications of MATHIST under study. Included in these factors would be births outside legal marriages, seasonal separations of spouses, and the dissolution of marriages for a variety of reasons. Despite the readily apparent difficulties inherent in comparing period rates with simulated cohort rates, qualitative comparisons among the two types of rates are, nevertheless, of interest. From such comparisons, it should be possible to make some informed judgments, regarding the plausibility of the mathematical assumptions underlying MATHIST as well as the appropriateness of the numerical specifications used in the computer runs under study. It should be emphasized that even if simulated and observed age-specific fertility rates resemble each other fairly closely, this resemblance does not imply that the mechanisms underlying the simulation are those producing the observed rates in a real population.

Presented in Table 7.7.6 are selected Swedish age-specific fertility rates based on nineteenth and twentieth century data recorded in Keyfitz and Flieger (1968). Table 5.7.7 contains the corresponding rates from selected countries with high and low fertility as reported in *World Fertility*, a 1981 publication of the Population Reference Bureau.[1] Those countries selected for high period fertility rates, during the late sixties and early seventies, were Afghanistan, Algeria, Mexico, and Bangladesh. Belgium, France, Britain (the United Kingdom – U.K.), and the United States (U.S.) were selected for low period fertility in 1970.

When making comparisons among age-specific fertility rates of several countries, the sources of data used in their computation should be kept in mind. Rates for Afghanistan, Mexico, and Bangladesh were based on surveys; while those for Algeria, Belgium, Fance, the U.K., and U.S. were based on birth registration data. Readers interested in further details on the nature of the data used in the computations may consult the publication, *World Fertility*,

---

[1] Population Reference Bureau, Inc., 1337 Connecticut Avenue, N.W., Washington, D.C. 20036.

mentioned above. The data for Afghanistan and Algeria were "adjusted"; but, no "adjustments" were applied to the other countries in Table 5.7.7. One would presume that unadjusted data is more reliable than adjusted data.

Cohort age-specific fertility rates from twelve of the twenty-four computer runs from MATHIST are presented in Tables 5.7.8 and 5.7.9. Rates from computer runs with early marriage and short anovulatory times appear in Table 5.7.8; Table 5.7.9 is devoted to rates from computer runs with late marriage and long anovulatory times. Observe that each column in these tables corresponds to one of the three levels of fecundabilities, D_1, D_2, and D_3.

Many comparisons may be made among the twenty-eight sets of age-specific fertility rates listed in Tables 5.7.6 to 5.7.9. To keep the exposition manageable, however, attention will be confined to comparisons among total fertility rates, birth rates for the age group [15, 20), and the age group with the maximum birth rate. Comparisons of observed period total fertility rates with simulated cohort rates will, in some measure, supply validity checks on the ability of MATHIST to estimate average family sizes for married women reaching age 50. Comparisons among birth rates for the age interval [15, 20) provides a basis for judging the applicability of the distributions for ages at first marriage used in the simulations. Finally, the age group with a maximum birth rate is a useful indicator of the age patterns of childbearing.

In what follows, descriptions of computer runs will be terse. Before reading the remarks given below, it might be helpful to consult the definitions of levels of each factor outlined in Table 5.7.4.

5.7.3.1 Comparisons of Total Fertility Rates

1. The simulated total fertility rate (TFR) of 10.33 for the computer run, $A_1 B_1 C_1 D_1$ in Table 5.7.8, lies outside any period TFR in Tables 5.7.6 and 5.7.7. This observation suggests that the run, $A_1 B_1 C_1 D_1$, probably represents a purely hypothetical cohort.

2. The cohort TFR's, 7.14 and 7.58, for computer runs, $A_1 B_1 C_1 D_2$ and $A_1 B_1 C_2 D_1$ in Table 5.7.8, are within the range of the TFR's for the high level fertility countries, Afghanistan, Algeria, Mexico, and Bangladesh, in Table 5.7.7. Observe that the computer run, $A_1 B_1 C_2 D_2$, differs from the run, $A_1 B_1 C_2 D_1$, only with respect to D_2, the level of fecundability with a high variance.

3. The TFR's 2.82 and 2.36, for the computer runs $A_2 B_2 C_2 D_2$ and $A_2 B_2 C_2 D_3$ in Table 5.7.9, lie within the range of the period TFR's for the low fertility countries in Table 5.7.7. But, as expected from the late marriage pattern used for the computer runs in Table 5.7.9, the age pattern of childbearing for these runs differs markedly from those for the low fertility countries in Table 5.7.7.

5.7.3.2 Comparisons of Birth Rates for the Age Group [15, 20)

1. The fertility rates for the age group [15, 20), in the high fertility countries of Table 5.7.7, greatly exceed the simulated rates for this age group in Table 5.7.8. Thus, maternity histories in the high fertility countries, as of about 1970, began

at a much earlier age, on the average, than those implied by the U.S. age at first marriage distribution used in the simulations presented in Table 5.7.8.

2. With the exception of the rate 48.62 for the age group $[15, 20)$ around 1965, all Swedish rates for this age group are less than those for the early marriage cohorts simulated in Table 5.7.8. Swedish maternity histories, as represented by the data in 5.7.6, thus began at a later age, on the average, than those implied by the U.S. marriage distribution used in the simulations.

3. With respect to the low fertility countries in the lower panel of Table 5.7.7, the birth rates, corresponding to the age $[15, 20)$ for Belgium and France, fell below any simulated rates in Table 5.7.8; while those for the U.K. and the U.S. exceeded those in Table 5.7.8. Like maternity histories in Sweden, see especially the rates for the age group $[15, 20)$ around 1945 and 1955 in Table 5.7.6, those in Belgium and France, around 1970, began at a later age, on the average, than those implied by the U.S. marriage distribution. The high U.S. birth rate of 69.2, for the age group $[15, 20)$, probably represents a substantial number of births outside legal marriages.

4. All simulated birth rates for the age group $[15, 20)$ in Table 5.7.9 are significantly smaller than any of the observed rates for this age group recorded in Tables 5.7.6 and 5.7.7. This observation suggests that the late marriage pattern for La Hulpe was an isolated phenomenon, at least to the extent that it was not wide spread in Sweden nor does it seem to occur widely in modern populations.

5.7.3.3 Comparisons of Age Groups with Maximum Birth Rates

1. For all the nineteenth century Swedish rates for the periods listed in the upper panel of Table 5.7.6, the age group $[30, 35)$ had the maximum birth rate. This age group also had the maximum birth rate for the computer run, $A_1 B_1 C_1 D_1$ with early marriage, listed in Table 5.7.8. As might be expected from the runs with late marriage in Table 5.7.9, the age group $[30, 35)$ had the maximum birth rate in all computer runs except the run, $A_2 B_2 C_1 D_2$ in which the high variance in fecundability (D_2) was in force.

2. With the exception of the rates for the period 1955, where the age groups $[20, 25)$ and $[25, 30)$ had essentially the same maximum rate of 132, the age group with the maximum birth rate for twentieth century Swedish data, in Table 5.7.6, was $[25, 30)$. A similar pattern emerged for the computer runs, $A_1 B_1 C_1 D_2$ and $A_1 B_1 C_2 D_2$, in Table 5.7.8, with early marriage. Both these runs had the fecundability level D_2 with a high variance.

3. In all the low fertility countries in Table 5.7.7, the age group $[20, 25)$ had a maximum birth rate. The same pattern occurred in computer runs, $A_1 B_1 C_1 D_3$ and $A_1 B_1 C_2 D_3$, in Table 5.7.8, with early marriage. In both these runs, level D_3 of fecundability was in force, indicating that contraception was used after two live births. Adding induced abortions (C_2) to contraception in run, $A_1 B_1 C_2 D_3$, had a depressing effect on fertility, when compared to run $A_1 B_1 C_1 D_3$, but the relative age pattern of childbearing was not changed.

5.7.3.4 On the Plausibility of the Mathematical Assumptions Underlying MATHIST

One approach to judging the plausibility of a set of mathematical assumptions is to test their implications against observations in the real world. To the extent that simulated patterns of cohort age-specific fertility rates, as well as cohort total fertility rates, closely resembled a variety of observed period age-specific and total fertility rates in historical and modern populations, the mathematical assumptions underlying MATHIST seem to have passed two plausibility tests at the qualitative level. Putting together sets of appropriate computer input for MATHIST, based on data from various sources, still remains a problem in implementing the computer programs. It seems likely that further software development will be required before this problem can be brought to a satisfactory resolution. In the meantime, a variety of hypothetical computer input, based on the current parametric specifications of MATHIST, may be generated with relative ease. Given such input, it would be feasible to conduct a variety of factorial experiments similar to the twenty-four sample runs under discussion.

Table 5.7.6. Selected Swedish age-specific fertility rates per thousand females

Age group	Nineteenth century periods			
	1800	1825	1850	1875
[15, 20)	14.16	11.78	5.14	8.29
[20, 25)	98.20	112.25	79.42	100.40
[25, 30)	191.62	221.09	191.09	208.65
[30, 35)	222.51	246.55	238.67	223.15
[35, 40)	186.67	210.63	208.23	215.73
[40, 45)	98.09	117.15	119.29	129.94
[45, 50)	21.83	21.76	19.22	19.00
TFR[a]	4.16	4.71	4.31	4.53
Age group	Twentieth century periods			
	1935	1945	1955	1965
[15, 20)	17.53	31.98	38.11	48.62
[20, 25)	75.05	121.47	132.44	140.90
[25, 30)	91.13	144.92	132.45	153.98
[30, 35)	78.27	112.39	84.98	89.33
[35, 40)	52.75	68.83	46.31	39.28
[40, 45)	23.19	24.25	14.57	9.89
[45, 50)	2.57	1.95	1.13	0.72
TFR[a]	1.70	2.52	2.25	2.41

[a] Total fertility rate per female.

Table 5.7.7. Recent period age-specific fertility rates per thousand females – selected countries

Age group	High fertility			
	Afghanistan 1972–1973	Algeria 1970	Mexico 1967–1969	Bangladesh 1974
[15, 20)	166.0	132.4	126.1	198.3
[20, 25)	356.0	347.6	315.2	337.3
[25, 30)	354.0	363.6	336.0	310.9
[30, 35)	311.0	330.3	278.6	261.5
[35, 40)	238.0	258.1	207.1	197.0
[40, 45)	136.0	122.3	114.0	95.4
[45, 50)	78.0	17.0	42.0	13.5
TFR[a]	8.28	7.86	7.10	7.07

Age group	Low fertility			
	Belgium 1970	France 1970	U.K. 1970	U.S. 1970
[15, 20)	31.1	37.7	49.9	69.2
[20, 25)	149.4	164.7	154.1	166.3
[25, 30)	143.1	149.6	151.9	143.5
[30, 35)	78.1	87.5	79.0	73.3
[35, 40)	36.4	41.5	34.4	31.6
[40, 45)	9.9	11.4	8.7	8.2
[45, 50)	0.7	0.9	0.6	0.5
TFR[a]	2.24	2.47	2.39	2.46

[a] Total fertility rate per female.

Table 5.7.8. Simulated cohort age-specific fertility rates per thousand females – early marriage and short anovulatory times

Age group	No induced abortion Computer run		
	$A_1 B_1 C_1 D_1$	$A_1 B_1 C_1 D_2$	$A_1 B_1 C_1 D_3$
[15, 20)	56.68	46.83	55.89
[20, 25)	345.11	265.10	258.59
[25, 30)	481.21	340.99	213.32
[30, 35)	493.53	328.89	154.32
[35, 40)	446.56	285.62	120.60
[40, 45)	234.70	152.56	62.44
[45, 50)	16.735	13.93	5.90
TFR[a]	10.33	7.14	4.34

Continued

Table 5.7.8 (contd.)

Age group	Induced abortions Computer run		
	$A_1B_1C_2D_1$	$A_1B_1C_2D_2$	$A_1B_1C_2D_3$
[15, 20)	56.26	46.51	55.77
[20, 25)	299.60	236.62	241.74
[25, 30)	349.04	254.25	165.17
[30, 35)	335.54	222.33	101.74
[35, 40)	300.12	186.89	75.37
[40, 45)	165.65	99.45	38.54
[45, 50)	16.12	9.52	3.63
TFR[a]	7.58	5.26	3.41

[a] Total fertility rate per female.

Table 5.7.9. Simulated cohort age-specific fertility rates per thousand females – late marriage and long anovulatory times

Age group	No induced abortions Computer run		
	$A_2B_2C_1D_1$	$A_2B_2C_1D_2$	$A_2B_2C_1D_3$
[15, 20)	0.28	0.24	0.27
[20, 25)	40.67	34.30	40.36
[25, 30)	154.75	127.41	136.64
[30, 35)	235.03	187.39	159.67
[35, 40)	248.34	192.13	131.20
[40, 45)	145.94	110.21	65.12
[45, 50)	14.69	10.89	5.91
TFR[a]	4.17	3.30	3.73

Age group	Induced abortions Computer run		
	$A_2B_2C_2D_1$	$A_2B_2C_2D_2$	$A_2B_2C_2D_3$
[15, 20)	0.27	0.24	0.27
[20, 25)	40.51	34.19	40.31
[25, 30)	145.95	121.47	132.41
[30, 35)	201.83	162.68	140.99
[35, 40)	200.61	153.82	103.56
[40, 45)	114.69	84.05	47.30
[45, 50)	11.41	8.08	4.08
TFR[a]	3.56	2.82	2.36

[a] Total fertility rate per female.

5.7.4 Computer Generated Graphs of Selected Output from MATHIST

Another approach to the study of computer output from MATHIST is to view computer generated graphs. Among the several functions in the numerical output of MATHIST that could be graphed, only two were chosen for display in this section. One of these functions was the cohort gross maternity function $b(t)$, which is defined mathematically by Eq. (4.3.5). This function could also be called a cohort fertility density function, see Eq. (4.3.1). A second function chosen for display was $p_n(t)$, $n \geq 0$, giving the distribution of live births in a cohort of females as a function $t \geq 0$ of their age. Mathematically, this function is defined by Eq. (4.2.13). By evaluating this function at age $t = 50$ years, the distribution of the number of live births experienced in a cohort of women surviving to age 50 is obtained.

Graphs of the cohort gross maternity functions for the computer runs, $A_1 B_1 C_1 D_1$, $A_1 B_1 C_1 D_2$, and $A_1 B_1 C_1 D_3$, are presented in Fig. 5.7.1. Compared in this figure are the three levels of fecundability, D_1, D_2, and D_3 as summarized in Table 5.7.4, against a background of early marriage (A_1), short anovulatory periods (B_1), and no induced abortions (C_1). As can be seen from Fig. 5.7.1, the graphs of the three functions differ markedly, indicating that fecundability alone can affect fertility substantially. From Table 5.7.5, it can be seen that the total fertility rates associated with these functions are 10.33, 7.14, and 4.50.

Compared in Fig. 5.7.2 are the graphs of the cohort gross maternity functions for the computer runs, $A_1 B_1 C_1 D_2$, $A_1 B_1 C_2 D_2$, $A_1 B_2 C_1 D_2$, and $A_1 B_2 C_2 D_2$, represented in the second panel in Table 5.7.5. In these computer runs early marriage (A_1) and high but variable fecundability (D_2) were held constant. Differences in the graphs are thus attributable to changes in the levels of the factors B and C, representing lengths of postpartum sterile periods and pregnancy outcome probabilities. Graphs corresponding to computer runs in which induced abortions were in force (C_2) after two live births have shapes differing rather sharply, especially after 25 years of age, from those in which no

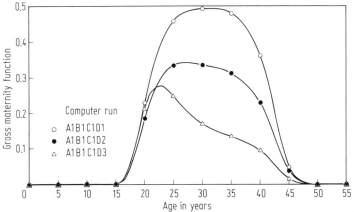

Fig. 5.7.1. Effects of the distribution of fecundability on the cohort gross maternity function – early marriage, short postpartum sterile period, and no induced abortion

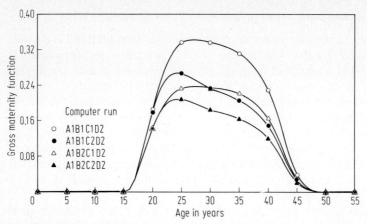

Fig. 5.7.2. Effects of induced abortion and length of postpartum sterile period on the cohort gross maternity function – early marriage, high and variable fecundability

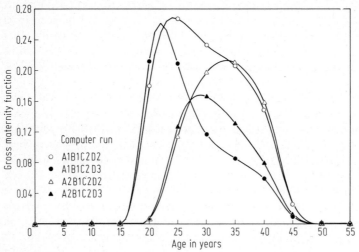

Fig. 5.7.3. Effects of early vs. late marriage, high vs. low fecundability on the cohort gross maternity function – induced abortion, short postpartum sterile period

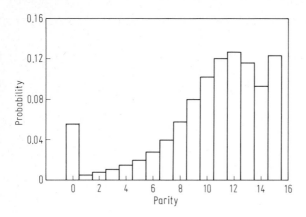

Fig. 5.7.4. Distribution of number of live births in cohort of women surviving to age 50 – computer run $A_1 B_1 C_1 D_1$

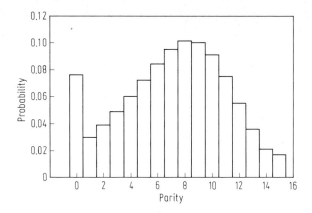

Fig. 5.7.5. Distribution of number of live births in cohort of women surviving to age 50 – computer run $A_1 B_1 C_1 D_2$

induced abortions (C_1) were in force. A low level of fecundability after two live births (D_3) resulted in a cohort gross maternity function with a similar graph in Fig. 5.7.1 for the computer run $A_1 B_1 C_1 D_3$.

Figure 5.7.3 contains graphs of the cohort gross maternity functions for the computer runs $A_1 B_1 C_2 D_2$, $A_1 B_1 C_2 D_3$, $A_2 B_1 C_2 D_2$, and $A_2 B_1 C_2 D_3$, with total fertility rates of 5.26, 3.41, 3.56, and 2.63, respectively. In these runs, short anovulatory periods (B_1) and induced abortions (C_2) after two live births are held constant and the interaction of ages at marriage (A_1 and A_2) with two fecundability levels (D_2 and D_3) were studied. Early (late) marriage is represented by the curves with left (right) skewness. As was expected, from the shapes of the graphs in Fig. 5.7.3, it can be seen that the acceptance of contraception after two live births had a greater effect in decreasing fertility in the early level of age at marriage than in the late level.

Displayed in Figs. 5.7.4 and 5.7.5 are the histograms, corresponding to computer runs $A_1 B_1 C_1 D_1$ and $A_1 B_1 C_1 D_2$, for the distribution of the number of live births experienced in a cohort of women surviving to age 50. The histogram for the run, $A_1 B_1 C_1 D_1$, is skewed to the right with a mode of 12 births; while that for run, $A_1 B_1 C_1 D_2$, is more symmetrically distributed around a mode of 8 births. As can be seen from Table 5.7.5, the means of these two distributions differ markedly, 10.33 versus 7.14, but their variances, 15.39 and 14.73, are nearly the same. All differences between the two histograms in Figs. 5.7.4 and 5.7.5 attributable to levels D_1 and D_2 of fecundability.

5.8 A Stochastic Model of Anovulatory Sterile Periods Following Live Births

As has been demonstrated in the computer simulations reported in Sect. 5.7, the distribution of the lengths of anovulatory sterile periods following childbirth can have a significant effect on the flow of births into a population. Two factors affecting the lengths of these periods are breastfeeding practices and

Table 5.8.1. State space for model of waiting times to the resumption of ovulation following childbirth

O_1	Mother ovulatory
O_2	Mother anovulatory, infant weaned
O_3	Mother anovulatory, infant partially breastfed
O_4	Mother anovulatory, infant fully breastfed
O_5	Mother anovulatory, infant dead

infant mortality. Presented in this section is a stochastic model, taking these two factors into account, for the distribution of waiting times to the resumption of ovulation following childbirth. The model will be formulated as an age-dependent semi-Markov process in continuous time. In what follows, the stochastic process under consideration will be treated intuitively; in a subsequent chapter, the mathematical foundations of age-dependent semi-Markov processes will be discussed in more detail.

Listed in Table 5.8.1 are the five states of the model for the distribution of waiting times to the resumption of ovulation following childbirth. As an aid in interpreting these states, think of a live birth occurring at $t=0$. When a mother resumes ovulation after childbirth and is again at risk of pregnancy, she is said to be in state O_1. The other states of the model represent types of interactions between a mother and her infant; interactions that influence waiting times till entrance into O_1. If a mother is still anovulatory but her infant has been weaned, she is said to be in state O_2; if a mother is anovulatory and her infant is partially nursing, i.e., the infant consumes food other than his mother's milk, she is said to be in state O_3. The state O_4 is defined similarly, except that the only food consumed by the infant is his mother's milk, i.e., the infant is on full breastfeeding. Lastly, a mother is said to be in state O_5 if she is anovulatory but her infant is dead.

One-step transitions among the states listed in Table 5.8.1 are governed by the transition densities of the process. For the model under consideration, the 5×5 matrix of transition densities takes the form

$$\mathbf{a}(x,t)=\begin{bmatrix} 0 & 0 & 0 & 0 & 0 \\ a_{21}(x,t) & 0 & 0 & 0 & 0 \\ a_{31}(x,t) & a_{32}(x,t) & 0 & 0 & a_{35}(x,t) \\ a_{41}(x,t) & 0 & a_{43}(x,t) & 0 & a_{45}(x,t) \\ a_{51}(x,t) & 0 & 0 & 0 & 0 \end{bmatrix}, \tag{5.8.1}$$

where $x,t\geq0$. The variable x, in all the functions displayed in the matrix in (5.8.1), is the number of time units that have elapsed since the birth of an infant at some time origin $t=0$. Given that state O_i is entered at time $x\geq0$, the conditional density $a_{ij}(x,t)$ governs the time taken for the one-step transition $O_i\to O_j$, where $i,j=1,2,\ldots,5$.

From now on the function $A_{ij}(x,t)$ will represent the integral

$$A_{ij}(x,t)=\int_0^t a_{ij}(x,s)ds, \quad t\geq0. \tag{5.8.2}$$

Observe that if state O_i is entered at time $x \geq 0$, then $A_{ij}(x, t)$ is the conditional probability that the one-step transition, $O_i \to O_j$, occurs sometime during the time interval $(x, x+t]$, $t > 0$.

Suppose state O_i, $i \neq 1$, is entered at time x. Given this event, let $P_{i1}(x, t)$ be the conditional probability that the absorbing state O_1 is reached sometime during the time interval $(x, x+t]$, $t > 0$. If, for example, newly born infants are put on full breastfeeding, then $P_{41}(0, t)$ is the distribution function of the waiting times till their mothers resume ovulation, i.e., enter state O_1.

If state O_4 is entered at $t = 0$, then, according to row 4 of the density matrix in (5.8.1), only three transitions, $O_4 \to O_1$, $O_4 \to O_3$, and $O_4 \to O_5$, are possible on the first step. The probability that the first step transition, $O_4 \to O_1$, occurs sometime during the time interval $(0, t]$, $t > 0$, is $A_{41}(0, t)$. A second possibility is that the first step transition, $O_4 \to O_3$, occurs after s, $0 \leq s < t$, time units, with $a_{43}(0, s)ds$, ds small, as an approximate probability. Given this event, $P_{31}(s, t-s)$ is the conditional probability that state O_1 is entered sometime during the $t-s$ time units remaining in the interval $(0, t]$. Integrating over all possible values of s yields the formula

$$\int_0^t a_{43}(0, s) P_{31}(s, t-s)ds \tag{5.8.3}$$

for the conditional probability of a first transition to state O_3 followed by one or more transitions to state O_1 during the time interval $(0, t]$, given that state O_4 is entered at $t = 0$. A formula similar to (5.8.3) may be written down if state O_5 is entered at the first step. Finally, adding probabilities for the three mutually exclusive events just described, leads to the formula

$$P_{41}(0, t) = A_{41}(0, t) + \int_0^t a_{43}(0, s) P_{31}(s, t-s)ds$$

$$+ \int_0^t a_{45}(0, s) P_{51}(s, t-s)ds \tag{5.8.4}$$

for determining the distribution function $P_{41}(0, t)$, $t > 0$.

Contained in the integrals on the right in (5.8.4) are the unknown functions $P_{31}(x, t)$ and $P_{51}(x, t)$. From the density matrix in (5.8.1), it is easy to see that $P_{51}(x, t) = A_{51}(x, t)$ for all $x, t \geq 0$. Furthermore, a renewal argument similar to that outlined in the derivation of (5.8.4) may be used to derive the equation

$$P_{31}(x, t) = A_{31}(x, t) + \int_0^t a_{32}(x, s) P_{21}(x+s, t-s)ds$$

$$+ \int_0^t a_{35}(x, s) P_{51}(x+s, t-s)ds, \tag{5.8.5}$$

which is valid for all $x, t \geq 0$. It has already been noted that $P_{51}(x, t) = A_{51}(x, t)$; it also follows, from the second row of the density matrix in (5.8.1), that $P_{21}(x, t) = A_{21}(x, t)$ for all $x, t \geq 0$. Therefore, the functions $P_{41}(0, t)$ and $P_{31}(x, t)$ in (5.8.4) and (5.8.5) are completely determined by the functions in the one-step transition matrix defined in (5.8.1). If infants are only partially breastfed immediately following their birth, then $P_{31}(0, t)$ is the distribution function of the waiting times till their mothers resume ovulation. The function $P_{31}(0, t)$ is determined by letting $x=0$ in (5.8.5).

Considered next is the problem of specifying the densities in the matrix $\mathbf{a}(x, t)$ set forth in (5.8.1). One approach to specifying the densities in each row of the matrix $\mathbf{a}(x, t)$, with more than one nonzero density, is to apply the classical theory of competing risks with independent sojourn times described in Sects. 3.2 and 3.3. Suppose, for example, the process begins in state O_4 at $t=0$; let $F_{4j}(t)$, $j=1, 3$, or 5, be latent distribution functions, governing the timing of transitions out of state O_4. Moreover, suppose the p.d.f. of $F_{4j}(t)$ is $f_{4j}(t)$. Then, the latent risk function associated with $F_{4j}(t)$ is

$$\theta_{4j}(t)=\frac{f_{4j}(t)}{1-F_{4j}(t)}, \quad t>0, \tag{5.8.6}$$

where $j=1, 3$, or 5.

All densities in the first row of the matrix $\mathbf{a}(0, t)$ may be expressed in terms of the latent risk functions in (5.8.6). For, let

$$A_4(0, t)=A_{41}(0, t)+A_{43}(0, t)+A_{45}(0, t)$$

and $\tag{5.8.7}$

$$\theta_4(t)=\theta_{41}(t)+\theta_{43}(t)+\theta_{45}(t),$$

when $t>0$. Then, because latent sojourn times are assumed to be independent, it follows that

$$1-A_4(0, t)=\prod_j (1-F_{4j}(t))=\exp\left[-\int_0^t \theta_4(s)ds\right], \tag{5.8.8}$$

where $j=1, 3$, or 5 and $t>0$. It would be instructive to compare formula (5.8.8) with (3.2.12) and (3.2.13). An application of formula (3.2.15), then leads to the conclusion that the densities in question have the formulas

$$a_{4j}(0, t)=\exp\left[-\int_0^t \theta_4(s)ds\right]\theta_{4j}(t), \tag{5.8.9}$$

for $t>0$. When the process starts in state O_3 at $t=0$, similar formulas may be written down for densities $a_{3j}(0, t)$ in terms of risk functions associated with the latent distribution functions $F_{3j}(t)$, for $j=1, 2$, or 5.

In order to completely determine the one-step transition densities of the process, all functions in the matrix $\mathbf{a}(x, t)$ must be determined for every $x>0$ and $t\geq0$. Suppose state O_i is entered at x time units following the start of the process. Given this event, let $F_{ij}(x, t)$ be the conditional latent distribution function, governing the timing of the transition $O_i \to O_j$, $i \neq j$. It will be assumed that there are basic latent distribution functions $F_{ij}(t)$ such that

$$F_{ij}(x, t)=\frac{F_{ij}(x+t)-F_{ij}(x)}{1-F_{ij}(x)}, \tag{5.8.10}$$

for all $x \geq 0$ and $t \geq 0$. In other words, it is assumed that aging of the process is taken into account through the basic latent distribution functions and Eq. (5.8.10).

An interesting implication of Eq. (5.8.10) is that all densities in the matrix $\mathbf{a}(x, t)$ may be expressed in terms of the risk functions $\theta_{ij}(t)$ of the basic latent distribution functions $F_{ij}(t)$. For, let $\eta_{ij}(x, t)$ be the risk function of the conditional distribution function $F_{ij}(x, t)$. Then, by definition,

$$\eta_{ij}(x, t) = \frac{f_{ij}(x, t)}{1 - F_{ij}(x, t)}, \qquad (5.8.11)$$

for $x \geq 0$ and $t \geq 0$, where $f_{ij}(x, t)$ is the partial derivative of $F_{ij}(x, t)$ with respect to t. By using the definition of $F_{ij}(x, t)$ given in (5.8.10), it follows that

$$\eta_{ij}(x, t) = \theta_{ij}(x + t). \qquad (5.8.12)$$

Therefore, the conditional risk functions in (5.8.11) are determined by translating the basic risk functions as in (5.8.12).

Now let

$$\theta_i(t) = \sum_j \theta_{ij}(t), \qquad j \neq i. \qquad (5.8.13)$$

Another application of the reasoning applied in the derivation of formula (5.8.9), leads to the conclusion that

$$a_{ij}(x, t) = \exp\left[-\int_0^t \theta_i(x + s) ds \right] \theta_{ij}(x + t), \qquad (5.8.14)$$

for $x \geq 0$ and $t > 0$, is a general formula for the density functions in the matrix $\mathbf{a}(x, t)$. If a one-step transition from state O_i to state O_j is impossible, then the latent risk function $\theta_{ij}(t)$ is identically zero. It is interesting to observe that $a_{ij}(x, t)$ is a translation of the density $a_{ij}(0, t)$ in the sense of Eq. (5.8.14).

A computer implementation of the model just described has been carried out by Pickens (1982), using APL as the programming language. The algorithms used by Pickens in his computer implementation of the model are described in Problem 6.

Considerable attention has been given to investigating the impact of levels of nursing and infant mortality on the duration of anovulation following childbirth. In this connection, the papers – Cantrelle and Leridon (1971), Perez et al. (1971), Ginsberg (1973), Masnick (1979), and Page and Lesthaeghe (1980) – may be consulted. In some societies, a general recognition of the association between infant deaths and too early weaning seems to have led to sexual taboos following childbirth. Santow (1978), Chap. 5, has provided an interesting review of the subject. Much attention has also been devoted to fertility regulation during human lactation, see, for example, Parkes et al. (1977).

Example 5.8.1. *Numerical Examples Based on a Parametric Model.*

In many data sets, partial and full nursing are lumped into a class labeled "nursing" and infant mortality is ignored. Under such circumstances, a three state age-dependent semi-Markov process may be considered. States O_1 and O_2 are again defined as in Table 5.8.1; but, a mother is said to be in state O_3, if she is anovulatory and her infant is nursing. The three-state stochastic process, to be considered in these examples, has a one-step transition density matrix of the form,

$$\mathbf{a}(x,t) = \begin{bmatrix} 0 & 0 & 0 \\ a_{21}(x,t) & 0 & 0 \\ a_{31}(x,t) & a_{32}(x,t) & 0 \end{bmatrix}, \tag{5.8.15}$$

where $x, t \geq 0$. Three basic latent risk functions, $\theta_{21}(t)$, $\theta_{31}(t)$, and $\theta_{32}(t)$, completely determine the process.

Ginsberg (1973) considered a model in which latent distributions, governing the durations of anovulation after childbirth, followed a gamma type law with index parameter 2 and scale parameters λ_{21} and λ_{32}. When the latent distribution functions $F_{i1}(t)$, $i=2$ or 3, are of the gamma type with index parameter 2 and scale parameters λ_{i2}, then their p.d.f.'s have the form

$$f_{i1}(t) = \lambda_{i1}^2 t \exp[-\lambda_{i1}t], \tag{5.8.16}$$

where $t \geq 0$. It is well known that the survival function associated with the p.d.f. in (5.8.16) has the form

$$1 - F_{i1}(t) = \exp[-\lambda_{i1}t][1 + \lambda_{i1}t], \tag{5.8.17}$$

for $t \geq 0$. Therefore, the basic latent risk functions $\theta_{21}(t)$ and $\theta_{31}(t)$ have the simple form

$$\theta_{i1}(t) = \frac{\lambda_{i1}^2 t}{1 + \lambda_{i1}t}, \tag{5.8.18}$$

for $i=2$ or 3 and $t \geq 0$.

Two forms of the latent risk function $\theta_{32}(t)$ were used in the numerical examples under discussion. In one form, corresponding to short durations of nursing, it was assumed that $\theta_{32}(t) = \lambda_{32}$, a positive constant, for all $t \geq 0$. As is well known, assuming the risk function $\theta_{32}(t)$ is the constant λ_{32} is equivalent to assuming that the latent distribution function $F_{32}(t)$ is that of an exponential distribution with scale parameter λ_{32}. In a second form, it was assumed that $\theta_{32}(t)$ was the risk function determined by the empirical distribution of long durations of nursing practiced by women in the rural zone of Senegal, as reported by Cantrelle and Leridon (1971). It is generally recognized that long periods of nursing are associated with long periods of anovulation.

Two assignments of the parameters λ_{21} and λ_{31}, determining the latent distributions for durations of anovulation following childbirth, were used in the numerical examples under discussion. A latent distribution for short periods of anovulation was determined by using estimates of parameters reported by Ginsberg (1973). More precisely, let λ_2 and λ_3 be estimates of the parameters in Ginsberg's Table I. Then, the parameter values for short durations of nursing were given by $\lambda_{21} = 30\lambda_3$ and $\lambda_{31} = 30\lambda_2$, a month being taken as thirty days. To determine latent distributions for long periods of anovulation, infor-

Table 5.8.2. Parameter values for latent gamma distributions governing short and long periods of anovulation

Parameter	Short	Long
λ_{21}	2.2536	0.4867
	(0.8875)[a]	(4.1093)
λ_{31}	0.4296	0.0885
	(4.6555)	(22.6)

[a] Mean times in months of 30 days.

mation from rural Senegal was used. By allowing for at least one month of anovulation following childbirth, the mean anovulatory period of Senegalese women, whose child died after two years, was estimated at about 22.6 months. Solving the equation, $2/\lambda_{31} = 22.6$, for the mean of a gamma distribution with index parameter 2 and scale parameter λ_{31}, yields the estimate $\lambda_{31} = 0.0885$. According to a hypothesis suggested by Ginsberg, $\lambda_{21} = 5.5\,\lambda_{31} = 0.4867$. Listed in Table 8.5.2 are the values of the parameters corresponding to short and long periods of anovulation following childbirth. Also listed are the means of the gamma distributions with these parameter values.

Selected values of the two latent distributions, governing durations of nursing, are presented in Table 5.8.2. Short durations of nursing were assumed to follow an exponential distribution with the parameter value $\lambda_{32} = 0.6$. Long durations of nursing were assumed to follow an empirical distribution for women in the rural zone of Senegal as reported by Cantrelle and Leridon (1971) and are listed in the third and fourth columns of Table 5.8.3.

Selected values of computer output, determined by the values in Tables 5.8.2 and 5.8.3 and Eq. (5.8.5) with $x=0$, are presented in Table 5.7.1. Whenever an empirical distribution was listed only at selected values as in Table 5.8.3, a monotone interpolation procedure was used to compute values down to a monthly time scale. Level B_1 in Table 5.7.1 was determined by short

Table 5.8.3. Latent distribution functions for short and long durations of nursing

Short nursing durations		Long nursing durations	
age of child (months)	distribution function	age of child (months)	distribution function
0	0.000	0	0.000
6	0.973	6	0.000
12	0.999	12	0.020
18	1.000	18	0.076
24	1.000	24	0.418
30	1.000	30	0.861
36	1.000	36	0.970
Mean	1.667[a]	Mean	24.0[a]

[a] Expressed in months.

anovulatory and short nursing durations; level B_2 was determined by long anovulatory and long nursing durations. The two sets of computer output listed in Table 5.7.1 are merely two of eight computer runs described by Pickens (1982) in his Chap. 2.

5.9 A Semi-Markovian Model for Waiting Times to Conception Under Contraception

Apart from selecting the parameters of the Beta-Geometric distribution to represent a hypothetical pattern of contraceptive use, no attention has been paid to the construction and computer implementation of models, involving several methods of contraception in which switching among methods may occur. Two highly effective methods of contraception, with low risks of pregnancy, are intrauterine devices (IUD's) and oral contraceptive (pills), taken by women to prevent ovulation. Other less effective methods, with higher risks of pregnancy, include condoms, diaphragms, foams, and jellies. In some cases, one partner in a marriage may accept a sterilization to prevent unwanted pregnancies.

Formulated in this section is a model of waiting times to conception, based on the concept of an absorbing semi-Markov process in discrete time presented in Sect. 5.2. Chosen as absorbing states for this process were: E_1 – pregnant and E_2 – contraceptively sterilized. Four transient states were also included in the model. These states were: E_3 – the null state, E_4 – using an IUD, E_5 – using a pill, and E_6 – using some other method or methods of contraception. For ease of reference, the six states of the process are presented in Table 5.9.1.

As an aid to understanding the nature of the model under consideration, it will be useful to consider the general form of the matrix of densities for the one-step transitions of the process, from which the p.d.f. of the waiting times to conception may be computed. Just as in Sect. 5.2, let ϑ represent the state space and let ϑ_1 and ϑ_2 be disjoint subsets of absorbing and transient states. Because all elements in rows corresponding to absorbing states $i\in\vartheta_1$ are identically zero, it suffices to consider a 4×6 matrix, $(a_{ij}(t);\ i\in\vartheta_2,\ j\in\vartheta)$ of density functions, governing one-step transitions among transient states and to absorbing states. Given that transient state $i\in\vartheta_2$ is entered at $t=0$, let $p_{i1}(t)$, $t\geq1$, be the p.d.f. of the waiting times to conception. Then, by using a renewal argument, it follows that the set of four functions, $(p_{i1}(t),\ i\in\vartheta_2)$, satisfies the

Table 5.9.1. Six states of a semi-Markov model of waiting times to conception under contraception

Absorbing states		Transient states	
E_1	Pregnant	E_3	The null state
E_2	Contraceptively sterilized	E_4	Using an IUD
		E_5	Using a pill
		E_6	Using other methods

renewal equation

$$p_{i1}(t)=a_{i1}(t)+\sum_{s=1}^{t}\sum_{v}a_{iv}(s)p_{v1}(t-s),\qquad(5.9.1)$$

where i and v run over the set $\mathcal{9}_2$ of transient states. If $b_i\geq0$ is the fraction of a cohort in transient state i at $t=0$, then

$$p_1(t)=\sum_i b_i p_{i1}(t),\qquad i\in\mathcal{9}_2,\qquad(5.9.2)$$

is the p.d.f. of the waiting times to conception.

When numerical specifications of the one-step transition densities are given, numerical solutions of Eq. (5.9.1) may be computed, using the methods discussed in Problem 2. Furthermore, if the fractions $b_i\geq0$ with the property, $b_4+b_5+b_6=1$, are given, then computer output, based on Eq. (5.9.1), may be used to calculate the p.d.f. $p_1(t)$ in (5.9.2). Depending on the assumptions a user wishes to make, computer generated versions of either (5.9.1) or (5.9.2) may be used as input to MATHIST for the p.d.f. of the waiting times to conception.

There are at least three approaches to numerically specifying the one-step transition densities of the process. One approach is to estimate these functions from survey or other data on maternity histories. An example of this approach is the work of Pickens and Mode (1981) on similar models, using data from the 1973 U.S. National Survey of Family Growth.[1] Two other approaches involve constructing parametric formulas for the one-step transition densities. In a second approach, concepts from the theory of competing risks, as outlined in Sect. 3.3, are used; and, a third approach involves only the concept of conditional probabilities to derive parametric formulas for the one-step transition densities of the process.

Considered first is the case, based on the theory of competing risks, with independent latent sojourn times and constant latent risk functions. For every transient state $i\in\mathcal{9}_2$ and state $j\in\mathcal{9}$ such that $i\neq j$, let θ_{ij} be a constant latent risk function such that $\theta_{ij}>0$ for at least one j. Let the positive constant θ_i be defined by

$$\theta_i=\sum_j\theta_{ij},\qquad i\neq j.\qquad(5.9.3)$$

Then, if $A_{ij}^{(c)}(t)$ is the continuous time analogue of the function $A_{ij}(t)$ defined in (5.2.3), it follows from Example 3.4.1 that

$$A_{ij}^{(c)}(t)=p_{ij}(1-\exp[-\theta_i t]),\qquad(5.9.4)$$

where $t\geq0$ and $p_{ij}=\theta_{ij}/\theta_i$, $i\neq j$. The distribution function of the sojourn time in state i is, therefore,

$$A_i^{(c)}(t)=\sum_j A_{ij}^{(c)}(t)=1-\exp[-\theta_i t],\qquad(5.9.5)$$

for $t\geq0$.

---

[1] U.S. National Center for Health Statistics (1978). National Survey of Family Growth, Cycle I. United States Vital and Health Statistics, Series 2, No. 36. Government Printing Office, Washington, D.C.

According to Eq. (5.2.7), p_{ij} is the conditional probability that $j \neq i$ is the next state visited, given that state i is entered at $t=0$. If j is the next state visited, then $A_{ij}^{(c)}(t)/p_{ij}$ is the conditional distribution function of the sojourn time in state i. From Eq. (5.9.4), it follows that this distribution function is, $1 - \exp[-\theta_i t]$, for every state $j \neq i$.

Not having the conditional distribution function, $A_{ij}^{(c)}(t)/p_{ij}$, depend on j is, in some respects, counter intuitive. For it seems plausible that the sojourn time in state i could depend on $j \neq i$, the next state visited. This observation leads to a third approach for constructing the function $A_{ij}^{(c)}(t)$ based on the concept of conditional probabilities. Let p_{ij} be the conditional probability of a transition to state j, given that state i is entered at $t=0$. Given this event, suppose the sojourn time in state i is exponentially distributed with positive parameter θ_{ij}. Then, the function $A_{ij}^{(c)}(t)$ in (5.9.4) takes the form

$$A_{ij}^{(c)}(t) = p_{ij}(1 - \exp[-\theta_{ij}t]), \tag{5.9.6}$$

for $t \geq 0$. Unlike the exponential distribution in (5.9.5), the distribution function of the sojourn times in state $i \in \mathcal{I}_2$ is now the mixture

$$A_i^{(c)}(t) = \sum_j p_{ij}(1 - \exp[-\theta_{ij}t]) \tag{5.9.7}$$

of exponential distributions, where

$$\sum_j p_{ij} = 1, \quad i \neq j. \tag{5.9.8}$$

It is clear that the exponential distribution functions on the right in (5.9.7) may be replaced by other parametric forms.

An important consideration in the computer implementation of parametric models is the number of parameters that must be specified in order to generate numerical forms of the one-step transition densities. Also of importance in differentiating among models as a basis for statistical inference is the number of parameters that must be estimated. Consideration of the maximum number of parameters needed to determine the functions $A_{ij}^{(c)}(t)$, according to formula (5.9.4) or formulas (5.9.6) and (5.9.8), is, therefore, of interest. For each of the four transient states $i \in \mathcal{I}_2$, the maximum number of one-step transitions out of state i is five. If the functions $A_{ij}^{(c)}(t)$ are specified according to formula (5.9.4), then a parameter θ_{ij} is associated with each one-step transition. Hence, twenty is the maximum number of parameters that must be specified in this case. On the other hand, if the functions $A_{ij}^{(c)}(t)$ are to be determined according to formulas (5.9.6) and (5.9.8), then a maximum number of five θ-parameters must be associated with each state. Because of condition (5.9.8), a maximum of only four p-parameters must be specified for each transient state. Thus, altogether, a maximum of thirty-six parameters would be needed to specify the model in the second case.

It is recognized that if a continuous time version of an absorbing semi-Markov process were considered, the explicit formulas for the densities in (5.9.1) could be derived if the functions were determined by either formula

(5.9.4) or formulas (5.9.6) and (5.9.8). For each of these two cases, however, the formulas for the densities in (5.9.1) could be quite different so that separate computer programs may be required. Moreover, any programs implementing the model in either case would also be dependent upon the underlying exponential distributions and could not accommodate models based on other assumptions. Rather than writing programs for special models, from the point of view of software design, it is much more efficient to develop general purpose programs for the discrete time formulation underlying formulas (5.9.1) and (5.9.2). When such software is available, discrete sections of continuous time models determined by the formula

$$a_{ij}(t) = A_{ij}^{(c)}(t) - A_{ij}^{(c)}(t-1), \qquad t \geq 1, \tag{5.9.9}$$

may be used as computer input, where the functions on the right can be arbitrary. For the models of waiting times to conception under discussion, the t-variable in (5.9.9) would represent months.

The need to specify twenty or more parameters in a computer simulation model can be a formidable barrier to making many computer runs. Models requiring the specification of fewer parameters are, thus, of practical interest. In some cases, the number of parameters needed to specify a model may be greatly reduced, if the motivations underlying the use of contraception are taken into account. When contraceptives are used to delay or space births in populations in which infants are not nursed for long periods of time, married women (couples) will be classified as spacers; when contraceptives are used to prevent further pregnancies, women (couples) will be classified as limiters. One would expect the structure of the matrices of one-step transition densitites to differ among spacers and limiters. Two plausible examples of these different structures are given below. Discrete time analogues of the continuous time models considered in this section are discussed in Problem 7.

Example 5.9.1. *A One-Step Transition Matrix of Density Functions for Spacers*

Since spacers want more children, they would not be expected to seek a contraceptive sterilization; but, they would be expected to use some method, or methods, of contraception in attempting to space their children. A scenario of contraceptive use for spacers might be as follows.

Suppose a physician advises a couple to try an IUD. For some couples an IUD may prove to be suitable; whereas, for other couples, undesirable side effects may be associated with its use. Among those couples for whom an IUD is unsuitable but who still wish to use an effective method of contraception, a pill could be the next choice. For some, a pill may also have undesired side effects so that they finally resort to use of a diaphragm, condom, or perhaps a combination of both. Suppose all methods of contraception have some risk of failure, i.e., a pregnancy may occur while using a method. Furthermore, if a pregnancy has not occurred while a method is being used, a switch to the null state occurs when a pregnancy is desired. The following one-step transition matrix of densities,

$$\begin{bmatrix} a_{31}(t) & 0 & 0 & 0 & 0 \\ a_{41}(t) & 0 & a_{43}(t) & 0 & a_{45}(t) & 0 \\ a_{51}(t) & 0 & a_{43}(t) & 0 & 0 & a_{56}(t) \\ a_{51}(t) & 0 & a_{63}(t) & 0 & 0 & 0 \end{bmatrix}, \tag{5.9.10}$$

for transitions among and out of transient states, describes the scenario just outlined. Implicit in the structure of the matrix in (5.9.10) is the assumption that once the null state E_3 is entered, no method of contraception is re-accepted.

Some spacers, who view the pill as the only acceptable method of contraception, may wish to avoid the use of an IUD altogether. If a pill produces undesirable side effects, they will resort to less effective methods of contraception. The last two rows of the matrix in (5.9.10) describes such a scenario.

When specifying explicit forms of the densities in (5.9.10), it would seem that transition functions of the type given in (5.9.6) would be preferable to those in (5.9.4); for, if two parameters, p_{ij} and θ_{ij}, can be associated with each possible transition, then there would be greater opportunities for realistic descriptions of contraceptive failures, switches among methods of contraception, and desires to switch to the null state when a pregnancy is wanted. Observe that the density matrix in (5.9.10) provides an example of an absorbing semi-Markov process in discrete time in which all transient states do not communicate.

Example 5.9.2. *A One-Step Transition Matrix of Density Functions for Limiters*

When couples have attained a desired family size, some may be willing to accept a contraceptive sterilization, involving either the male or female. A sterilization could, moreover, coincide with the termination of some method of contraception.

To make the discussion concrete, suppose that patterns of contraceptive use follow those in Example 5.9.1 except that users of other methods may accept a pill as a more effective method of preventing unwanted pregnancies. In addition, suppose that a contraceptive sterilization may be accepted while using any method of contraception or while in the null state. Let it also be supposed that limiters avoid the null state except for the possibility of brief periods while waiting for a contraceptive sterilization. Then, the following one-step transition matrix of densities,

$$\begin{bmatrix} a_{31}(t) & a_{32}(t) & 0 & 0 & 0 & 0 \\ a_{41}(t) & a_{42}(t) & 0 & 0 & a_{45}(t) & 0 \\ a_{51}(t) & a_{52}(t) & 0 & 0 & 0 & a_{56}(t) \\ a_{61}(t) & a_{62}(t) & 0 & 0 & a_{65}(t) & 0 \end{bmatrix}, \tag{5.9.11}$$

may be used to describe the scenario just outlined.

It is of interest to observe that, according to the structure of the matrix in (5.9.11), a couple may accept a pill but then discontinue this method to accept some other method of contraception. After using this method for a while, it is then possible to reaccept a pill. Clearly, other versions of the matrix in (5.9.11) could easily be considered.

5.10 Notes on Cohort and Period Projections of Fertility

MATHIST has a capability for computing cohort fertility density functions very efficiently. With such a capability available, it is of interest to explore ways in which computer output from MATHIST may be used to implement some version of the evolutionary potential birth process discussed in Sect. 4.9. To make the discussion more concrete, suppose a population is undergoing a fertility change in response to modernization, as reflected by increasing acceptance of contraceptives. Without going into details, let it also be supposed that computer input to MATHIST can be selected in such a way that the cohort fertility density function $b_c(t, y)$ in (4.9.15), calculated under the assumptions prevailing in MATHIST, reflect this fertility change for successive cohorts born at some given set of epochs designated by values of the variable t. Explored in this section is the problem of using this computer output to project period total fertility rates at some designated epochs $t = t_0, t_0 + 1, \ldots, t_0 + h$, where h is a positive integer.

Consider a population at some epoch $t \in R$. Then, from Eq. (4.9.19), connecting period and cohort fertility density functions, it follows that,

$$B_p(t, x, h) = \int_x^{x+h} b_c(t - y, y) dy,\qquad (5.10.1)$$

is the mean number of births at epoch t born to women in the age interval $(x, x+h]$. If $h = 5$ and ages 15 and 50 are the lower and upper limits of the childbearing ages, then,

$$v_p(t, \infty) = \sum_{x=15}^{45} B_p(t, x, 5),\qquad (5.10.2)$$

is the period total fertility rate at epoch t. In principle, a period total fertility rate is easy to determine from cohort fertility density functions; but, there are practical limitations, regarding the number of cohort fertility density functions that can be computed.

A feasible approach to reducing the number of cohort fertility density functions needing calculation is to take $h = 1$ in (5.10.1) and approximate $B_p(t, x, 1)$ by the integral

$$B_p^*(t, x, 1) = \int_x^{x+h} b_c(t - x, y) dy.\qquad (5.10.3)$$

When time is measured in years, it seems reasonable to suppose that the fertility density functions $b_c(t - y, y)$ for cohorts born at epochs $t - y$, $y \in (x, x+1]$, would differ little from the function $b_c(t - x, y)$. One would expect, therefore, that $B_p^*(t, x, 1)$ would be a close approximation to $B_p(t, x, 1)$. If it is again assumed that the 15 and 50 are the lower and upper limits of the childbearing ages, then,

$$v_p^*(t, \infty) = \sum_{x=15}^{49} B_p^*(t, x, 1),\qquad (5.10.4)$$

is an approximation to the period total fertility rate at epoch t.

Considered next is the following question. If the period total fertility rate in (5.10.4) is to be calculated at each of the epochs $t = t_0, t_0 + 1, \ldots, t_0 + h$, where h is a positive integer, then how many cohort fertility density functions (c.f.d.f.'s) need to be calculated? According to formula (5.10.4), a c.f.d.f. must be calculated for each age $x = 15, \ldots, 49$ at epoch t_0, making a total of 35. For each epoch (year) of the projection, one cohort advances beyond the childbearing ages and another advances into the childbearing ages. Hence, if a projection is carried out for $h \geq 1$ epochs, then a total of $35 + h$ c.f.d.f.'s need to be calculated.

Ten-year projections are frequently carried out; when $h = 10$, 45 c.f.d.f.'s need to be calculated, a number in the range of feasibility for MATHIST if only c.f.d.f.'s were required. If $h = 10$, then the cohorts in the projection would be born at epochs $t_0 - 49, t_0 - 48, \ldots, t_0 - 5$, ordered from left to right in the order of their births. For $h \geq 1$, cohorts born at the epochs $t_0 - 49, t_0 - 48, \ldots, t_0 + h - 15$, would be needed to make a projection spanning h epochs.

Special software, using some of the subroutines making up MATHIST, would be required to implement the above ideas. Among the more difficult problems that must be approached in designing this software would be that of structuring computer input to reflect the evolution of fertility in successive cohorts. Indeed, even describing such an evolution in terms of component functions of the model, as outlined in Sect. 5.4, is beyond the scope of this section. Given the level of development in MATHIST, as described in Sect. 5.5, it appears that the ideas outlined above could actually be implemented within a reasonable span of time.

A case, requiring the calculation of relatively few c.f.d.f.'s, is worthy of mention. For simulations pertaining to a population in the beginning of a transition to low fertility, it may be reasonable to assume that all cohorts in the initial population may be characterized by the same c.f.d.f. Under this assumption, a projection spanning ten years would require the calculation of only ten c.f.d.f.'s. Using MATHIST to calculate ten c.f.d.f.'s would not be a difficult task.

5.11 Further Reading

Computer models of human reproduction have been used quite extensively in demography for about two decades. Sheps and Menken (1973) compiled a very useful list of the literature on such models up to about 1971. Included in this list were papers on micro-models, involving the use of Monte Carlo methods to compute realizations of some stochastic process, and macro-models, involving the computation of summary measures, such as means and variances, for some stochastic process. Contained in the volume of Dyke and MacCluer (1974), reporting the proceedings of a conference, are a set of papers on applying computer simulation methods to the study of human populations.

A problem of importance in evaluating family planning programs is that of developing model relationships between birth rates and birth control practices.

Bogue et al. (1973) and Nortman et al. (1977) have designed and implemented computer models to deal with this problem. Wachter (1978), in collaboration with E.A. Hammel and P. Laslett, has reported on the applications of Monte Carlo methods to the study of historical social structure. Inoue (1977) used a combination of micro and macro simulation methods in his study of policy measures which affect fertility. Some authors have approached the problem of forecasting births by applying statistical time series methodology. Two examples of this approach are the papers of Lee (1978) and McDonald (1979).

Problems and Miscellaneous Complements

1. (a) To show the sequence of random variables $\langle X_v \rangle$, $v \geq 0$, has stationary transition probabilities observe that from (5.2.1) and (5.2.3), it follows that

$$P[X_n = j, \, Y_n \leq t \,|\, X_{n-1} = i] = A_{ij}(t)$$

for all $n \geq 1$. Consequently, the marginal p.d.f. of X_n, given that $X_{n-1} = i$, is given by

$$P[X_n = j \,|\, X_{n-1} = i] = \lim_{t \uparrow \infty} A_{ij}(t) = p_{ij},$$

if $i \neq j$, and by

$$P[X_n = j \,|\, X_{n-1} = i] = \lim_{t \uparrow \infty} (1 - A_i(t)) = p_{ii}$$

if $i = j$, see (5.2.7) and (5.2.8). Because these limits hold for all $n \geq 1$, the sequence, $\langle X_v \rangle$, has stationary transition probabilities.

(b) To show that the sequence of random variables, $\langle X_v \rangle$, is a Markov chain, it is sufficient to demonstrate that the joint conditional p.d.f. of the random variables $X_1, X_2, \ldots, X_n$, given that $X_0 = i_0$, is

$$\prod_{v=1}^{n} p_{i_{v-1} i_v}$$

for every $n \geq 1$. Derive the above p.d.f. from the joint conditional p.d.f. in (5.2.2).

(c) Show that Eq. (5.2.1) implies that the conditional distribution function of the random variable Y_n, given that $X_{n-1} = i$, is

$$A_i(t) = P[Y_n \leq t \,|\, X_{n-1} = i],$$

see (5.2.4). Observe that this conditional distribution depends on the last state visited but not on $n \geq 1$. Moreover, it does not depend on the values of the random variable in the sequence $\langle Y_v \rangle$, $v \geq 1$, either preceding or following Y_n. Hence, $\langle Y_v \rangle$, $v \geq 1$, is a sequence of independent random variables. Nondependence on n imparts a kind of stationarity in the distributions of the random variables in the sequence $\langle Y_v \rangle$.

2. Discussed in this set of problems are numerical algorithms for solving systems of renewal equations of the type given in (5.2.11) and (5.2.12). In what follows, it will be supposed there are $r_1 \geq 0$ absorbing and $r_2 > 0$ transient states in a state space consisting $r = r_1 + r_2$ states.

(a) The operation of matrix convolution is basic to solving systems of renewal type equations. Let $\mathbf{q}(t) = (a_{ij}(t))$ be a $r_2 \times r_2$ matrix of density functions, where i and j are transient states in $\mathcal{G}_2$. Let $\mathbf{q}^{(1)}(t) = \mathbf{q}(t)$; and, let $\mathbf{q}^{(2)}(t)$ be a $r_2 \times r_2$ matrix whose ij-th element $a_{ij}^{(2)}(t)$ is

$$a_{ij}^{(2)}(t) = \sum_k \sum_{s=0}^{t} a_{ik}(s) a_{kj}(t-s), \quad k \in \mathcal{G}_2.$$

In general, for any pair of transient states i and j in $\mathcal{G}_2$, the ij-th function in the $r_2 \times r_2$ matrix $\mathbf{q}^{(n)}(t)$ is defined recursively by

$$a_{ij}^{(n)}(t) = \sum_k a_{ik}^{(n-1)}(s) a_{kj}(t-s)$$

$$= \sum_k a_{ik}(s) a_{kj}^{(n-1)}(t-s), \quad k \in \mathcal{G}_2.$$

Show that this equation may be written in the compact matrix form,

$$\mathbf{q}^{(n)}(t) = \sum_{s=0}^{t} \mathbf{q}^{(n-1)}(s)\mathbf{q}(t-s) = \sum_{s=0}^{t} \mathbf{q}(s)\mathbf{q}^{(n-1)}(t-s).$$

(b) Let $\mathbf{q}^{(0)}(t)$ be a $r_2 \times r_2$ matrix-valued function such that $\mathbf{q}^{(0)}(0) = \mathbf{I}$, a $r_2 \times r_2$ identity matrix, and $\mathbf{q}^{(0)}(t) = \mathbf{0}$, a $r_2 \times r_2$ zero matrix, for $t \geq 1$. Show that $\mathbf{q}^{(0)}(t)$ is the identity function with respect to the operation of matrix convolution. It suffices to show that

$$\sum_{s=0}^{t} \mathbf{q}^{(0)}(s)\mathbf{q}(t-s) = \sum_{s=0}^{t} \mathbf{q}(s)\mathbf{q}^{(0)}(t-s) = \mathbf{q}(t)$$

for all $t \geq 0$.

(c) The following "geometric series"

$$\mathbf{m}(t) = \mathbf{q}^{(0)}(t) + \mathbf{q}^{(1)}(t) + \dots, \quad t \geq 0,$$

of matrix convolutions arises in solving system of renewal type equations. It will be assumed that $\mathbf{q}(0) = \mathbf{0}$, a $r_2 \times r_2$ zero matrix. Show that the series defining the $r_2 \times r_2$ matrix-valued function $\mathbf{m}(t)$ converges for all $t \geq 0$. Then, use an argument analogous to that used in proving Theorem 4.8.1 to show that $\mathbf{m}(t)$, $t \geq 0$, may be calculated by using the recursive system $\mathbf{m}(0) = \mathbf{I} = \mathbf{q}^{(0)}(0)$ and

$$\mathbf{m}(t) = \sum_{s=0}^{t-1} \mathbf{m}(s)\mathbf{q}(t-s)$$

for $t \geq 1$.

(d) Define a $r_2 \times r_2$ diagonal matrix of functions $\mathbf{D}(t) = (d_{ij}(t))$ by $d_{ij}(t) = \delta_{ij}(1 - A_i(t))$; let $\mathbf{P}(t) = (P_{ij}(t))$ be the $r_2 \times r_2$ matrix of current state probabilities in Eqs. (5.2.11). Show that Eqs. (5.2.11) may be written in the compact matrix form

$$P(t) = D(t) + \sum_{s=0}^{t} q(s) P(t-s), \qquad t \geq 0.$$

(e) To gain insights into solving the equation in part (d), it will be helpful to drop t and write the equation in the even more succint form,

$$P = D + q * P,$$

where $*$ stands for the operation of matrix convolution. Show that if the expression $D + q * P$ is substituted on the right for P, then

$$P = (q^{(0)} + q^{(1)}) * D + q^{(2)} * P.$$

In general, if this process is repeated n times, then an equation of the form

$$P = (q^{(0)} + q^{(1)} + \ldots + q^{(n-1)}) * D + q^{(n)} * P$$

results. This last equation suggests that the matrix function $m(t)$ defined in part (c) plays a fundamental role in solving the equation, $P = D + q * P$, for P.

(f) In fact, the results in part (e) suggest that

$$P(t) = \sum_{s=0}^{t} m(s) D(t-s), \qquad t \geq 0,$$

provided that the function $q^{(n)} * P$ converges to a zero matrix uniformly in $t \geq 0$ as $n \to \infty$. Observe that if the equation, $P = D + q * P$, is written in the form, $(q^{(0)} - q) * P = D$, then, formally, m is the inverse of $q^{(0)} - q$.

(g) For every pair of transient states i and j in ϑ_2, $n \geq 2$, and $t \geq 0$, show that

$$a_{ij}^{(n)}(t) \leq \sum p_{ik}^{(n-1)}, \qquad k \in \vartheta_2,$$

where $(p_{ij}^{(n)})$, $n \geq 1$, is the n-th power of the one-step transition matrix (p_{ij}) of the Markov chain $\langle X_v \rangle$, $v \geq 0$, see (5.2.7) and (5.2.8). When ϑ_2 is a set of communicating transient states in a Markov chain with absorbing states, then $(p_{ij}^{(n)})$, $i, j \in \vartheta_2$, approaches a zero matrix as $n \to \infty$, see Kemeny and Snell (1976), Chap. III. Conclude from this result that the matrix valued function $q^{(n)} * P$ converges to a zero matrix uniformly in $t \geq 0$ as $n \to \infty$.

(h) Let $R(t) = (A_{ij}(t))$ and $(P_{ij}(t))$, where $i \in \vartheta_2$ and $j \in \vartheta_1$, be $r_2 \times r_1$ matrices of functions, with $r_1 > 0$. Show that the solution of Eq. (5.2.12) is

$$P(t) = \sum_{s=0}^{t} m(s) R(t-s), \qquad t \geq 0.$$

3. With reference to Eq. (5.3.4), let $h_{41}(v, t)$ and $h_{51}(v, t)$ be the densities on the right hand side.

(a) Show that

$$h_{41}(v, t) = \sum_{s=0}^{t} a_{42}(v, s) h_{21}(v, t-s)$$

and

$$h_{51}(v, t) = \sum_{s=0}^{t} a_{52}(v, s) h_{21}(v, t-s)$$

for all $v \geq 0$ and $t \geq 0$.

(b) Define the density $f_{22}(v, t)$ by

$$f_{22}(v, t) = \sum_{j=4}^{5} \sum_{s=0}^{t} f_{2j}(v, s) a_{j2}(v, t-s),$$

for $v \geq 0$ and $t \geq 0$. Then, use the results in part (a) and Eq. (5.3.4) to show that $h_{21}(v, t)$ satisfies the renewal equation

$$h_{21}(v, t) = f_{21}(v, t) + \sum_{s=0}^{t} f_{22}(v, s) h_{21}(v, t-s)$$

for all $v \geq 0$ and $t \geq 0$.

(c) The results in parts (a) and (b) provide an efficient algorithm, because the last equation in part (b) can be solved by an application of a scalar version of Littman's algorithm. Note that solving Eqs. (5.3.8) requires a matrix version of Littman's algorithm.

(d) Taken together, the algorithm for calculating $h_{21}(v, t)$ outlined in Sect. 5.3 and the results of this problem are of didactic value, because they illustrate two different but mathematically equivalent approaches to deriving algorithms. Observe that part (b) above could be used as a point of departure. Then, if so desired, the formulas in part (a) could be used to calculate $h_{41}(v, t)$ and $h_{51}(v, t)$.

4. The p.d.f. $f(x)$ of a random variable X having a negative binomial distribution with parameters $\rho > 0$ and θ, $0 < \theta < 1$, is defined by (5.4.4) and (5.4.5).

(a) Let $\eta = 1 - \theta$. Show that the generating function of the random variable X is

$$g(s) = \theta^{\rho}(1 - \eta s)^{-\rho},$$

where $|s| \leq 1$. Recall that $g(s) = E[s^X]$.

(b) Use the generating function in part (a) to show that

$$E[X] = \frac{\rho \eta}{\theta}$$

and

$$\text{Var}[X] = \frac{\rho \eta}{\theta^2}.$$

Then, deduce the results in (5.4.7) from the relationship $T = \tau + X$.

(c) Show that

$$\frac{f(t+1)}{f(t)} = \frac{(\rho + t)(1 - \theta)}{t + 1}$$

for all $t \geq 0$.

(d) The p.d.f. $f(t)$ decreases for all t such that $f(t+1)<f(t)$; the p.d.f. $f(t)$ increases for all t such that $f(t)<f(t+1)$. Use the result in part (c) to show that $f(t)$ decreases for all t such that

$$t>\frac{\rho(1-\theta)-1}{\theta}.$$

For what values of t is the p.d.f. $f(t)$ increasing?

(e) The p.d.f. $f(t)$ has a double mode if there is a positive integer t_0 such that $f(t_0)=f(t_0+1)$ and $f(t)<f(t_0)$ for all $t\neq t_0$ and $t\neq t_0+1$. Use the results in parts (c) and (d) to show that

$$t_0=\frac{\rho(1-\theta)-1}{\theta},$$

provided that the expression on the right is a positive integer. If this expression is not a positive integer, how would you interpret t_0?

5. Developed in this problem are some results stemming the reparameterization, $\gamma=\alpha+\beta$ and $\delta=\alpha/\gamma$, of the Beta distribution introduced in Sect. 5.7.

(a) If $\gamma=\alpha+\beta$ and $\delta=\alpha/\gamma$, show that $\alpha=\gamma\delta$ and $\beta=\gamma(1-\delta)$.

(b) Use formulas (1.5.2) and (1.5.3) to verify that the mean and variance of the Beta distribution are given by (5.7.1).

(c) Verify that the coefficient of variation for the Beta distribution is given by formula (5.7.2).

(d) Verify that if $\alpha=1$ and $\beta>0$, then the p.d.f. of the Beta distribution is given by (5.7.3).

(e) Let $\eta>0$ be the mode of the Beta distribution. Use formula (1.5.4) to show that

$$\eta=\frac{\gamma\delta-1}{\gamma-2},$$

provided that $\gamma>2$.

(f) When values of $\gamma>2$ and $\eta>0$ are assigned in (e), the value of δ is determined by

$$\delta=\frac{1+(\gamma-2)\eta}{\gamma}.$$

Investigate the behavior of the righthand side when $\eta>0$ is fixed and $\gamma\to\infty$. Interpret the result in terms of the mean and mode of the Beta distribution.

6. Outlined in this problem is a multiple decrement life table algorithm used in the computer implementation of the models for the anovulatory period following childbirth described in Sect. 5.8. Even if the densities $a_{ij}(x,s)$ in (5.8.2) are specified in terms of parametric formulas, the calculation of the integrals $A_{ij}(x,t)$ in (5.8.2), for all i and j and selected values of x and t, by quadrature formulas, can lead to heavy computations in large scale computer studies. The multiple decrement life table algorithm described below not only makes the computations manageable but it is also amenable to nonparametric input

based on observational data. Because the computation of $A_{ij}(x, t)$ is the same for every i and x, it will suffice to consider m functions of the form $A_j(t)$ for $1 \le j \le m$.

(a) Consider the problem of computing values of the functions $A_1(\cdot), \ldots, A_m(\cdot)$ on some finite number of points: $0 = t_0 < t_1 < \ldots < t_n$. Let $A(t) = A_1(t) + \ldots + A_m(t)$ and define the conditional probabilities q_{jk} by

$$q_{jk} = \frac{A_j(t_k) - A_j(t_{k-1})}{1 - A(t_{k-1})}$$

for $1 \le j < m$ and $1 \le k \le n$. Then, if $q_k = \sum_j q_{jk}$, $p_k = 1 - q_k$, $P_0 = 1$ and $P_k = p_1 p_2 \ldots p_k$ for $k \ge 1$, it follows that

$$A_j(t_k) = \sum_{v=1}^{k} P_{v-1} q_{jv},$$

for $1 \le j \le m$ and $1 \le k \le n$. Thus, if the q_{jk} can be computed, the last equation provides a solution to the problem. Discretized forms of the densities may be determined by $a_j(t_k) = A_j(t_k) - A_j(t_{k-1}) = P_{k-1} q_{jk}$.

(b) The conditional probabilities q_{jk} may be expressed in terms of latent risk functions $\theta_1(t), \ldots, \theta_m(t)$. Let $\theta(t) = \theta_1(t) + \ldots + \theta_m(t)$. Use the formula

$$A_j(t) = \int_0^t (1 - A(s)) \theta_j(s) ds,$$

and the assumption of independent latent sojourn times to show that

$$A_j(t_k) - A_j(t_{k-1}) = \exp\left[- \int_0^{t_{k-1}} \theta(x) dx \right] \int_{t_{k-1}}^{t_k} \exp\left[- \int_{t_{k-1}}^{s} \theta(x) dx \right] \theta_j(s) ds.$$

Then, conclude from the definition of q_{jk} that

$$q_{jk} = \int_{t_{k-1}}^{t_k} \exp\left[- \int_{t_{k-1}}^{s} \theta(x) dx \right] \theta_j(s) ds.$$

(c) The expression for q_{jk} can be further simplified if it is assumed that there are constants c_{jk} such that $\theta(t) c_{jk} = \theta_j(t)$ for all $t \in (t_{k-1}, t_k]$. Show that under this proportionality assumption, the conditional probability may be expressed in the form

$$q_{jk} = c_{jk}[1 - \exp[-\eta_k]],$$

where

$$\eta_k = \int_{t_{k-1}}^{t_k} \theta(s) ds.$$

(d) In the computer implementation of this last formula for q_{jk}, continuous risk functions were approximated by the step functions, $\theta_j^*(t) = \theta_j(t_k)$ for all $t \in (t_{k-1}, t_k]$, so that the integral η_k took the simple form $\eta_k = \theta_j(t_k)(t_k - t_{k-1})$. Furthermore, the proportionality constants were chosen as $c_{jk} = \theta_j(t_k)/\theta(t_k)$. A further simplification results when $t_k - t_{k-1} = 1$, a unit of time, for all k. The unit of time used in implementing the models described in Sect. 5.8 was a month of thirty days. Nonparametric latent risk functions were estimated from empirical distribution or survival functions by numerical interpolation and differentiation. The above approximation procedures seem to have yielded reasonable numerical results when applied to the 36×36 time lattice needed to implement the models described in Sect. 5.8.

7. Sometimes the densities in (5.9.9), defined on the discrete time points $t = 1, 2, ...$, can be recognized as well-known discrete distributions.

(a) Let X be a continuous type random variable T such that $T = t$ if, and only if, $t - 1 < X \le t$ for $t = 1, 2, ...$. Then.

$$P[T = t] = P[t - 1 < X \le t].$$

Let $\eta = \exp[-\theta]$ and suppose X has on exponential distribution with parameter θ. Show that

$$P[T = t] = (1 - \eta)\eta^{t-1}, \quad t = 1, 2, ...$$

Thus, T follows a geometric distribution.

(b) Suppose the function $A_{ij}^{(c)}(t)$ is determined according to formula (5.9.4). Conclude from formula (5.9.9) that there exists a positive parameter η_i such that

$$a_{ij}(t) = p_{ij}(1 - \eta_i)\eta_i^{t-1}, \quad t = 1, 2, ...$$

(c) If the function $A_{ij}^{(c)}(t)$ is determined by formula (5.9.6), show that there exists a positive parameter η_{ij} such that

$$a_{ij}(t) = p_{ij}(1 - \eta_{ij})\eta_{ij}^{t-1}, \quad t = 1, 2, ...$$

(d) Situations, in which the one-step transition densities, $a_{ij}(t)$, are defined directly in terms of a discrete distribution, may arise. For example, densities of the type

$$a_{ij}(t) = p_{ij}f_{ij}(t),$$

could be considered, where $f_{ij}(t)$ is an arbitrary p.d.f. on the positive integers $1, 2, ...$

References

Bogue, D.J., Edmonds, S. and Bogue, E.J. (1973): *An Empirical Model for the Demographic Evaluation of Impact of Contraception and Marital Status on Birth Rates*. Community and Family Study Center. University of Chicago

Brigham, E.O. (1974): *The Fast Fourier Transform*. Prentice-Hall, Inc., Englewood Cliffs, New Jersey

Busby, R.C. and Mode, C.J. (1982): Theory and Applications of a Computationally Efficient Cohort Simulation Model of Human Reproduction. *Mathematical Biosciences* 64:45–74

Cantrelle, P. and Leridon, H. (1971): Breast Feeding, Mortality in in Childhood, and Fertility in a Rural Zone of Senegal. *Population Studies* 25:505–533

Çinlar, E. (1969): Markov Renewal Theory. *Advances in Applied Probability* 1:123–187

Çinlar, E. (1975): *Introduction to Stochastic Processes*. Prentice-Hall, Inc., Englewood Cliffs, N.J.

Doob, J.L. (1953): *Stochastic Processes*. John Wiley and Sons, Inc., New York and London

Dyke, B. and MacCluer, J.W. (1974): *Computer Simulation in Human Population Studies.* Academic Press, New York

Eaton, J.W. and Mayer, A.J. (1953): The Social Biology of Very High Fertility Among Hutterites – The Demography of a Unique Population. *Human Biology* 25: 206–264

French, F.E. and Bierman, J.E. (1962): Probabilities of Fetal Mortality. *Public Health Rep.* 177:835–847

Ginsberg, R.B. (1973): The Effect of Lactation on the Length of the Post Partum Anovulatory Period – An Application of a Bivariate Stochastic Model. *Theoretical Population Biology* 4:276–299

Hostetler, J.A. (1974): *Hutterite Society.* The Johns Hopkins University Press. Baltimore, Maryland

Inoue, S. (1977): Choice of Policy Measures to Affect Fertility – A Computer Micro-Simulation Study. *Population Bulletin of the United Nations* 10:14–35

Kemeny, J.G. and Snell, J.L. (1976): *Finite Markov Chains.* Springer-Verlag, New York, Heidelberg, Berlin

Keyfitz, N. and Flieger, W. (1968): *World Population.* The University of Chicago Press, Chicago and London

Lee, R.D. (1978): New Methods for Forecasting Fertility – An Overview. *Population Bulletin of the United Nations* 11:6–11

Leridon, H. (1977): *Human Fertility – The Basic Components.* The University of Chicago Press, Chicago and London

Masnick, G.S. (1979): The Demographic Impact of Breastfeeding – A Critical Review. *Human Biology* 51:109–125

McDonald, J. (1979): A Time-Series Approach to Forecasting Australian Total Live-Births. *Demography* 16:675–601

Nortman, D.L., Potter, R.G., Kirmeyer, S.W. and Bongaarts, J. (1977): *Model Relations Between Birth Rates and Birth Control Practices.* Volumes I and II. The Population Council, New York, N.Y.

Page, H. and Lesthaeghe, R. (1980): The Postpartum Period – Development and Application of Model Schedules. *Population Studies* 34:143–169

Parkes, A.S., Thomson, A.M., Potts, J. and Herbertson, M.A. (1977): *Fertility Regulation During Human Lactation – Proceedings of Sixth IPPF Biomedical Workshop.* Journal of Biosocial Science, Supplement No. 4

Perez, A., Vela, P., Potter, R. and Masnick, G. (1971): Timing and Sequence of Resuming Ovulation and Menstruation After Childbirth. *Population Studies* 17:155–166

Perrin, E.B. and Sheps, M.C. (1964): Human Reproduction: A Stochastic Process. *Biometrics* 20:28–45

Perrin, E.B. and Sheps, M.C. (1965): A Mathematical Model for Human Fertility Patterns. *Archives of Environmental Health* 10:694–698

Pickens, G.T. (1982): *A Computerized Stochastic Model of the Postpartum Period.* Ph.D. Thesis, University of Pennsylvania, Philadelphia, Pa.

Pickens, G.T. and Mode, C.J. (1981): Sequence of Events Following Adoption of Contraception – An Exploratory Analysis of 1973 Fertility History Data. *Social Biology* 28:111–125

Pittenger, D.B. (1973): An Exponential Model of Female Sterility. *Demography* 10:113–124

Popkin, B.M., Billsborrow, R.E. and Akin, J.S. (1982): Breast-Feeding Patterns in Low-Income Countries. *Science* 218:1088–1093

Ross, J.A. and Madhavan, S. (1981): A Gompertz Model for Birth Interval Analysis. *Population Studies* 35:439–454

Santow, G. (1978): *A Simulation Approach to the Study of Human Fertility.* Martinus Nijhuff Social Sciences Division. Leiden-Boston

Sheps, M.C. and Menken, J.A. (1973): *Mathematical Models of Conception and Birth.* The University of Chicago Press, Chicago and London

Tietze, C. (1957): Reproductive Span and Rate of Reproduction Among Hutterite Women. *Fertility and Sterility* 8:89–97

Tietze, C. (1981): *Induced Abortion – A World Review.* Population Council, New York

Wachter, K.W. (1978): *Statistical Studies of Historical Social Structure.* Academic Press, New York

Watkins, S.C. and McCarthy, J. (1980): The Female Life Cycle in a Belgian Commune – La Hulpe, 1847–1866. *Journal of Family History* 5:167–179

Chapter 6. Age-Dependent Models of Maternity Histories Based on Data Analyses

6.1 Introduction

It is widely agreed that simulation models should be closely linked to data analyses. Indeed, a justifiable criticism of the software design described in Chap. 5 would be that its underlying mathematical models need to be tested further in the analysis of data. Accordingly, the purpose of this chapter is to give an account of some research directed toward providing firm links between data analyses and the construction and implementation of computer simulation models. Problems arising in evaluating and measuring the impact of family planning programs on fertility provided the substantive motivation for the research reported in this chapter.

Age is a variable that is frequently taken into account in the analysis of survey data used in evaluating family planning programs. Consequently, the chapter begins with an outline of the mathematical foundations underlying age-dependent semi-Markov processes in discrete time. A new age-dependent model of maternity histories, based on the theory of age-dependent semi-Markov processes, is then constructed. Following the development of this model and the algorithms needed to implement it, several sections are devoted to a discussion of computer input derived from the analysis of survey data collected in Taiwan during the sixties and early seventies. Four sample computer runs are then discussed in detail. The primary purpose of these runs was to provide comparative estimates of the fertility of women in Taiwan around 1970, if contraception and induced abortion were not used to avert unintended births.

Estimates of births experienced by women after contact with a family planning program are frequently based on assumptions that are difficult to validate. As part of the quest for valid models of maternity histories when women interact with a family planning program, a mathematical model of a longitudinal family planning experiment, conducted in Taichung, Taiwan, during the sixties, was constructed by linking age-dependent semi-Markov models together in a kind of time series. It is also shown that the proposed model of the Taichung experiment passed several validation tests. Analyzing large longitudinal data sets presents special problems, which often require new methods of data analysis. It seems likely that the methods developed in analyzing the Taichung data will be useful in analyzing other large longitudinal data sets, arising in the biological and social sciences. Following the development and validation of the Taichung model, the chapter concludes with suggested implications for evaluating other family planning programs and a simulation experiment designed to measure the impact of the Taichung experiment on fertility.

6.2 Age-Dependent Semi-Markov Processes in Discrete Time with Stationary Transition Probabilities

Some concepts, taken from the theory of age-dependent semi-Markov processes in continuous time, have already been used in Sect. 5.8, where a model describing anovulatory sterile periods following live births was discussed. As promised in that section, the purpose of this section is to outline the mathematical foundations underlying age-dependent semi-Markov processes. Like the discussion in Sect. 5.2, attention will be focused on processes in discrete time, in order to bring the mathematics into closer alignment with the arithmetic done by computers. A translation of the theory to continuous time should be straightforward.

Just as in Sect. 5.2, let ϑ be a finite set of states and let $\omega = \langle \omega_0, \omega_1, \omega_2, \ldots \rangle$ be a sample path of the process. In an age-dependent process, the age of an individual when entering an initial state is taken into account; consequently, the initial element ω_0 in a sample path will be the pair $\langle i_0, x \rangle$, when an individual is of age x when entering an initial state i_0. For every $v \geq 1$, the pair $\omega_v = \langle i_v, y_v \rangle$ represents the case state i_v is entered at the v-th step after a sojourn time of y_v in state i_{v-1}. The truncated sample path $\langle \omega_0, \omega_1, \ldots, \omega_n \rangle$ will again be denoted by $\omega^{(n)}$, $n \geq 0$.

In terms of random variables, the pair $\langle X_n, Y_n \rangle$ describes the process at the n-th step, $n \geq 0$. For $n \geq 1$, the random variable X_n indicates the state entered at the n-th step after a random sojourn time Y_n in state X_{n-1}. The pair of initial random variables $\langle X_0, Y_0 \rangle$ indicates the initial state and the age of an individual when entering the initial state. For every $n \geq 0$, a particular realization of the pairs of random variables $\langle X_0, Y_0 \rangle, \ldots, \langle X_n, Y_n \rangle$ will be denoted by a truncated sample path $\omega^{(n)}$. Observe that if an individual is of age x when entering the initial state i_0 and his history is represented by $\omega^{(n)}$, $n \geq 1$, then he is of age $x + y_1 + y_2 + \ldots + y_n = t_n$ when entering state i_n at the n-th step.

A fundamental object underlying the theory of age-dependent semi-Markov processes is a one-step transition matrix $\mathbf{a}(x, t) = (a_{ij}(x, t))$ of density functions defined for all pairs of non-negative integers $x, t = 0, 1, 2, \ldots$ and every pair of states i and j in ϑ. All functions in the matrix $\mathbf{a}(x, t)$ satisfy the inequality, $0 \leq a_{ij}(x, t) \leq 1$, for every pair of integers $x, t \geq 0$. Given the matrix $\mathbf{a}(x, t)$ of one-step transition densities, the basic assumption characterizing an age-dependent semi-Markov process in discrete time may be stated as follows. Let $\omega^{(n-1)}$, $n \geq 1$, represent a realization of the random variables $\langle X_0, Y_0 \rangle, \ldots, \langle X_{n-1}, Y_{n-1} \rangle$. If $X_{n-1} = i$, then it will be assumed that

$$P[X_n = j, Y_n = y | \omega^{(n-1)}] = a_{ij}(t_{n-1}, y), \tag{6.2.1}$$

for all $n \geq 1$, states i and j in $\mathcal{I}$, and nonnegative integers t_{n-1} and y. The transition probabilities of the process are stationary, since the right-hand side of (6.2.1) depends on n only through the variable t_{n-1}.

Assumption (6.2.1) is Markovian in the sense that the conditional probability that state j is entered at the n-th step depends only on the state i visited at step $n-1$; assumption (6.2.1) implies an age-dependent property in the sense that the conditional distribution of the sojourn time in state i depends on the age t_{n-1} of an individual when state i is entered. Hence, the name, age-dependent semi-Markov process, is used to describe the stochastic structure under consideration. Unlike the semi-Markov process in discrete time discussed in Sect. 5.2, the random variables in the sequence $\langle Y_v \rangle$, $v \geq 1$, representing sojourn times in states, are not independent, because for every $n \geq 1$ the conditional density on the right in (6.2.1) depends on the values of the random variables Y_v, $0 \leq v \leq n-1$.

By well-known rules for dealing with conditional distributions, it follows that $P[X_v = i_v, Y_v = y_v, 1 \leq v \leq n | \omega^{(0)}]$, the joint conditional p.d.f. of the pairs of random variables $\langle X_1, Y_1 \rangle$, ..., $\langle X_n, Y_n \rangle$ given that $\omega^{(0)} = \langle X_0, Y_0 \rangle = \langle i_0, x \rangle$, may be represented in the form

$$\prod_{v=1}^{n} P[X_v = i_v, Y_v = y_v | \omega^{(v-1)}] \qquad (6.2.2)$$

for all states $i_0, i_1, ..., i_n$ in $\mathcal{I}$ and nonnegative integers $y_1, y_2, ..., y_n$. But, under assumption (6.2.1), it follows that (6.2.2) is equal to

$$\prod_{v=1}^{n} a_{i_{v-1} i_v}(t_{v-1}, y_v). \qquad (6.2.3)$$

Further insights into the implications of assumption (6.2.1) may be gained by considering the random variables

$$T_v = x + Y_1 + ... + Y_v, \qquad 1 \leq v \leq n. \qquad (6.2.4)$$

Note the random variable T_v represents the age of an individual when entering state i_v at the v-th step. Let $T_0 = x$. Then, from (6.2.4), it follows that there is a one-to-one correspondence between the random variables $T_1, T_2, ..., T_n$ and $Y_1, Y_2, ..., Y_n$. In fact, the inverse of the transformation in (6.2.4) is

$$Y_v = T_v - T_{v-1}, \qquad 1 \leq v \leq n. \qquad (6.2.5)$$

Because of the one-to-one correspondence represented by (6.2.4) and (6.2.5), formula (6.2.3) implies that the conditional joint p.d.f. of the pairs of random variables $\langle X_1, T_1 \rangle$, ..., $\langle X_n, T_n \rangle$, given that $\langle X_0, T_0 \rangle = \langle i_0, x \rangle$, is

$$\prod_{v=1}^{n} a_{i_{v-1} i_v}(t_{v-1}, t_v - t_{v-1}), \qquad (6.2.6)$$

where $i_0, i_1, ..., i_n$ are states in $\mathcal{I}$ and $t_0 \leq t_1 \leq t_2 \leq ... \leq t_n$ are nonnegative integers. But, (6.2.6) implies that the conditional p.d.f. of the pair of random variables $\langle X_n, T_n \rangle$, $n \geq 1$, given that $\langle X_v, T_v \rangle = \langle i_v, t_v \rangle$, $0 \leq v \leq n-1$, is

$a_{i_{n-1}i_n}(t_{n-1}, t_n - t_{n-1})$. Therefore, the sequence of pairs $\langle X_v, T_v \rangle$, $v \geq 1$, enjoys the Markov property, a property that is useful in deriving age-dependent renewal equations. It should be noted that the conditioning just described applies only to paths of positive probability.

Before deriving age-dependent renewal equations, which are analogues of Eqs. (5.2.11) and (5.2.12), it will be necessary to briefly consider the concepts of absorbing and transient states, within the structure under consideration. For all pairs of states i and j in $\mathbf{9}$ and nonnegative integers x and t, define a function $A_{ij}(x, t)$ by

$$A_{ij}(x, t) = \sum_{s=0}^{t} a_{ij}(x, s); \tag{6.2.7}$$

and let

$$A_i(x, t) = \sum_j A_{ij}(x, t), \quad j \in \mathbf{9}. \tag{6.2.8}$$

Because the sequence of pairs $\langle X_v, T_v \rangle$, $v \geq 0$, enjoys the Markov property and the transition probabilities are stationary, $A_i(x, t)$ is the conditional distribution function of the sojourn time in state i, given that an individual enters state i at age x.

A state $i \in \mathbf{9}$ will be called transient if, and only if,

$$\lim_{t \uparrow \infty} A_i(x, t) = 1 \tag{6.2.9}$$

for all $x \geq 0$. No matter what the age x of an individual when entering transient state i, movement out of the state occurs eventually with probability one. A state $i \in \mathbf{9}$ will be called absorbing if

$$A_i(x, t) = 0 \tag{6.2.10}$$

for all $x, t \geq 0$. Once entered an absorbing state is never left, from the perspective of an age-dependent semi-Markov process.

Just as in Sect. 5.2, when there are transient states in $\mathbf{9}$, it will be useful to partition $\mathbf{9}$ into a set $\mathbf{9}_1$ of absorbing states and a disjoint set $\mathbf{9}_2$ of transient states. Communication among the transient states in $\mathbf{9}_2$ is defined just as Sect. 5.2. All other conventions regarding the specification of the matrix $\mathbf{a}(t)$ outlined in Sect. 5.2 continue to hold for matrix $\mathbf{a}(x, t)$ for every $x \geq 0$. In principle, the finite dimensional conditional marginal distributions for the sequences of random variables $\langle X_v \rangle$ and $\langle Y_v \rangle$, $v \geq 1$, given that $\langle X_0, Y_0 \rangle = \langle i_0, x \rangle$, could be worked out from (6.2.3); but, these distributions are not of central interest to the developments that follow.

A random function of interest is $Z(t)$, indicating the state in $\mathbf{9}$ an individual occupies at time $t > 0$. This random function is defined exactly as in the discussion preceding (5.2.9). The probabilities of all statements concerning $Z(t)$ are, however, determined by the assumptions underlying the stochastic process under discussion. Of interest in the developments to follow in subsequent sections of this chapter are the current state probabilities

$$P_{ij}(x, t) = P[Z(t) = j | X_0 = i, Y_0 = x], \tag{6.2.10a}$$

defined for all pairs of states i and j in $\mathcal{I}$ and nonnegative integers x and t. Special cases of these probabilities have already been used in Sect. 5.8 for the case of continuous time; a more general treatment will be needed, however, for the developments to follow in subsequent sections.

For the case i and j are transient states in $\mathcal{I}_2$, renewal-type arguments, completely analogous to those used in deriving Eq, (5.2.11), may be invoked to show that the current state probabilities in (6.2.11) satisfy the age-dependent renewal type equations

$$P_{ij}(x,t)=\delta_{ij}(1-A_i(x,t))+\sum_v\sum_{s=0}^{t}a_{iv}(x,s)P_{vj}(x+s,t-s), \tag{6.2.11}$$

where $v\in\mathcal{I}_2$, δ_{ij} is the Kronecker delta, and x and t are nonnegative integers. It should be noted that the renewal argument used to derive the summands $a_{iv}(x,s)P_{vj}(x+s,t-s)$ in the double sum on the right in (6.2.11) is valid for every s and v, because the sequence of pairs of random variables $\langle X_v,T_v\rangle$, $v\geq0$, enjoys the Markov property. When $i\in\mathcal{I}_2$ but j is an absorbing state in $\mathcal{I}_1$, then another renewal type argument may be used to derive the equation

$$P_{ij}(x,t)=A_{ij}(x,t)+\sum_v\sum_{s=0}^{t}a_{iv}(x,s)P_{vj}(x+s,t-s), \tag{6.2.12}$$

where $v\in\mathcal{I}_2$ and x and t are nonnegative integers. In Eqs. (6.2.11) and (6.2.12), x is the age of an individual when entering state i.

Equations (5.8.4) and (5.8.5), which could easily be solved within the structure of the model discussed in Sect. 5.8, are actually special cases of Eqs. (5.2.12). Whenever it is possible, some special structure of a model should be exploited to solve renewal equations in a simple way. Many models of interest, however, may not have a simple structure, making it necessary to consider a general method for finding numerical solutions to systems of age-dependent renewal equations.

Equations (6.2.11) and (6.2.12) are of the same mathematical form and may thus be considered within a common framework. Let $\mathbf{q}(x,t)=(a_{ij}(x,t))$ be a $r_2\times r_2$ matrix corresponding to $r_2>0$ transient states; and, let $\mathbf{P}(x,t)$ be a $r_2\times r_1$ or a $r_2\times r_2$ matrix of current state probabilities, depending on whether the columns of $\mathbf{P}(x,t)$ correspond to r_1 absorbing or r_2 transient states. If there are no absorbing states, then $r_1=0$ and $\mathbf{P}(x,t)$ is a $r_2\times r_2$ matrix of functions. Similarly, let $\mathbf{D}(x,t)=(d_{ij}(x,t))$ be either a $r_2\times r_1$ or $r_2\times r_2$ matrix of functions. Then, Eqs. (6.2.11) and (6.2.12) are of the general form

$$\mathbf{P}(x,t)=\mathbf{D}(x,t)+\sum_{s=0}^{t}\mathbf{q}(x,s)\mathbf{P}(x+s,t-s), \tag{6.2.13}$$

where x and t are nonnegative integers.

Upon consulting Problem 2 (d) of Chap. 5, it can be seen that (6.2.13) is an age-dependent version of the matrix renewal equation discussed in that problem. One would expect, therefore, that a matrix version of Littman's algorithm, see Theorem 4.8.1, could be used to find numerical solutions to Eq. (6.2.13). Indeed, it is shown in Problem 1, that when the matrix-valued functions in Eq. (6.2.13) are defined on a finite lattice of $\langle x,t\rangle$-points, then a calculable matrix-valued function $\mathbf{m}(x, t)$ may be found such that

$$\mathbf{P}(x, t) = \sum_{s=0}^{t} \mathbf{m}(x, s) \mathbf{D}(x+s, t-s), \qquad (6.2.14)$$

on this finite lattice of $\langle x, t \rangle$-points.

Age-dependent semi-Markov processes of the type discussed in this chapter were introduced in Mode (1976, 1977) in connection with the formulation of models of human reproduction and their applications in family planning evaluation. Hoem (1972) also discussed a class of Markov processes in connection with actuarial tables and duration dependence in demography. As explained in Mode (1982), the approach used by Hoem, in determining the basic probabilities underlying a stochastic process, differs fundamentally from the sample path approach used in this section and elsewhere. In contrast to the approach used by Hoem, the sample path approach is not only easy to understand but also readily suggests algorithms for implementing either micro- or macro-simulation models on a computer.

6.3 An Age-Dependent Semi-Markovian Model of Maternity Histories

Formulated in Sect. 5.3 was a model for waiting times among live births based on linking absorbing semi-Markov processes together according to parity. For each parity, the states of the semi-Markov process were the five physiological states defined in Table 5.3.1. The idea of linking stochastic processes together to build a model of maternity histories had been developed earlier in a more general framework based on absorbing, age-dependent, semi-Markov processes, see Mode and Soyka (1977). Not only did this paper contain algorithms for computing distributions of waiting times among live births but there were also algorithms for computing current state probabilities, making it possible to obtain a more complete resolution of maternity histories than that based on the algorithms of Sect. 5.3. During the course of implementing the algorithms developed in Mode and Soyka (1977) on a computer, a new and more computationally efficient version of the model was discovered. Developed in this section is the theory underlying this new version of the model.

Just as in Sect. 5.3, a set of physiological states women may occupy during the course of their maternity histories is a fundamental component of the new formulation. A basic feature of the new formulation, differentiating it from that in Sect. 5.3, is that permanent physiological sterility will be treated as an absorbing state rather than a stopping time. Whether to treat permanent physiological sterility as an absorbing state or a stopping time, as in Sect. 5.3, is a difficult question to resolve when formulating stochastic models of human reproduction. A useful approach to finding answers to the question is to compare the performance of two alternative formulations, with respect to ease

Table 6.3.1. Definitions of the six physiological states in maternity histories

E_1	Permanent physiological sterility
E_2	Fecundable – fertile and capable of becoming pregnant
E_3	Pregnant
E_4	Physiologically sterile period following live birth
E_5	Physiologically sterile period following induced abortion
E_6	Physiologically sterile period following other types of pregnancy outcomes

of computer implementation and desirable numerical properties. Listed in Table 6.3.1 are definitions of the six physiological states used in formulating the new version of the model.

Another problem encountered in formulating models of maternity histories is the choice of state space for the stochastic process underlying the model. Unfortunately, there are no clear-cut guidelines to follow when making this choice and one must proceed by trial and error. At first sight, it would appear that a reasonable guiding principle to follow would be that of choosing a minimal number of states so that the essentials of a formulation could easily be grasped by the human mind. This guiding principle was, in fact, used in constructing the model developed in Mode and Soyka (1977); for, instead of considering a large set of states, a small state space was chosen and simple stochastic processes were linked together, according to parity, to build a model of maternity histories. Subsequent developments, motivated primarily by numerical problems to be discussed in a later section of this chapter, actually led to a formulation with an elarged state space. The two formulations are mathematically equivalent but that based on an enlarged state space is not only more efficient computationally but also has superior numerical properties.

In the new formulation, each state in the state space ϑ is represented by the ordered pair $\langle v, E_j \rangle$, $1 \leq j \leq 6$, corresponding to a married woman of parity $v \geq 0$ in physiological state E_j. For every $v \geq 0$, let $\vartheta(v)$ be the set of six states defined by

$$\vartheta(v) = [\langle v, E_j \rangle, 1 \leq j \leq 6]. \tag{6.3.1}$$

Then, the state space ϑ of the process may be represented by the disjoint union of sets

$$\vartheta = \sum_{v=0}^{\infty} \vartheta(v). \tag{6.3.2}$$

Only finitely many states in ϑ will be realized in any maternity history. For every $v \geq 0$, the state $\langle v, E_1 \rangle$, indicating a woman enters the permanently sterile state E_1 when she is of parity $v \geq 0$, will be absorbing; and, all non-absorbing states will be transient.

Turning next to the one-step transition densities of the process, let $a_{ij}(v, \xi; x, t)$ be the conditional density governing the timing of the transition from state $\langle v, E_i \rangle$ to state $\langle \xi, E_j \rangle$, given that at $t=0$ state $\langle v, E_i \rangle$ was entered at age x. Then, $\mathbf{a}(x, t) = (a_{ij}(v, \xi; x, t))$ is the matrix of one-step transition densities for the process. Because women must advance linearly from parity v to v

+1, the matrix $\mathbf{a}(x, t)$ has a special structure. Let $\mathbf{a}(v, \xi; x, t) = (a_{ij}(v, \xi; x, t))$, $i, j = 1, 2, \ldots, 6$, be a 6×6 submatrix of $\mathbf{a}(x, t)$, governing transitions from states in the set $\mathcal{I}(v)$ to states in the set $\mathcal{I}(\xi)$. It follows that if women can only advance linearly from parity v to parity $v+1$, then $\mathbf{a}(v, \xi; x, t) = \mathbf{0}$, a 6×6 zero matrix, unless $\xi = v$ or $\xi = v+1$. Therefore, if the sets of states $\mathcal{I}(v)$, $v \geq 0$, are ordered according to the nonnegative integers $v = 0, 1, 2, \ldots$, then the matrix $\mathbf{a}(x, t)$ has a simple partitioned form. Upon dropping the variables x and t to ease the notation, it can be seen that the one-step transition density matrix for the process may be represented in the simple form

$$\mathbf{a} = \begin{bmatrix} \mathbf{a}(0,0) & \mathbf{a}(0, 1) & \mathbf{0} & \mathbf{0} & \cdot & \cdot & \cdot \\ \mathbf{0} & \mathbf{a}(1, 1) & \mathbf{a}(1, 2) & \mathbf{0} & \cdot & \cdot & \cdot \\ \mathbf{0} & \mathbf{0} & \mathbf{a}(2, 2) & \mathbf{a}(2, 3) & \cdot & \cdot & \cdot \\ \cdot & \cdot & \cdot & \cdot & \cdot & \cdot & \cdot \\ \cdot & \cdot & \cdot & \cdot & \cdot & \cdot & \cdot \\ \cdot & \cdot & \cdot & \cdot & \cdot & \cdot & \cdot \end{bmatrix}. \tag{6.3.3}$$

From (6.3.3) it is readily apparent that the matrix $\mathbf{a}$ of one-step transition densities is completely determined as soon as the 6×6 matrices $\mathbf{a}(v, v)$ and $\mathbf{a}(v, v+1)$ are specified for every $v \geq 0$. For every $v \geq 0$, let the states $\langle v, E_j \rangle$ in the set $\mathcal{I}(v)$ be ordered according to the positive integers $1, 2, \ldots, 6$. Then, it will be assumed that the matrix $\mathbf{a}(v, v)$ has the form

$$\mathbf{a}(v, v) = \begin{bmatrix} 0 & 0 & 0 & 0 & 0 & 0 \\ a_{21} & 0 & a_{23} & 0 & 0 & 0 \\ 0 & 0 & 0 & 0 & a_{35} & a_{36} \\ a_{41} & a_{42} & 0 & 0 & 0 & 0 \\ a_{51} & a_{52} & 0 & 0 & 0 & 0 \\ a_{61} & a_{62} & 0 & 0 & 0 & 0 \end{bmatrix}, \tag{6.3.4}$$

for every $v \geq 0$. In order to lighten the notation, the four-variable argument, $\langle v, v; x, t \rangle$, has been dropped from each element in $\mathbf{a}(v, v)$.

Some basic assumptions underlying the process have been made explicit in the structure of the matrix in (6.3.4). The first row of the matrix consists of all zeros, because the state $\langle v, E_1 \rangle$ is absorbing for every $v \geq 0$. From the second row, it follows that, within each set $\mathcal{I}(v)$ of states, the only transitions out of the fecundable state E_2 are either to the permanently sterile state E_1 or the pregnant state E_3. Once in the pregnant state E_3, the only transition possible, while remaining in the set $\mathcal{I}(v)$, are to states E_5 and E_6, representing temporary sterile periods following an induced abortion or some other type of pregnancy outcome. Finally, from the last three rows of the matrix in (6.3.4), it can be seen that once states E_4, E_5, or E_6 are entered, then the only transitions possible, within the set $\mathcal{I}(v)$ of states, are to the states E_1 and E_2.

When a pregnant women of parity v experiences a live birth, then she moves from state $\langle v, E_3 \rangle$ to state $\langle v+1, E_4 \rangle$ in $\mathcal{I}(v+1)$. It will be assumed that this is the only transition possible from any state in $\mathcal{I}(v)$ to any state in the set $\mathcal{I}(v+1)$. The 6×6 matrix $\mathbf{a}(v, v+1)$ will thus consist of all zeros, except for the density $a_{34}(v, v+1)$. This concludes the discussion, pertaining to the specification of the matrix $\mathbf{a}(x, t)$ of one-step transition densities for the process.

Before algorithms can be developed for calculating current state probabilities, the square matrix $\mathbf{m}(x, t)$ on the right in (6.2.14), corresponding to transient states, will need to be described explicitly in terms of the notation for the stochastic process under discussion. For every pair of transient states $\langle v, E_i \rangle$ and $\langle \xi, E_j \rangle$, let $m_{ij}(v, \xi; x, t)$ be the corresponding element in the matrix $\mathbf{m}(x, t)$ for the process under consideration. Then, for every $x \geq 0$, $m_{ij}(v, \xi; x, 0) = \delta_{ij}\delta_{v\xi}$; and, for $t \geq 1$, the recursion relation defined in Problem 1 (b) of this chapter takes the form

$$m_{ij}(v, \xi; x, t) = \sum_{k} \sum_{\lambda} \sum_{s=0}^{t-1} m_{ik}(v, \lambda; x, s)a_{kj}(\lambda, \xi; x+s, t-s), \qquad (6.3.5)$$

where i, j, and k run over the integers $2, \ldots, 6$ and λ runs over the nonnegative integers $0, 1, 2, \ldots$.

When a woman marries at age x, transient state $\langle 0, E_2 \rangle$ is entered. It is, therefore, of special interest to study the evolution of the process when the initial state is $\langle 0, E_2 \rangle$. If $\langle 0, E_2 \rangle$ is the initial state, then, from the structure of the matrices in (6.3.3) and (6.3.4), it follows that if $j=2$, Eq. (6.3.5) reduces to

$$m_{22}(0, \xi; x, t) = \sum_{j=5}^{6} \sum_{s=0}^{t-1} m_{2j}(0, \xi; x, s)a_{j2}(\xi, \xi; x+s, t-s). \qquad (6.3.6)$$

Similarly, if $j=3$, then Eq. (6.3.5) becomes

$$m_{23}(0, \xi; x, t) = \sum_{s=0}^{t-1} m_{22}(0, \xi; x, s)a_{23}(\xi, \xi; x+s, t-s). \qquad (6.3.7)$$

Replace ξ by $\xi+1$ in (6.3.5). Then, if $j=4$, Eq. (6.3.5) reduces to

$$m_{24}(0, \xi+1; x, t) = \sum_{s=0}^{t-1} m_{23}(0, \xi; x, s)a_{34}(\xi, \xi+1; x+s, t-s). \qquad (6.3.8)$$

Finally, if $j=5$ or 6, then Eq. (6.3.5) becomes

$$m_{2j}(0, \xi; x, t) = \sum_{s=0}^{t-1} m_{23}(0, \xi; x, s)a_{3j}(\xi, \xi; x+s, t-s). \qquad (6.3.9)$$

Equations (6.3.6) through (6.3.9) are clearly a recursive system in $t \geq 1$ for every pair of nonnegative integers x and ξ. For if the elements in the matrix $\mathbf{a}(x, t)$ are given in advance, then the initial values $m_{2j}(0, \xi; x, 0) = \delta_{2j}\delta_{0\xi}$, $j = 5, 6$, may be used to determine $m_{22}(0, \xi; x, 1)$ from Eq. (6.3.6). From the values of $m_{22}(0, \xi; x, s)$ for $s=0, 1$, the values of $m_{23}(0, \xi; x, t)$ may be determined from Eq. (6.3.7) for $t=0, 1$, and 2. Then, the values of $m_{23}(0, \xi; x, s)$ for $s=0, 1$, may be used to determine the values of $m_{24}(0, \xi+1; x, t)$ and $m_{2j}(0, \xi; x, s)$, $j=5, 6$, for $t=0, 1, 2$. After the values of $m_{2j}(0, \xi; x, s)$ have been calculated for $j=5, 6$ and $s=0, 1$, the value of $m_{22}(0, \xi; x, 2)$ may be calculated from (6.3.6). The process just described can, obviously, be continued indefinitely.

Having set down an algorithm for calculating the elements of the matrix $\mathbf{m}(x, t)$ for the case the state $\langle 0, E_2 \rangle$ is entered at $t=0$ when a woman is age x, the stage has been set for the derivation of algorithms for calculating the current state probabilities by specializing formulas (6.2.11), (6.2.12), and (6.2.14) to the process under consideration. As a first step, define the functions $A_{ij}(v, \xi; x, t)$ and $A_i(v; x, t)$ by

$$A_{ij}(v, \xi; x, t) = \sum_{s=0}^{t} a_{ij}(v, \xi; x, t) \tag{6.3.10}$$

and

$$A_i(v; x, t) = \sum_{j} \sum_{\xi} A_{ij}(v, \xi; x, t). \tag{6.3.11}$$

Equations (6.3.10) and (6.3.11) are merely analogues of Eqs. (6.2.7) and (6.2.8) for the process under consideration.

Given that state E_2 is entered at $t=0$ when a woman marries at age x, let $P_{2j}(0, \xi; x, t)$ be the conditional probability of being in state $\langle \xi, E_j \rangle$ at time $t>0$. Then, if $\langle \xi, E_j \rangle$ is the absorbing state $\langle v, E_1 \rangle$, it follows that

$$P_{21}(0, \xi; x, t) = \sum_{s=0}^{t} m_{22}(0, \xi; x, s) A_{21}(\xi, \xi; x+s, t-s)$$

$$+ \sum_{k=4}^{6} \sum_{s=0}^{t} m_{2k}(0, \xi; x, s) A_{k1}(\xi, \xi; x+s, t-s), \tag{6.3.12}$$

for $t \geq 0$. Observe that Eq. (6.3.12) has been derived by adapting Eqs. (6.2.12) and (6.2.14) to the stochastic process under consideration, taking into account structure of the matrices in (6.3.3) and (6.3.4). Similarly, if $\langle v, E_j \rangle$ is a transient state so that $2 \leq j \leq 6$, then by adapting Eqs. (6.2.11) and (6.2.14) to the present notation, it follows that

$$P_{2j}(0, \xi; x, t) = \sum_{s=0}^{t} m_{2j}(0, \xi; x, s) D_{jj}(\xi; x+s, t-s) \tag{6.3.13}$$

for $t \geq 0$, where $D_{jj}(\xi; x, t) = 1 - A_j(\xi; x, t)$.

The probability of being in one of the states $\langle \xi, E_j \rangle$, $1 \leq j \leq 6$, at time $t>0$ may not be of particular interest. Rather than focusing on parity, in some cases, it may be of more interest to know the probability of being in some physiological state E_j at time $t>0$. For example, managers of family planning programs would be interested in knowing the fraction of women in a cohort in the fecundable state E_2, who are at risk of pregnancy. Given that state $\langle 0, E_2 \rangle$ is entered at $t=0$ when a woman marries at age x, let $P_{2j}(0; x, t)$ be the probability she is in physiological state E_j at time $t>0$. Then,

$$P_{2j}(0; x, t) = \sum_{\xi=0}^{\infty} P_{2j}(0, \xi; x, t) \tag{6.3.14}$$

for $1 \leq j \leq 6$, $x \geq 0$, and $t \geq 0$.

Even if there are only finitely many non-zero terms on the right in (6.3.14), it can be seen from Eqs. (6.3.12) and (6.3.13), that Eqs. (6.3.6) through (6.3.9) would have to be iterated in t for many values of ξ before the function in (6.3.14) could be calculated. Depending on the assumptions in force, regarding the maximum parity women may attain, this iterative procedure would have to be carried out for as many as fifteen or twenty values of ξ. As expected, the number of iterations required can greatly be reduced, if it is assumed that the process is homogeneous after some parity $\rho \geq 1$. Mathematically, the homogeniety assumption is expressed by the conditions that $\mathbf{a}(\xi, \xi) = \mathbf{a}(\rho, \rho)$ and $\mathbf{a}(\xi, \xi + 1) = \mathbf{a}(\rho, \rho + 1)$ for all $\xi \geq \rho \geq 1$.

For every pair of states $\langle v, E_i \rangle$ and $\langle \xi, E_j \rangle$, let

$$m_{ij}(v, \rho^+; x, t) = \sum_{\xi = \rho}^{\infty} m_{ij}(v, \xi; x, t) \tag{6.3.15}$$

at the point $\langle x, t \rangle$. Then, from Eqs. (6.3.6) and (6.3.9), it follows that $m_{2j}(0, \rho^+; x, t)$, $2 \leq j \leq 6$, satisfy the recursion relations

$$m_{22}(0, \rho^+; x, t) = \sum_{j=5}^{6} \sum_{s=0}^{t-1} m_{2j}(0, \rho^+; x, s) a_{j2}(\rho, \rho; x+s, t-s), \tag{6.3.16}$$

$$m_{23}(0, \rho^+; x, t) = \sum_{s=0}^{t-1} m_{23}(0, \rho^+; x, s) a_{23}(\rho, \rho; x+s, t-s), \tag{6.3.17}$$

$$m_{24}(0, (\rho+1)^+; x, t) = \sum_{s=0}^{t-1} m_{23}(0, \rho^+; x, s) a_{34}(\rho, \rho+1; x+s, t-s), \tag{6.3.18}$$

and

$$m_{2j}(0, \rho^+; x, t) = \sum_{s=0}^{t-1} m_{23}(0, \rho^+; x, s) a_{3j}(\rho, \rho; x+s, t-s), \tag{6.3.19}$$

for $j = 5, 6$ and $t \geq 1$. The initial values $m_{ij}(v, \rho^+; x, 0)$ are completely determined from (6.3.15) and the initial values $m_{ij}(v, \xi; x, 0) = \delta_{ij} \delta_{v\xi}$.

Under the assumption the process is homogeneous after some parity $\rho \geq 1$, the current state probabilities in (6.3.14) may be represented in the form

$$P_{2j}(0; x, t) = \sum_{\xi=0}^{\rho-1} P_{2j}(0, \xi; x, t) + P_{2j}(0, \rho^+; x, t), \tag{6.3.20}$$

for $1 \leq j \leq 6$ and points $\langle x, t \rangle$. For the physiologically sterile state E_1, the function on the extreme right in (6.3.20) takes the form

$$P_{21}(0, \rho^+; x, t) = \sum_{s=0}^{t} m_{22}(0, \rho^+; x, s) A_{21}(\rho, \rho; x+s, t-s)$$

$$+ \sum_{k=4}^{6} \sum_{s=0}^{t} m_{2k}(0, \rho^+; x, s) A_{k1}(\rho, \rho; x+s, t-s). \tag{6.3.21}$$

For the other physiological states E_j, $2 \leq j \leq 6$, the function has the form

$$P_{2j}(0, \rho^+; x, t) = \sum_{s=0}^{t} m_{2j}(0, \rho^+; x, s) D_{jj}(\rho; x+s, t-s). \tag{6.3.22}$$

Not only are the functions $m_{ij}(v, \xi; x, t)$, determined by Eq. (6.3.5) useful in calculating the current state probabilities of the process but they also have interesting interpretations in their own right. For every pair of transient states $\langle v, E_i \rangle$ and $\langle \xi, E_j \rangle$, let

$$M_{ij}(v, \xi; x, t) = \sum_{s=0}^{t} m_{ij}(v, \xi; x, s) \qquad (6.3.23)$$

at all points $\langle x, t \rangle$. Then, it is shown in Problem 2 that the function in (6.3.23) has the following interpretation. Suppose transient state $\langle v, E_i \rangle$ is entered by an individual of age x at $t=0$. Then, $M_{ij}(v, \xi; x, t)$ is the mean number of visits to transient state $\langle \xi, E_j \rangle$ during the time interval $(0, t]$, $t > 0$. The matrix $\mathbf{M}(x, t) = (M_{ij}(v, \xi; x, t))$ of functions is sometimes referred to as the renewal matrix; thus, it makes sense to refer to $\mathbf{m}(x, t)$ as the renewal density matrix. It is also of interest to note that the matrix $\mathbf{M}(x, t)$ is an analogue of the fundamental matrix for finite absorbing Markov chains as defined by Kemeny and Snell (1976).

Because of the structure of the matrix in (6.3.3), state $\langle \xi, E_4 \rangle$ can be visited once, and only once, during a maternity history. Visits to states in the sequence $\langle \xi, E_4 \rangle$, $\xi \geq 1$, are thus a discretized version of the age-dependent potential birth process discussed in Sect. 4.8. In the notation of that section, let $H(n, x, t)$ be the conditional distribution function of the waiting times to the $(n+1)$-th live birth, given that parity n was attained at age x. Then,

$$M_{44}(n, n+1; x, t) = H(n, x, t). \qquad (6.3.24)$$

Of fundamental importance in an age-dependent potential birth process is $H^{(n)}(x, t)$, the conditional distribution function of waiting times to the n-th live birth, $n \geq 1$, given marriage at age x. According to (4.8.11), this distribution function can be computed iteratively from the functions $H(n, x, t)$ and their densities $h(n, x, t)$, a task requiring many computations. Within the framework under consideration marriage at age x is equivalent to the event state $\langle 0, E_2 \rangle$ is entered at $t=0$; and, given this event $H^{(n)}(x, t)$ is the conditional mean number of visits to state $\langle n, E_4 \rangle$ during the time interval following marriage at age x. Therefore,

$$M_{24}(0, n; x, t) = H^{(n)}(x, t), \qquad (6.3.25)$$

for all $n \geq 1$ and pairs of nonnegative integers $\langle x, t \rangle$.

Hence, once values of the renewal density $m_{24}(0, n; x, t)$ have been calculated, many functions of interest in utilizing the theory of an age-dependent potential birth process are given as a bonus. For, in the notation of Sect. 4.8, (6.3.25) implies that the p.d.f. of the number of live births at t time units following marriage at age x is

$$q_n(x, t) = P[K(t) = n \mid T_0 = x]$$
$$= M_{24}(0, n; x, t) - M_{24}(0, n+1; x, t), \qquad (6.3.26)$$

for $n \geq 0$, provided that $M_{24}(0, 0; x, t) = 1$ for all points $\langle x, t \rangle$. Furthermore, the mean of the distribution in (6.3.26) is

$$v(x, t) = \sum_{n=1}^{\infty} M_{24}(0, n; x, t), \qquad (6.3.27)$$

see (4.8.13). The equation

$$b(x, t) = \sum_{n=1}^{\infty} m_{24}(0, n; x, t),\tag{6.3.28}$$

also connects renewal densities with the gross maternity function $b(x, t)$ as defined in Sect. 4.8. When the process is homogeneous after some parity $\rho \geq 1$, then (6.3.28) takes the form

$$b(x, t) = \sum_{n=0}^{\rho-1} m_{24}(0, n; s, t) + m_{24}(0, \rho^{+}; x, t).\tag{6.3.29}$$

The gross maternity function in (6.3.29) may thus be computed by $\rho - 1$ passes through the algorithm defined by Eqs. (6.3.6) to (6.3.9) and one pass through Eqs. (6.3.16) to (6.3.19). If, however, it is desired to compute the p.d.f. in (6.3.26), a pass through Eqs. (6.3.6) to (6.3.9) is required for each term $q_n(x, t)$, $n \geq 0$. The distribution of age at marriage in a cohort is taken into account just as before. For example, a discretized version of formula (4.8.17) for calculating $b(t)$, the cohort birth rate function, continues to apply.

Two other conditional means are of interest in summarizing computer output suggested by the formulation under consideration. Consider again those women who enter state $\langle 0, E_2 \rangle$ at $t=0$ when they marry at age x. Given this event, let $M_{2j}(0; x, t)$ be the conditional mean number of visits to physiological state E_j, $2 \leq j \leq 6$, during the time interval $(0, t]$, $t > 0$. Similarly, let $S_{2j}(0; x, t)$ be the conditional mean length of time spent in state E_j, $1 \leq j \leq 6$, during the time interval $(0, t]$. When the process is homogeneous after parity $\rho \geq 1$, then

$$M_{2j}(0; x, t) = \sum_{v=0}^{\rho-1} M_{2j}(0, v; x, t) + M_{2j}(0, \rho^{+}; x, t),\tag{6.3.30}$$

for $2 \leq j \leq 6$ and all points $\langle x, t \rangle$. Moreover, if $P_{2j}(0; x, t)$ is the current state probability defined in (6.3.14), then

$$S_{2j}(0; x, t) = \sum_{s=1}^{t} P_{2j}(0; x, s)\tag{6.3.31}$$

for all $1 \leq j \leq 6$ and points $\langle x, t \rangle$. At every point $\langle x, t \rangle$,

$$\sum_{j=1}^{6} P_{2j}(0; x, t) = 1;\tag{6.3.32}$$

therefore,

$$\sum_{j=1}^{6} S_{2j}(0; x, t) = t\tag{6.3.33}$$

for all points $\langle x, t \rangle$. Equation (6.3.33) provides a useful check on the numerical accuracy of the calculations. Further details on the derivation of Eqs. (6.3.30) and (6.3.31) are given in Problem 2.

6.4 On Choosing Computer Input for an Age-Dependent Model of Maternity Histories

As suggested in Sect. 6.3, ease of computer implementation is an important factor to take into account when comparing two models of maternity histories based on alternative formulations. But, before the question of ease of computer implementation can be fruitfully discussed, it will be necessary to consider what computer inputs are needed to implement the model of maternity histories developed in Sect. 6.3. A comparison of the models formulated in Sects. 5.3 and 6.3 suggests that much of the discussion of Sect. 5.4, pertaining to component functions for models of human reproduction will continue to apply, provided that these components can be adapted to the age-dependent case.

In addition to the age-dependent property, a fundamental property differentiating the model discussed in Sect. 6.3 from that in Sect. 5.3 is that entrance into the permanently sterile state E_1 is treated as a competing risk, whenever a woman is in the fecundable state E_2 or a temporary sterile state following a live birth (E_4), an induced abortion (E_5), or some other type of pregnancy outcome (E_6); see rows two, four, five and six of the matrix in (6.3.4). Berman (1963) seems to have been the first to clearly point out a relationship between a model of competing risks with independent latent sojourn times and the one-step transition functions of a semi-Markov process in continuous time. A method of constructing one-step transition functions for age-dependent semi-Markov processes, based in part on the observations of Berman, has already been discussed in Sect. 5.8, in connection with a model of anovulatory sterile periods following live births.

Provided that explicit forms of the latent risk functions were known, the method of Sect. 5.8 could, in principle, be used to calculate the one-step transition densities in the submatrices $\mathbf{a}(v, v)$ and $\mathbf{a}(v, v+1)$, $v \geq 0$, in (6.3.3). To carry out these calculations an adaptation of formula (5.9.9) to the age-dependent case could be used to discretize continuous time formulas. But, finding explicit forms of latent risk functions, validating them, and estimating parameters when required are research problems in their own right, demanding an extensive effort in data analysis and computer programming. Faced with such problems when desiring to implement a model on a computer, an investigator may find it expedient to use a combination of parametric forms from the archives and nonparametric forms based on the analysis of recent data. Discussed in the remainder of this section is such a combination of forms actually used in generating numerical input to the model formulated in Sect. 6.3.

Because theories of competing risks with independent latent sojourn times are easiest to deal with in continuous time, let $A_{ij}^{(c)}(v, \xi; x, t)$ be a continuous time version of the function defined in (6.3.10). For the case $i=2$ and $j=1$ or 3, let $F_{21}(v; x, t)$ and $F_{23}(v; x, t)$ be the continuous latent distribution functions determining the function $A_{2j}^{(c)}(v, v; x, t)$, $j=1, 3$. Then, by applying formula (3.2.5) to the theory of competing risks with independent latent sojourn times, it follows that

$$A_{21}^{(c)}(v, v; x, t) = \int_0^t (1 - F_{23}(v; x, s)) f_{21}(v; x, s) ds$$

and   (6.4.1)

$$A_{23}^{(c)}(v, v; x, t) = \int_0^t (1 - F_{21}(v; x, s)) f_{23}(v; x, s) ds,$$

where f_{21} and f_{23} are the densities of the distribution functions F_{21} and F_{23}. Problem 3 may be consulted for further details on the derivation of the formulas in (6.4.1). Similar formulas corresponding to the functions in rows four, five, and six of the matrix in (6.3.4) could also be written down.

According to the formulas in (6.4.1), a knowledge of the latent distribution functions $F_{21}(v; x, t)$ and $F_{23}(v; x, t)$ for every $v \geq 0$ and age $x \geq 0$ is required to determine the functions on the left in (6.4.1). In practice, such knowledge is at best difficult, or even impossible, to acquire so that it is natural to look for simplifying assumptions. One of these assumptions is to suppose that the latent distribution function $F_{21}(v; x, t)$, representing risks of permanent physiological sterility, does not depend on v, a woman's parity. Mathematically, this assumption is equivalent to stating that for each age x, there is a latent distribution function $F_{21}(x, t)$ such that

$$F_{21}(v; x, t) = F_{21}(x, t),$$ (6.4.2)

for all $v \geq 0$ and points $\langle x, t \rangle$.

Implicit in Eq. (6.4.2) is the assumption that childbearing *per se* does not affect the onset of permanent sterility. This assumption seems to be plausible for most women, although there are situations in which a pregnancy may prevent further pregnancies. An example is an ectopic pregnancy, so named because the embryo becomes implanted outside the uterus, usually in a Fallopian tube. Should both Follopian tubes be rendered unfunctional as a result of ectopic pregnancies, further pregnancies would be precluded.

Even if assumption (6.4.2) is in force, the distribution function on the right must be specified for every $x \geq 0$. A simple approach to specifying this latent distribution function is to assume that there is a latent distribution function $F_{21}(t)$, governing the onset of permanent sterility as a function t of age in a cohort of women. The function $F_{21}(x, t)$ is then determined by conditioning so that

$$F_{21}(x, t) = \frac{F_{21}(x + t) - F_{21}(x)}{F_{21}(x)}.$$ (6.4.3)

For every $x \geq 0$, $F_{21}(x, t)$ is clearly a distribution function in $t \geq 0$. A similar argument was used in Sect. 5.8, see formula (5.8.10).

Having determined the function $F_{21}(x, t)$ according to formula (6.4.3), it remains to choose an explicit form of the function $F_{21}(t)$. As mentioned in Sect. 5.4, in connection with the Weibull model of sterility displayed in (5.4.12), considerable effort by several investigators has been devoted to research on this function. Used in the computer runs based on the model developed in

Sect. 6.3 was the parametric model proposed by Pittenger (1973), a model that may be described as follows.

Let $t_0 < t_1 < t_2 < \ldots$ be the ages at which the values of the function $F_1(t)$ are to be calculated; and, let q_v be the conditional probability that a fertile woman of age t_{v-1}, $v \geq 1$, becomes permanently sterile during the age interval $(t_{v-1}, t_v]$. Pittenger postulated that this conditional probability q_v had the exponential form $q_v = \alpha \beta^{x_v}$, where $x_v = t_v - t_0$ for $v \geq 1$. By trial and error, the values of the parameters, $\alpha = 0.0002$ and $\beta = 1.251242$, were judged by Pittenger to yield good results. From the statistical point of view, this method of choosing parameter values leaves much to be desired. Simple parametric formulas taken from the archives do, however, have expository advantages when describing computer input.

In carrying out the calculations, the value of t_0 was chosen as 12.5 years and the other x-values were in increments of $1/12$, corresponding to years composed of twelve months of equal length. The values of the function $F_{21}(t)$ were computed by an obvious modification of the algebra in the life table algorithm outlined in Sect. 2.4. Sample values of the latent distribution function at selected values of x and t are presented in Table 6.4.1.

The other component needed to specify the one-step transition functions in (6.4.1) is $F_{23}(v, x, t)$, the latent distribution function governing waiting times to pregnancy following entrance into the fecundable state E_2 by a woman of parity v and age x. According to the formulas in (6.4.1), this function is to be specified by age and parity. A possible approach to making this specification would be that of using continuous time analogues of the parametric distributions developed in Chap. 1 on fecundability. Rarely, however, are estimates of fecundability by age and parity of women available in the archives. Even if such estimates were available, they would probably be based on data from historical, noncontracepting populations and there would be some question as to whether such estimates would be applicable to recent or current populations.

Ideally, it would be nice to have estimates of natural fecundability based on either recent or current maternity histories in which contraception was used to space and limit births. For in attempting to measure the demographic impact of family planning programs, it is sometimes useful to have estimates of

Table 6.4.1. Selected values of the latent distribution function $F_{21}(x, t)$ for permanent sterility based in Pittenger's exponential model

Values of t in years	Age – values of x					
	24	29	34	39	43	46
1	0.004	0.011	0.035	0.106	0.261	0.510
2	0.008	0.025	0.077	0.225	0.502	0.823
3	0.014	0.043	0.127	0.354	0.705	0.964
4	0.021	0.064	0.186	0.489	0.856	1.000
5	0.030	0.090	0.255	0.622	0.948	1.000
6	0.041	0.121	0.334	0.745	0.990	1.000

fertility in a population, if contraception were not being used. At first sight, it might appear impossible to obtain estimates of natural fecundability in contracepting populations; but, many maternity histories are punctuated by null segments, i.e., episodes in which no contraception was being used. Null segments may end in various ways, including end of survey, pregnancy, the acceptance of a method of contraception, or, perhaps, a contraceptive sterilization if no more children were desired. A question arises, therefore, as to whether null segments may be used to estimate natural fecundability.

The mutually exclusive ways in which a null segment might end may be viewed as competing risks. It is thus clear that multiple decrement life table procedures may be used to describe the timing and "causes" of termination for null segments, just as if mortality and causes of death were being considered. From the statistical point of view, the columns of a multiple decrement life table are estimates of the functions $G_1(t), \ldots, G_m(t)$, $m \geq 2$, described in Sect. 3.2. Under the assumption of independent latent sojourn times (lifespan), the i-th latent distribution function $F_i(t)$ is determined by the formula

$$F_i(t) = 1 - \exp\left[-\int_0^t \frac{g_i(s)\,ds}{1-G(s)}\right], \tag{6.4.4}$$

where $1 \leq i \leq m$, $t \geq 0$, $G(t) = G_1(t) + \ldots + G_m(t)$, and $g_i(t)$ the density of $G_i(t)$. Problem 3 may be consulted for technical details on the derivation of (6.4.4). The basic idea underlying the estimation of the latent distribution function $F_{23}(v; x, t)$ was to classify null segments by age and parity, estimate the G-function by multiple decrement life table methods, and then exploit formula (6.4.4). A more detailed discussion of results obtained, using the estimation procedure just described, will be provided in the next section.

Continuous time formulas of exactly the same form as that in (6.4.1) may also be written down for transitions out of the states E_4, E_5, and E_6, see the last three rows of the matrix in (6.3.4). To determine these one-step transition functions, six latent distribution functions, $F_{i1}(v; x, t)$ and $F_{i2}(v; x, t)$, $i = 4, 5$, and 6, must be specified by age and parity. All assumptions regarding transitions into the permanently sterile state E_1 outlined above will continue to be in force; each latent distribution function $F_{i1}(v; x, t)$ is thus a copy of the function $F_{21}(x, t)$ determined by (6.4.3). It remains to specify the three latent distribution functions $F_{i2}(v; x, t)$. $i = 4, 5$, and 6. All these distribution functions were assumed to be independent of age and parity. The distribution $F_{42}(t)$, governing waiting times to the resumption of ovulation following childbirth, was chosen as the Pascal distribution; consult the discussion in Sect. 5.4 on the mathematical properties of this distribution. Following a suggestion by Barrett (1969), the mean of the Pascal distribution was taken as eleven months; in other words, women resumed ovulation eleven months, on the average, after childbirth. The latent distribution functions $F_{52}(t)$ and $F_{62}(t)$ were determined by the densities listed in Table 5.4.2. For ease of reference, the latent distribution functions under discussion are partially listed in Table 6.4.2. Observe that, according to the Pascal distribution, only thirty percent of the women resume ovulation six months following childbirth.

Table 6.4.2. Selected values of latent distribution functions governing waiting times to the resumption of ovulation following childbirth, induced abortion, or other type of pregnancy outcome

Month	Live birth F_{42}	Induced abortion F_{52}	Other F_{62}
1	0.000	0.250	0.000
2	0.033	0.750	0.100
3	0.087	1.000	0.350
4	0.154	1.000	0.650
5	0.226	1.000	0.900
6	0.300	1.000	1.000

Given numerical forms of the latent distribution functions just described, which are only partially represented in Tables 6.4.1 and 6.4.2, the one-step transition functions corresponding to the last three rows of the matrix in (6.3.4) may be computed by evaluating integral formulas of the type displayed in (6.4.1). Because of the uncertainty surrounding the latent distribution functions under discussion, no special numerical procedures were used to approximate these integrals. In fact, all integrals were evaluated by their approximating sums as determined by the monthly values of the functions and subintervals of one month in length. Whenever they were needed, density values on the monthly intervals were computed by differencing distribution functions. On the whole, these approximations seem to have worked quite well and were regarded as values of the discretized functions, such as $A_{41}(x, t)$ and $A_{42}(x, t)$, on a lattice of $\langle x, t \rangle$-points. Arrays were calculated for each of the last three rows in the matrix (6.3.4); but, for purposes of illustration only selected values of the functions $A_{41}(x, t)$ and $A_{42}(x, t)$ are presented in Table 6.4.3. By a detailed

Table 6.4.3. Selected values of one-step transition functions governing transitions to physiologically sterile and fecundable states following childbirth

Months	Values of A_{41}			Values of A_{42}		
	Age group			Age group		
	20–24	30–34	40–44	20–24	30–34	40–44
1	0.000	0.003	0.022	0.000	0.000	0.000
2	0.001	0.006	0.043	0.033	0.033	0.032
3	0.001	0.008	0.063	0.087	0.086	0.082
4	0.001	0.011	0.081	0.153	0.152	0.143
5	0.001	0.013	0.098	0.226	0.223	0.207
6	0.002	0.015	0.113	0.300	0.296	0.272
12	0.002	0.023	0.174	0.669	0.656	0.567
24	0.003	0.029	0.208	0.946	0.921	0.746
36	0.003	0.030	0.213	0.990	0.961	0.765
48	0.003	0.030	0.213	0.996	0.966	0.766
60	0.003	0.030	0.213	0.996	0.967	0.766
Limit [a]	0.003	0.030	0.213	0.996	0.967	0.766

[a] Value at age fifty.

study of the numerical forms of the latent distribution functions $F_{41}(t)$ and $F_{42}(x, t)$, it could be shown that the sum $A_{41}(x, t) + A_{42}(x, t)$ should be nearly equal to one at $t = 60$ months for all ages x. That this equality holds approximately can be seen from Table 6.4.3, suggesting that the approximations in question are performing quite well in that they meet mathematical expectations. All values of the one-step transition densities in (6.3.3) were obtained by numerically differencing arrays of the type partially represented in Table 6.4.3.

The last functions needed to specify the matrix of one-step transition densities in (6.3.3) are those governing the type and timing of pregnancy outcomes; see the third rows in the submatrices $\mathbf{a}(v, v)$ and $\mathbf{a}(v, v+1)$ in (6.3.3). Used in specifying these matrices were the assumptions and methods described in Sect. 5.4. More details of the actual numerical values used in the simulations will be presented in the next section.

6.5 Estimates of Fecundability Functions Based on Null Segments and Other Computer Input

Suggested in the previous section was a technique, based on applying formula (6.4.4), for estimating the latent distribution functions $F_{23}(v; x, t)$ in (6.4.1) from null segments in maternity histories. Outlined in this section are applications of this technique to data on maternity histories gathered in two Taiwan surveys that provided segment coding. Each segment was classified by age of wife, parity, and family planning status, i.e., spacer or limiter. Spacers are couples who, according to the wife's report, were practicing contraception in order to delay an eventually desired birth; limiters were practicing contraception to prevent further childbearing.

During 1973 an island wide survey was conducted on Taiwan in an attempt to gain information on the knowledge of, attitudes towards, and the practice of contraception for about 5,600 currently married women of reproductive age. Hereafter, this survey will be referred to as KAP-IV. Because the rate of refusal to respond to questions on maternity histories was very low, the data collected were thought to be representative of married Taiwanese women during the period covered by the survey. To enhance the credibility of the data, segment coding was confined to the last five years before interview or back to marriage, whichever came first. Detection of physiological sterility depended on a question concerning menopause. Freedman et al. (1974) have discussed the quality of the data.

A second source of segment coded data was the Taichung Medical IUD Follow-Up Study. In this longitudinal study, the initial sample consisted of 5,832 women from the city of Taichung, who had an IUD, intra-uterine device, inserted between July 1962 and May 1965. By 1971, six- to nine-year histories had been collected on the Taichung women based on annual contacts with 84% of the initial sample. Among those women lost to follow-up, three-quarters were known to have moved from Taichung. Both clinical and survey methods were used in the study. If, for example, women were known to be wearing an IUD, they were asked to return to the clinic for an examination; otherwise, they were interviewed at their homes on a yearly basis.

At the time of the original IUD insertion, the sample of wives averaged 4.79 births; and, within this sample, 82% accepted their first device for purposes of family limitation. According to Moots (1975), the Taichung acceptors of IUDs were characteristically of high fertility, who had less than average use of contraceptives prior to insertion. Before coming to the family planning clinic, many of the wives had resorted to induced abortion, which suggests they were highly motivated with respect to preventing further births. Additional information on the Taichung experiment has been given by Freedman and Takeshita (1969).

In both the KAP-IV and Taichung data sets, null segments, whose length was measured in months, were grouped into two sets described as full and closed. Within the KAP-IV data set, the full set included all null segments except those over-lapping postpartum amenorrhea; whereas, in the Taichung data set, only null segments immediately following childbirth were excluded from the full set. Within both full sets, a closed subset consisted of those null segments that either ended with a pregnancy or those for which there was a pregnancy in the maternity history subsequent to the initiation of a null segment.

Briefly, the rationale for choosing these two sets of null segments was as follows. It was thought that the full set of null segments would produce a more representative estimate of the fecundability functions for a non-contracepting population. Estimates of fecundability functions based on the closed sets of null segments would, on the other hand, provide some insights into the level of biases entailed by selecting the apparently more fecund women in the sample. Potter et al. (1979) have given a more complete discussion of the biases that might ensue from using the sets of null segments just described.

Within each set of full and closed null segments, multiple decrement life table procedures were carried out to provide estimates of the G-functions on the right in (6.4.4). The actual estimator used in making these calculations was the actuarial one described in Mode et al. (1977). After the G-functions had been estimated, they were substituted on the right in (6.4.4) to produce estimates of the latent fecundability functions according to an age, parity, and a limiter-spacer classification. The age groups used in this classification were: 20–24, 25–29, 30–34, and 35–39; the parity classes were 0, 1, and 2. Presented in Table 6.5.1 are selected values of the estimates of the latent fecundability functions for KAP-IV spacers based on a full set of null segments; the corresponding estimates, based on the closed set of null segments for KAP-IV spacers, are presented in Table 6.5.2. Table 6.5.3 contains selected values of the fecundability functions based on the full and closed sets of null segments for Taichung limiters.

Only some tentative interpretations of the estimates in Tables 6.5.1, 6.5.2, and 6.5.3, regarding how parity and age affect fecundability will be offered. For those estimates based on samples of fifty or more null segments of KAP-IV

Table 6.5.1. Estimates of selected values of latent fecundability functions for KAP-IV spacers – full set of null segments

Months	Age groups			
	20–24	25–29	30–34	35–39
	Parity zero			
6	0.681	0.644	0.273	0.000
12	0.829	0.756	0.580	0.333
24	0.904	0.871	0.842	0.333
72	0.938	0.925	1.000	0.333
Sample size	620	152	11	3
	Parity one			
6	0.578	0.527	0.430	0.200
12	0.790	0.723	0.587	0.200
24	0.913	0.837	0.670	0.600
72	0.938	0.896	0.759	0.600
Sample size	987	309	46	7
	Parity two			
6	0.498	0.510	0.535	0.516
12	0.766	0.735	0.696	0.516
24	0.918	0.900	0.782	0.677
72	0.985	0.938	0.878	0.839
Sample size	730	531	75	9

Table 6.5.2. Estimates of selected values of latent fecundability functions for KAP-IV spacers – closed set of null segments

Months	Age groups			
	20–24	25–29	30–34	35–39
	Parity zero			
6	0.749	0.746	–[a]	–
12	0.895	0.854	–	–
24	0.968	0.961	–	–
48	0.998	0.990	–	–
Sample size	447	111		
	Parity one			
6	0.669	0.661	–	–
12	0.863	0.848	–	–
24	0.965	0.953	–	–
48	0.998	0.994	–	–
Sample size	659	194		
	Parity two			
6	0.566	0.622	0.646	–
12	0.824	0.838	0.814	–
24	0.960	0.978	0.907	–
48	1.000	0.966	1.000	–
Sample size	498	282	56	

[a] Fewer than fifty segments.

Table 6.5.3. Estimates of selected values of latent fecundability functions for Taichung limiters

Months	Age groups			
	20–29	30–34	35–39	40–49
	Full set[a]			
6	0.525	0.422	0.338	0.177
12	0.710	0.605	0.509	0.231
24	0.844	0.752	0.678	0.332
72	0.960	0.915	0.777	0.509
Sample size	842	886	645	281
	Closed set[b]			
6	0.660	0.590	0.554	0.510
12	0.855	0.799	0.784	0.636
24	0.957	0.927	0.945	0.830
72	0.984	0.991	1.000	0.985
Sample size	603	531	326	85

[a] Based on full set of null segments.
[b] Based on closed set of null segments.

women presented in Table 6.5.1, the highest values of the fecundability function at six months was for women of parity zero. Most parity zero conceptions, for the period covered by the data, were the first following marriage, which suggests that waiting times to conception following marriage were shorter than those for subsequent entrances into the fecundable state during maternity histories. Within each parity class in Table 6.5.1, differences among the estimated values of the latent fecundability functions for the age groups 20–24 and 25–29 do not appear to be consistent or significant. The lower estimates of the values of the fecundability function at 12, 24, and 72 months for women of parity two in the age group 30–34 suggest, however, that the fecundability of women in this age group is indeed lower than that of the younger age groups.

As one would expect, the estimates in Table 6.5.2, based on the closed sets of null segments in the KAP-IV data, are almost always higher than those in Table 6.5.1, indicating that women contributing the closed set of null segments exhibited higher fertility, on the average, than those in the sample as a whole. Similar differences between the estimates in the full and closed sets of null segments for Taichung limiters are also manifest in Table 6.5.3. Differences in fecundability among the age groups in Table 6.5.3 also appear to be significant, with fecundability decreasing as age increases. It would be nice to have tests of statistical significance to aid in interpreting the estimates in Tables 6.5.1, 6.5.2, and 6.5.3; but, decisions regarding the construction and computer implementation of such tests would not be straightforward.

Given the computer arrays for the values of the latent fecundability functions partially represented in Tables 6.5.1, 6.5.2, and 6.5.3, the one-step transition functions A_{21} and A_{23}, determined by the integrals in (6.4.1) could be computed according to an age, parity, and limiter-spacer classification, using the approximation procedures described in Sect. 4.3. When implementing the model developed in Sect. 6.3 on a computer, each one of these functions would be stored in the computer as numerical arrays on a monthly time scale and a prescribed age grouping. Arrays of the type under consideration have already been partially represented in Table 6.4.3. It would be of interest to exhibit some of the numerical arrays in question. But, in order to keep the exposition manageable and conserve space, none of these arrays will even be partially represented here.

The last one-step transition functions needed to implement the model are those governing transitions out of the pregnant state E_3; inspect row three in the 6×6 matrices $\mathbf{a}(v, v)$ and $\mathbf{a}(v, v+1)$, $v \geq 0$, in (6.3.3) and (6.3.4). All of these transition densities were assumed to be independent of parity and were estimated, according to an age, parity, and limiter-spacer classification, by using a multiple decrement life table estimator on data from the KAP-IV and Taichung surveys. No attempt will be made to list the numerical arrays generated by using this procedure. Rather, attention will be confined to estimates of the pregnancy outcome probabilities, the π-probabilities in (5.4.1). Listed in Table 6.5.4 are the estimates of these π-probabilities.

As can be seen from Table 6.5.4, when compared to KAP-IV spacers, Taichung limiters exhibited relatively high rates of induced abortion as well as a more complete reporting of their spontaneous pregnancy losses. In fact, the

Table 6.5.4. Estimates of pregnancy outcome probabilities based on the KAP-IV and Taichung surveys

Age	Sample size	Live birth	Induced abortion	Other
KAP-IV parity zero spacers				
20–24	586	0.936	0.004	0.060
25–29	148	0.903	0.000	0.097
30–34	14	0.857	0.000	0.143
35–39	3	1.000	0.000	0.000
KAP-IV parity one spacers				
20–24	742	0.924	0.022	0.054
25–29	262	0.926	0.019	0.055
30–34	38	0.892	0.027	0.081
35–39	3	1.000	0.000	0.000
KAP-IV parity two spacers				
20–24	442	0.909	0.032	0.059
25–29	417	0.918	0.048	0.034
30–34	74	0.905	0.027	0.068
35–39	10	0.700	0.100	0.200
Taichung limiters all parities				
20–29	1,068	0.468	0.417	0.114
30–34	1,272	0.394	0.509	0.097
35–39	801	0.312	0.594	0.094

incidence of live births for Taichung limiters is below one-half for all three age groups and is actually exceeded by the reported incidence of induced abortions in the age groups 30–34 and 35–39. These high rates of induced abortions, occurring mostly in the first trimester of pregnancy, indicate that an appreciable fraction of potential spontaneous pregnancy terminations are anticipated by abortion procedures, see Potter et al. (1975). Thus, the reported rates of pregnancy losses, varying from 0.09 to 0.11, are high enough to suggest relatively complete reporting, although the possibility cannot be excluded that some induced abortions are being disguised as spontaneous ones. For, induced abortion was illegal throughout the period of the Taichung study, but the relevant statute was never strictly enforced. Almost without exception, the estimates of other pregnancy outcomes for KAP-IV spacers are well below the 13–23 % of spontaneous fetal wastage reported in specialized surveys, see Leridon (1977), Table 4.5. It appears, therefore, that women interviewed in the KAP-IV survey under-reported spontaneous fetal wastage.

6.6 Numerical Specifications of Four Computer Runs with Inputs Based on Survey Data

In Sects. 6.4 and 6.5, the discussion was centered around the definition and estimation of some latent distribution functions needed to implement the age-dependent semi-Markovian model of maternity histories developed in Sect. 6.3. Attention was also given to procedures for calculating one-step transition densities from the latent distribution functions. As can be seen from the discussion surrounding the matrices in (6.3.3) and (6.3.4), the densities in the 6 $\times 6$ submatrices $\mathbf{a}(v, v; x, t) = (a_{ij}(v, v; x, t))$ and $\mathbf{a}(v, v+1; x, t) = (a_{ij}(v, v+1; x, t))$ are to be calculated according to an age and parity classification. It is clear that if values of these densities were calculated for several parities $v \geq 0$ at all integral ages x throughout the childbearing years, very large arrays would have to be stored and manipulated in the computer. Few computer systems do, in fact, have capabilities for handling such large arrays, making it necessary to introduce age-dependence through a coarser grouping of the childbearing ages. Described in this section are the actual procedures used to generate values of the one-step transition densities for four computer runs based on the latent distribution functions considered in Sects. 6.4 and 6.5.

Six age groups were used in calculating values of the one-step transition densities for the computer runs described below. These age groups were: [20, 25), [25, 30), [30, 35), [40, 45), and [45, 50). For every parity v, the densities were assumed to be constant in x on each of these age intervals. For example, the combinations of latent distribution functions used to calculate the one-step transition densities $a_{21}(v; x, t)$ and $a_{23}(v; x, t)$ for parity $v = 0$ on each of the age intervals are displayed in Table 6.6.1. The function $F_{21}(x, t)$ in the panel on the right are those for the indicated values of x; namely, $x = 24, 29, 34, 39, 43$, and 46. Similarly, the functions $F_{23}(0; x, t)$ on the right in Table 6.6.1 for $x = 20, 25,$

30, and 35 are those for the age groups in either Table 6.5.1 or Table 6.5.2 for KAP-IV spacers, depending on whether estimates of fecundability function based on the full (open) or closed set of null segments were used as numerical input to a computer run. Within each age and parity class, the variable t was on a monthly time scale. Observe from Table 6.6.1 that the latent distribution function $F_{23}(0; 35, t)$ was used for all ages over 40. Similar conventions were followed in specifying numerical input for other parities, whenever the data were insufficient to estimate fecundability functions for ages over forty.

Some discussion of the rationale underlying the choice of six age groups and the particular values of the variable x in the latent distribution function $F_{21}(x, t)$ seems appropriate. Intuitively, it would seem natural to choose these values of x as the midpoints of the age intervals on the left in Table 6.6.1. This choice of x-values would probably tend to underestimate the effects of permanent sterility, since Pittenger's curve is believed to represent a minimum level of sterility. Moving x to the righthand side of the age intervals from 20 to 40 should tend to compensate for this underestimation. The values of x for ages over 40 were chosen in an attempt to avoid overcompensation for sterility in the last years of the childbearing ages. The reason for choosing six age intervals was to ensure that the model could be implemented on available computers.

The procedure used to calculate the one-step transition densities $a_{21}(v; x, t)$ and $a_{23}(v; x, t)$ for other parities $v \geq 1$ was the same as that described in Table 6.6.1, except that different estimates of the latent fecundability functions $F_{23}(v; x, t)$ were substituted in the panel on the right. It should also be mentioned that the latent distribution functions $F_{21}(x, t)$, $x = 24, 29, \ldots, 46$, in Table 6.6.1 were combined with those in Table 6.4.2 to compute the transition densities $a_{i1}(x, t)$ and $a_{i2}(x, t)$, $4 \leq i \leq 6$, in the last three rows of the matrix in (6.3.4) for the six age groups displayed in Table 6.6.1. As indicated earlier, some values of the transition functions A_{41} and A_{42} have been displayed in Table 6.4.3. Displays similar to Table 6.4.3 could be exhibited for all one-step transition functions under consideration but the space required for these displays would be prohibitive. Soyka (1983) may be consulted for further details.

Presented in Table 6.6.2 are the numerical specifications for four computer runs based on the survey data under discussion. In computer run I, for example, estimates of latent fecundability functions, based on the full (open) set

Table 6.6.1. Combinations of latent distribution functions used to calculate the one-step transition densities for women of parity zero

Age group	Latent distribution functions
$20 \leq x < 25$	$F_{21}(24, t)$ and $F_{23}(0; 20, t)$
$25 \leq x < 30$	$F_{21}(29, t)$ and $F_{23}(0; 25, t)$
$30 \leq x < 35$	$F_{21}(34, t)$ and $F_{23}(0; 30, t)$
$35 \leq x < 40$	$F_{21}(39, t)$ and $F_{23}(0; 35, t)$
$40 \leq x < 45$	$F_{21}(43, t)$ and $F_{23}(0; 35, t)$
$45 \leq x < 50$	$F_{21}(46, t)$ and $F_{23}(0; 35, t)$

of null segments for KAP-IV spacers, were used as computer input. As indicated by the symbolism in the third column of Table 6.6.2, the latent fecundability functions for parities 0, 1, and 2 were those for the corresponding parities among KAP-IV spacers, partially represented in Table 6.5.1. The symbol 2^+ indicates that the latent fecundability functions for KAP-IV spacers of parity two was used for all parities greater than two. Similar comments apply to the pregnancy outcome distributions specified in the fourth column of Table 6.6.2. Computer run II had the same pattern of input as that for run I, except that estimates of the latent fecundability functions were based on the set of closed null segments partially represented in Table 6.5.2.

Computer input for run III was the same as that for run I except that the latent fecundability functions for parities three or greater were estimated from open null segments for Taichung limiters, see Table 6.5.3. The input for computer run IV differs from that in run III only in that the pregnancy outcome distributions for parities three or greater were those for Taichung limiters, indicating that a high level of induced abortion was in force as shown in Table 6.5.4. Throughout all computer runs, transitions out of the temporary sterile states E_4, E_5, and E_6 were governed by the latent distribution functions partially represented in Tables 6.4.1 and 6.4.2.

Having discussed the inputs needed to implement the age-dependent semi-Markovian model of maternity histories described in Sect. 6.3, the question of ease of computer implementation can now be fruitfully discussed. Compared to MATHIST, the simulation system described in Sect. 5.5, the age-dependent property of the model under consideration requires that much larger arrays be manipulated in the computer. It is these large arrays that make the age-dependent model more difficult, and in many cases more expensive, to implement on a computer. Indeed, the arrays are so large that even with only six

Table 6.6.2. Numerical specifications for four computer runs

Computer run	Parity	Latent fecundability functions	Pregnancy outcome distributions
I	0	KAP-IV, parity 0, open	KAP-IV, parity 0
	1	KAP-IV, parity 1, open	KAP-IV, parity 1
	2^+	KAP-IV, parity 2, open	KAP-IV, parity 2
II	0	KAP-IV, parity 0, closed	KAP-IV, parity 0
	1	KAP-IV, parity 1, closed	KAP-IV, parity 1
	2^+	KAP-IV, parity 2, closed	KAP-IV, parity 2
III	0	KAP-IV, parity 0, open	KAP-IV, parity 0
	1	KAP-IV, parity 1, open	KAP-IV, parity 1
	2	KAP-IV, parity 2, open	KAP-IV, parity 2
	3^+	Taichung limiters, open	KAP-IV, parity 2
IV	0	KAP-IV, parity 0, open	KAP-IV, parity 0
	1	KAP-IV, parity 1, open	KAP-IV, parity 1
	2	KAP-IV, parity 2, open	KAP-IV, parity 2
	3^+	Taichung limiters, open	Taichung limiters

age groups the model cannot be implemented on many computer systems. When a computing system has a capability for handling large arrays, it would be entirely feasible to write a user-friendly menu similar to that for MATHIST in order to expedite feeding computer input into the algorithms developed in Sect. 6.3. Questions of computational efficiency and other numerical properties of the model will be discussed in subsequent sections in connection with computer output.

6.7 Computer Output Based on Survey Data

Discussed in this section are three sets of selected output from the four computer runs with numerical inputs based on Taiwanese survey data as described in Table 6.6.2. One of these sets was based on the means $M_{2j}(0; x, t)$, $2 \leq j \leq 6$, described in Eq. (6.3.30); a second set was based on the means $S_{2j}(0; x, t)$, $1 \leq j \leq 6$, described in Eq. (6.3.31). Finally, a third set was based on the current state probabilities $P_{2j}(0; x, t)$, $1 \leq j \leq 6$, defined in the discussion surrounding Eq. (6.3.20). Values of these three sets of functions were chosen for display, because they illustrate aspects of maternity histories not covered in the discussion of computer output from MATHIST in Sect. 5.7.

Presented in Table 6.7.1 are some calculated values of the mean functions $M_{2j}(0; x, t)$, $2 \leq j \leq 6$, for computer runs I and II; the corresponding values for computer runs III and IV are presented in Table 6.7.2. Recall that the means in these tables have the following interpretation. Consider women who marry at age $x < 50$ and remain married until age 50. For such women, the values in the tables are the mean number of visits ot state E_j, $2 \leq j \leq 6$, during the time interval $(x, 50]$ from x to age 50. In particular, the values in the rows corresponding to state E_4 represent the mean number of live births experienced by

Table 6.7.1. Mean number of visits to each transient physiological state by age fifty for given ages of marriage – computer runs I and II[a]

Computer run	State	Age at marriage					
		20	25	30	35	40	45
I	E_2	6.948	5.198	3.797	1.537	1.296	1.112
	E_3	6.183	4.433	3.042	0.624	0.401	0.215
	E_4	5.427	3.788	2.475	0.578	0.382	0.209
	E_5	0.256	0.174	0.131	0.018	0.006	0.000
	E_6	0.498	0.469	0.434	0.035	0.011	0.000
II	E_2	9.531	7.440	5.293	3.630	2.295	1.416
	E_3	9.054	6.960	4.801	3.146	1.794	0.831
	E_4	7.718	5.761	3.777	2.699	1.630	0.785
	E_5	0.453	0.354	0.278	0.146	0.049	0.003
	E_6	0.874	0.836	0.737	0.291	0.099	0.006

[a] All calculations carried out on a quarter-year time scale.

Table 6.7.2. Mean number of visits to each transient physiological state by age fifty for given ages of marriage – computer runs III and IV[a]

Computer run	State	Age at marriage					
		20	25	30	35	40	45
III	E_2	6.657	4.922	3.589	1.504	1.292	1.112
	E_3	5.839	4.100	2.775	0.580	0.393	0.215
	E_4	5.163	3.542	2.290	0.540	0.378	0.209
	E_5	0.229	0.146	0.104	0.013	0.005	0.000
	E_6	0.446	0.412	0.381	0.026	0.010	0.000
IV	E_2	7.219	5.178	3.650	1.508	1.293	1.112
	E_3	6.339	4.309	2.812	0.581	0.394	0.215
	E_4	4.023	2.985	2.139	0.530	0.376	0.209
	E_5	1.809	0.960	0.324	0.027	0.007	0.000
	E_6	0.507	0.364	0.349	0.025	0.010	0.000

[a] All calculations carried out on a quarter-year time scale.

these women for selected ages at marriage; while those corresponding to state E_5 represent the mean number of induced abortions experienced during the time interval from marriage to age 50.

Because the numerical input to computer runs I and II differed only with respect to fecundability functions, any differences among the means in the upper and lower panels of Table 6.7.1 are attributable to fecundability functions. Thus, in computer run I, based on fecundability functions estimated from the full (open) set of null segments for KAP-IV spacers, women marrying at ages 20 and 25 experienced 5.427 and 3.788 live births on the average. When fecundability functions based on the set of closed null segments from KAP-IV spacers were used in run II, these means rose to 7.718 and 5.761, respectively, see the lower panel in Fig. 6.7.1. In run I, women marrying at age 20 experienced 0.256 induced abortions on the average; but, when the fecundability functions estimated from closed null segments were used as input to run II, this mean rose to 0.453. Like MATHIST, the age-dependent model of maternity histories under consideration is very sensitive to choices of fecundability functions.

From Table 6.6.2, it can be seen that the input to computer run III differs from that of run I only in that the fecundability functions for parities three and higher were based on open null segments for Taichung limiters. According to the means presented in the upper panel of Table 6.7.2, this change in fecundability functions reduced the means 5.427 and 3.788, for run I in Table 6.7.1, to 5.163 and 3.542 live births for women marrying at ages 20 and 25. Differences among corresponding means for runs III and IV in Table 6.7.2 are attributable solely to the pregnancy outcome probabilities for Taichung limiters, who used high levels of induced abortion to limit births. Thus, women marrying at ages 20 and 25 in run IV would experience 4.023 and 2.985 live births on the average as compared to 5.163 and 3.542 for women in run I. Also observe that women marrying at ages 20 and 25 in run IV would experience 1.809 and 0.960

induced abortions on the average as compared to 0.229 and 0.146 in run III. In both Tables 6.7.1 and 6.7.2, the mean number of visits to the fecundable state E_2 exceeds all other means, because entrance into this state at marriage is counted as a visit.

Tables 6.7.3 and 6.7.4 contain calculated values of the means $S_{2j}(0; x, t)$, $1 \leq j \leq 6$, described in Eq. (6.3.31) for the four computer runs under discussion. According to Eq. (6.3.33), the column totals for these means for women marrying at the ages 20, 25, ..., 45 should be 360, 300, ..., 60 months. Due to the finite arithmetic operations done by computers, in which truncation and rounding errors are always present, there are no guarantees that the mathematical relationship in (6.3.33) would hold in the computer output. Close agreement between a mathematical prediction and computer output would provide an indication as to whether truncation and rounding errors are being controlled. As can be seen from the totals in Tables 6.7.3 and 6.7.4, there is very close agreement between the theoretical and calculated values in all cases, suggesting that truncation and rounding errors are indeed being controlled in the computer output under consideration.

Interesting substantive interpretation may be attached to the means in Tables 6.7.3 and 6.7.4. Consider, for example, the mean number of months spent in the permanently sterile state E_1 between marriage and age fifty for women marrying at age 20. In run I, see Table 6.7.3, this mean was 119.21 months; but, when higher levels of fecundability were used in run II, this mean decreased to 99.47, a difference of nearly 20 months. Even though the same latent risks of sterility were used in both computer runs, the higher level of fecundability used in run II significantly reduced the mean time spent in the state E_1.

Table 6.7.3. Mean number of months spent in each state between marriage and age fifty – computer runs I and II[a]

Computer run	State	Age at marriage					
		20	25	30	35	40	45
I	E_1	119.21	116.83	108.17	117.01	83.95	40.31
	E_2	119.27	97.02	74.05	50.34	27.98	15.79
	E_3	51.37	36.15	24.03	5.39	3.50	1.91
	E_4	66.27	46.15	30.56	7.03	4.47	1.98
	E_5	1.00	0.73	0.58	0.09	0.03	0.00
	E_6	2.88	2.75	2.61	0.23	0.07	0.00
	Total	360.00	300.00	240.00	180.00	120.00	59.99
II	E_1	99.47	97.64	93.68	81.95	61.51	31.02
	E_2	85.63	69.76	57.17	37.03	23.80	14.33
	E_3	73.82	55.62	37.16	25.66	15.21	7.30
	E_4	94.02	70.40	46.20	32.72	18.65	7.32
	E_5	1.85	1.53	1.26	0.71	0.23	0.01
	E_6	5.22	5.05	4.53	1.93	0.61	0.02
	Total	360.01	300.00	240.00	180.00	120.01	60.00

[a] All calculations carried out on a quarter-year time scale.

Table 6.7.4. Mean number of months spent in each state between marriage and age fifty – computer runs III and IV [a]

Computer run	State	Age at marriage					
		20	25	30	35	40	45
III	E_1	122.54	120.37	111.32	117.50	84.02	40.31
	E_2	122.13	99.36	75.42	50.56	28.02	15.79
	E_3	48.75	33.66	22.11	4.98	3.45	1.91
	E_4	63.15	43.62	28.43	6.71	4.43	1.98
	E_5	0.87	0.60	0.46	0.07	0.02	0.00
	E_6	2.55	2.39	2.26	0.18	0.06	0.00
	Total	359.99	300.00	240.00	180.00	120.00	59.99
IV	E_1	123.24	120.87	111.61	117.52	84.02	40.31
	E_2	133.27	104.86	76.79	50.56	28.03	15.79
	E_3	44.66	31.49	21.44	4.93	3.44	1.91
	E_4	49.08	36.72	26.61	6.60	4.42	1.98
	E_5	6.95	4.00	1.48	0.13	0.03	0.00
	E_6	2.80	2.07	2.06	0.17	0.06	0.00
	Total	360.00	300.01	239.99	180.00	120.00	59.99

[a] All calculations carried out on a quarter-year time scale.

Of course, the more pregnancies a woman experiences, the more time she spends in the temporary sterile state E_4, following a live birth. Thus, for women marrying at age 20 in run I, an average of 66.27 months was spent in state E_4; for women in run II, this mean rose to 94.02, a difference of about 28 months.

Another interesting observation that may be gathered from Table 6.7.4 is that induced abortions tend to increase the risk of subsequent pregnancies in maternity histories in the sense that the mean time spent in the fecundable state E_2 is increased. For example, in the low levels of induced abortion used in run III in Table 6.7.4, women on the average marrying at age 20 experienced 122.13 months in the fecundable state E_2; under the high levels of induced abortion used in run IV, see lower panel in Table 6.7.4, this mean increased to 133.27 months. Other interesting comparisons among the means in Tables 6.7.1 to 6.7.4 will be left to the reader.

Comparisons among simulated and observed total fertility rates are always of interest. Tuan (1958) has provided rather extensive information on the maternity histories of Chinese women living in rural Taiwan, based on a sample from a household registration system. At the time the survey was taken in the summer of 1953, nearly all the population of Taiwan had been covered by a system of household registration, dating back more than fifty years. Presented in the upper panel of Table 6.7.5 are cohort total fertility rates for women aged 45–64 at the time of the survey; contained in the middle panel of Table 6.7.5 are the cohort total fertility rates that resulted when the women in the sample were classified by age of marriage. Some period total fertility rates for Taiwan, taken from the same source as that for Table 5.7.7, are also presented in the lower panel of Table 6.7.5.

The only cohort total fertility rate in Table 6.7.5 that is directly comparable to those in Tables 6.7.1 and 6.7.2 is that in the middle panel for those women who were aged 20–24 at marriage. It is interesting to observe that the estimated value of 6.0 for this age group in Table 6.7.5 lies between the means 5.427 and 7.718 for computer runs I and II in Table 6.7.1 for women marrying at age 20. This observation suggests that estimates based on the full (open) set of null segments for KAP-IV spacers tended to underestimate fecundability; while, those based on closed null segments tended to overestimate it. It is also of interest to observe that the cohort total fertility rates in the upper panel of Table 6.7.5, classified by ages of women at the time of the survey, lie between the simulated rates 5.427 and 7.718 for runs I and II in Table 6.7.1 for women marrying at age 20.

That age at marriage also had a marked effect on cohort total fertility rates can be seen from the middle panel of Table 6.7.5. For women age 25 and over at marriage, the average number of births observed was 3.4, a number close to the simulated means of 3.788 and 3.542 for women marrying at age 25 in computer runs I and III as listed in Tables 6.7.1 and 6.7.2. Finally, it is of interest to observe from the lower panel of Table 6.7.5 that the period total fertility rates in Taiwan for 1965 and 1970 are closer to 5.163 and 4.023, the simulated rates reported in Table 6.7.2 for those women in computer runs III and IV marrying at age 20, than to the corresponding rates for computer runs I and II listed in Table 6.7.1.

In conclusion, computer runs I and II seem to be representative of maternity histories for rural Taiwanese women who completed childbearing prior to 1953; while, those histories represented in runs III and IV seem closer to those of Taiwanese women during the period 1965 to 1970. The low period total

Table 6.7.5. Historical cohort and period total fertility rates in Taiwan

	Age of rural women at survey[a]			
	45–49	50–54	55–59	60–64
CTFR[b]	7.3	7.2	6.9	6.9
	Age at marriage for women aged 45–64 at survey			
	Under 14	15–19	20–24	25 and over
CTFR	8.3	7.6	6.0	3.4
	Period			
	1965	1970	1978	
PTFR[c]	4.825	4.000	2.710	

[a] Survey conducted in 1953.
[b] Cohort total fertility rate.
[c] Period total fertility rate.

fertility rate of 2.710 for Taiwan in 1978 is indicative of the pronounced fertility changes that occurred on the island during the 1970s.

Computer generated graphs of the current state probability $P_{2j}(0; x, t)$, $1 \leq j \leq 6$, for women marrying at age $x = 20$ in computer run IV, are presented in Figs. 6.7.1 and 6.7.2. Figure 6.7.1 contains simultaneous plots of these probabilities as functions of time since marriage for the sterile state (E_1), the fecundable state (E_2), and the pregnant state (E_3). Contained in Fig. 6.7.2 are the corresponding simultaneous plots for the pregnant state (E_3) and the temporary sterile states following a live birth (E_4), an induced abortion (E_5), or some other type of pregnancy outcome (E_6).

As can be seen from the value $P_{22}(0; 20, 0) = 1$ in Fig. 6.7.1, all women in the cohort are in the fecundable state E_2 at $t = 0$, the time of marriage. Following marriage, the rapid decline in the graph of P_{22}, the fraction of women in the fecundable state E_2, is accompanied by a rapid rise in the graph of P_{23}, the fraction of women in the pregnant state E_3. The graph of P_{23} reaches a maximum sometime between the first and second years following marriage. After declining from a maximum, the graph of P_{23} reaches a relative maximum between three and four years after marriage and then steadily declines for the rest of the childbearing ages. In contrast to the graph of P_{23}, the graph of P_{22} is more parabolic in shape. As expected, the graph of the current state probability P_{21}, representing the fraction of women in the permanently sterile state E_1, is monotone increasing. That the value of P_{21} at 360 months is not one is a measure of the imperfect nature of both the computer input and the calculations, due in part to aggregating ages into six groups.

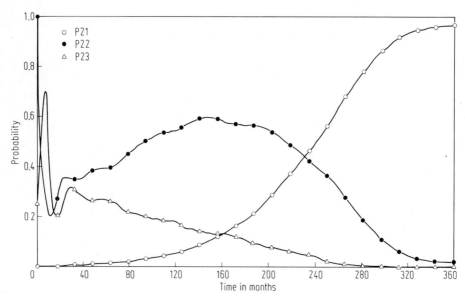

Fig. 6.7.1. Probability of being in the sterile (E_1), fecundable (E_2) or pregnant (E_3) states at time t since marriage at age 20

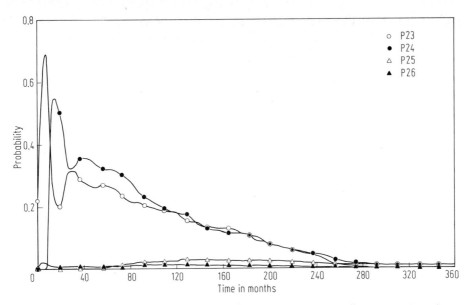

Fig. 6.7.2. Probability of being pregnant (E_3) or in a post-pregnancy sterile state states at time t since marriage at age 20

For purposes of making visual comparisons, the graph of the current state probability P_{23} for pregnancies is reproduced on a different scale in Fig. 6.7.2. As might be expected, the shape of the graph for the current state probability P_{24}, representing the fraction of women in a cohort in the sterile state E_4 following a live birth, is similar to that for P_{23}, the graph for pregnancies. Since entrance into state E_4 must always follow an entrance into state E_3, the graph of P_{24} is translated to the right of that for P_{23}. For most of the period three to ten years following marriage, there was a higher fraction of women in the state E_4 than in the pregnant state E_3. It is thus readily apparent, from the graphs in Fig. 6.7.2, that lengths of postpartum sterile periods can be a significant factor in protecting women from further conceptions in noncontracepting populations. Finally, it can be seen from Fig. 6.7.2 that, even when the high levels of induced abortion used by Taichung limiters are in force, at any time there is a relatively small fraction of women in a cohort who are in the sterile state E_5 following an induced abortion.

For the most part, attention in this section has been focused on substantive rather than numerical issues underlying the calculations. No mention, for example, has been made of the fact that all calculations reported in this section were done on a quarter-year time scale, see the footnotes to Tables 6.7.1 to 6.7.4. Questions regarding the choice of this time scale and the quality of the calculations presented here will be taken up in the next section.

6.8 Further Assessment of the Quality of Calculations in Sect. 6.7 and Conclusions

It is easy to understand why it is impractical to attempt to manipulate large two-dimensional arrays, in which both age and time are on a monthly scale within most computers, due to their limited memory capacity. As a practical measure, it is tempting to use some aggregation procedure on both the age and time dimensions, in order to reduce the size of the arrays used as input to the algorithms for the age-dependent model of maternity histories set forth in Sect. 6.3. A question that naturally arises is whether such aggregation would seriously bias the calculations and perhaps lead to a violation of mathematical laws, which, in turn, would lead to numerical uncertainties in the computer output. The purpose of this section is to provide some answers to this question and assess the quality of the calculations presented in Sect. 6.7. Some attention will also be paid to the reasons for deciding to implement the modules described in Chap. 5, rather than the age-dependent models described in this chapter.

During the course of the computer experiments discussed in the previous sections of this chapter, M.G. Soyka, a graduate student in mathematics at Drexel University, noticed that when calculations were done with the six age groupings used in Sect. 6.7 on a monthly time scale, the operation of age-dependent convolution was not associative. Thus, if A, B, and C were three two-dimensional arrays and $*$ represented the operation of age-dependent convolution, then the equation $A*(B*C)=(A*B)*C$ would not necessarily hold even to within rounding and truncation error. It can be shown that if the same aggregations are used on both the age and time dimensions, then the operation of age-dependent convolution is associative. When the law of associativity is violated, numerical uncertainties are introduced into the calculations; for, the computer output may depend fortuitously on the way a computer programmer chose to group the age-dependent convolutions.

To be brief, the model of maternity histories based on the renewal matrix as described in Sect. 6.3 is to be preferred to the modular model introduced in Mode and Soyka (1977), because its computer implementation depends on the associativity of age-dependent convolutions in a less crucial way. As will be shown by example, quite different numerical results may be obtained from two different computer implementations, although the algorithms underlying the two implementations determine functions that are mathematically equivalent.

Presented in Table 6.8.1 are estimates of the mean number of months spent in each state between marriage and age fifty as calculated by an implementation of the modular procedure outlined in Mode and Soyka (1977). Table 6.8.2 contains estimates of the same means as calculated by an implementation of the renewal matrix procedure described in Sect. 6.3. In both cases, calculations were carried out on a monthly time scale. As can be seen by comparing values in the tables, the two implementations lead to quite different numerical results. Because the numerical results in Table 6.8.2 depend less on the law of associativity for the operation of age-dependent convolution, they were judged more reliable. Upon comparing corresponding estimates in Tables 6.8.1 and 6.8.2, it would appear that the modular procedure has a tendency to under-estimate the mean time spent in the sterile state E_1 at the expense of over-estimating the mean times spent in states E_2, E_3, and E_4. It is also interesting to observe that the totals in Table 6.8.1 consistently underestimate the true values; in Table 6.8.2, however, where the renewal matrix procedure was used, the agreement between the theoretical and calculated totals is excellent.

Having discussed some evidence as to why the renewal matrix procedure produces numerical results that are more reliable than the modular procedure, attention will be turned to assessing biases that may result when the time scale is aggregated. In addition to estimates by single month, which resulted from the life table procedure used to generate computer input, two aggregations of the time scale, a bi-monthly and a quarter-year, were tested in computer experiments. A bi-monthly aggregation of the time scale was computed by selecting values of the monotone increasing distribution functions at multiples of two months; whenever densities were required, they were obtained by

Table 6.8.1. Mean number of months spent in each state between marriage and age fifty – calculations based on modular procedure[a]

State	Age at marriage					
	20	25	30	35	40	45
E_1	93.80	100.16	104.40	118.31	85.29	41.52
E_2	150.20	117.28	82.22	50.14	26.99	14.71
E_3	52.36	36.73	23.24	5.04	3.49	1.93
E_4	54.81	40.09	26.91	6.25	4.14	1.83
E_5	5.76	3.69	1.52	0.11	0.03	0.00
E_6	1.64	1.13	0.98	0.09	0.04	0.00
Total	358.57	299.08	239.27	179.94	119.98	59.99

[a] All calculations were carried out on a monthly time scale. Computer input was that for run IV in Table 6.6.2.

Table 6.8.2. Mean number of months spent in each state between marriage and age fifty – calculations based on renewal matrix procedure[a]

State	Age at marriage					
	20	25	30	35	40	45
E_1	126.48	123.88	114.68	119.21	85.52	41.52
E_2	133.13	104.27	75.81	49.46	26.84	14.71
E_3	44.62	31.43	21.30	4.93	3.45	1.93
E_4	47.65	35.23	25.12	6.14	4.10	1.83
E_5	5.57	3.38	1.33	0.11	0.03	0.00
E_6	2.55	1.81	1.76	0.15	0.06	0.00
Total	360.00	300.00	240.00	180.00	120.00	59.99

[a] All calculations were carried out on a monthly time scale. Computer input was that for run IV in Table 6.6.2.

differencing the values of the distribution functions at multiples of two months. A quarter-year aggregation of the time scale was computed in exactly the same way except that multiples of three months were considered.

Table 6.8.3 contains estimates of the means and variances of the numbers of live births experienced between marriage and age fifty, as computed according to the three aggregations of the time scale described above. All variances were computed using an algorithm not discussed here. When the bi-monthly and quarter-year values are compared to the single month values in Table 6.8.3, it can be seen that a coarser aggregation of the time scale leads to an under-estimation of both the mean and variance of the number of live births expe-rienced between marriage and age fifty. When reductions in computer costs that resulted from using the bi-monthly and quarter-year time scales are taken into account, however, the magnitude of the underestimation seems tolerable, especially in exploratory calculations.

In order to cut computer costs, all calculations presented in Sect. 6.7 were done on a quarter-year time scale. It is, therefore, of interest to assess these exploratory calculations in light of the results discussed above. Table 6.8.4 contains estimates of the mean number of visits to each transient physiological

Table 6.8.3. Means and variances of numbers of live births between marriage and age fifty by aggregation of time scale[a]

Age at marriage	Single month		Bi-monthly		Quarter-year	
	mean	variance	mean	variance	mean	variance
20	4.251	3.384	4.085	3.054	4.023	2.931
25	3.115	2.397	3.021	2.188	2.985	2.126
30	2.194	1.392	2.152	1.305	2.139	1.282
35	0.534	0.902	0.530	0.885	0.530	0.880
40	0.379	0.559	0.376	0.545	0.376	0.546
45	0.212	0.240	0.209	0.229	0.209	0.228

[a] Computer input was that for run IV in Table 6.6.2. Computer output was that from renewal matrix procedure.

Table 6.8.4. Mean number of visits to each transient physiological state by age fifty for given ages of marriage[a]

State	Age at marriage					
	20	25	30	35	40	45
E_2	7.708	5.469	3.772	1.519	1.299	1.117
E_3	6.828	4.599	2.929	0.590	0.398	0.218
E_4	4.251	3.115	2.194	0.534	0.379	0.212
E_5	2.013	1.098	0.378	0.029	0.008	0.001
E_6	0.563	0.386	0.357	0.026	0.010	0.001

[a] All calculations were carried out on a monthly time scale. Computer input was that for run IV in Table 6.6.2. Computer output was that from renewal matrix procedure.

state by age fifty for given ages of marriage. All means in Table 6.8.4 were calculated on the basis of a monthly time scale, using the computer input for run IV in Table 6.6.2. The values in Table 6.8.4 and those in the lower panel of Table 6.7.2 are, thus, approximations of the same means. From an inspection of the values in Tables 6.7.2 and 6.8.4, it can be seen that the pattern of under-estimation for the means in Table 6.8.3 continues to prevail in Table 6.7.2. This pattern of underestimation, however, does not seem to be sufficient to change any substantive conclusions in Sect. 6.7 based on comparing simulated means with the total fertility rates in Table 6.7.5.

Another perspective on the effects of aggregating the time scale may be obtained by comparing values in Table 6.8.2 with those in the lower panel of Table 6.7.4. Recall that a monthly time scale was used in calculating the values in Table 6.8.2; while, a quarter-year time scale was used in Table 6.7.4. Upon comparing corresponding values in these tables for women marrying at age 20, it appears that the quarter-year time scale used in doing the calculations of Table 6.7.4 tended to underestimate the mean number of months spent in the sterile state E_1. In part, this underestimation seems to have been compensated by over-estimating the mean times spent in state E_4 and E_5. Because of the differences between values in Tables 6.8.2 and 6.7.4 are not of sufficient magnitude to be of great substantive importance, doing exploratory calculations on a quarter-year time scale seems justified.

All of the material presented in Sects. 6.3 through 6.7, as well as that in the present section was based on a Ph.D. thesis, Soyka (1983). This thesis may be consulted for a detailed exposition of the problem of nonassociativity that results when age-dependent convolutions are done with arrays in which the age and time scales are aggregated differently. Many more tables, computer gener-ated graphs, and a listing of the important FORTRAN subroutines needed to implement the model of Sect. 6.3 may also be found in this thesis.

Whenever two alternative formulations are under consideration, some dis-cussion as to why the computer implementation of one model was offered for general use rather than another seems appropriate. Two problems, closely related to the large numerical arrays needed to implement age-dependent models, seemed to mitigate against offering the software implementing the model of Sect. 6.3 for general use. One of these problems was that of finding and processing the information needed to generate the two-dimensional arrays used in implementing the age-dependent model of maternity histories. As can be seen from the discussion in Sects. 6.4, 6.5, and 6.6, generating these arrays can require substantial labor. A second problem was that of non-associativity as discussed above, which could lead to numerical uncertainties in the com-puter output. In MATHIST, the simulation system described in Chap. 5, problems of generating and processing large two-dimensional arrays were side-tracked by considering only models based on ordinary convolutions and parity dependence. As mentioned before, the problem of developing menus to feed MATHIST was also much easier to solve than that of developing menus to feed the model of maternity histories in Sect. 6.3.

A problem that remains in comparing the two alternative formulations centers around the question as to how entrance into the permanently sterile state E_1 is to be treated. Should sterility be treated as an external stopping time, as in MATHIST; or should it be treated as an internal competing risk, as in the age-dependent model of Sect. 6.3? A similar question, regarding the treatment of mortality, could also be asked. More computer experimentation will be required to determine whether the two approaches to handling sterility lead to different substantive answers in forecasting fertility in cohorts. Age-dependence, as reflected by the onset of permanent sterility with increasing age, can be a crucial factor in forecasting fertility in many present-day populations, where childbearing is being delayed due to changing roles of women. Dealing with age-dependence is among the fundamental problems that must be faced in the formulation and computer implementation of models of maternity histories. The renewal matrix procedure in Sect. 6.3, as formulated by M.G. Soyka, seems to be a significant step forward in coming to grips with this problem. From the point of view of computational efficiency, it should also be mentioned that a non-age-dependent version of the model in Sect. 6.3 would compare favorably with MATHIST.

In Sect. 5.8, a stochastic model of anovulatory sterile periods following live births was developed. Like the one under consideration, this model also had an age-dependent property. But, in all calculations reported in Sect. 5.8, the same aggregations of the x and t scales were used so that numerical uncertainties, due to non-associativity of age-dependent convolutions, were not present in the computer output. Further development of age-dependent models of the type under consideration could be expedited by the availability of parametric formulas for generating a large variety of computer input by varying the parameters. Formulas for densities of the type displayed in (5.8.14), based on parametric latent risk functions, could be useful in generating such computer input.

6.9 A Non-Markovian Model for the Taichung Medical IUD Experiment

Semi-Markovian models for waiting times to conception, when contraceptives are used to delay or prevent a pregnancy, were outlined in Sect. 5.9. Although these models can be implemented on a computer with relative ease, there is no way of judging their validity unless they are used in analyzing real data. In Sect. 6.5, mention was made of the Taichung Medical IUD Follow-Up Study, which provided detailed longitudinal data on contraceptive-pregnancy histories. Such longitudinal data are well suited for developing and validating models describing the use of contraceptives to prevent and delay pregnancies. Sketched in this section are some aspects of a non-Markovian model of the Taichung medical IUD experiment based on linking absorbing, age-dependent semi-Markovian (AASM) processes in a kind of time series. Procedures used to validate the model will be discussed in a subsequent section. As will be seen in what follows, the model needed to describe and analyze the Taichung data was considerably more complicated than the ones described in Sect. 5.9.

Table 6.9.1. Set of states for model of Taichung medical IUD experiment

State	Definition
E_1	Menopausal[a]
E_2	Contraceptively sterilized
E_3	The null state
E_4	Using a pill
E_5	Using other methods
E_6	Using an IUD
E_7	IUD segment ends in expulsion
E_8	IUD segment ends in removal
E_9	Pregnant
E_{10}	Live birth[b]
E_{11}	Induced abortion[b]
E_{12}	Other[b]

[a] Used as an indicator of permanent sterility.
[b] Anovulatory sterile period following indicated type of pregnancy outcome.

During a preliminary analysis of the Taichung data, it was found that the set of six states defined in Table 5.9.1 was an inadequate framework for viewing the data. The primary reason for this inadequacy centered around the observation that events subsequent to an IUD segment in a contraceptive-pregnancy history depended, to some extent, on whether a segment ended in an *in situ* pregnancy, an expulsion, or a removal. A decision was thus made to include two states, representing expulsions and removals of IUD's, in the model. Listed in Table 6.9.1 are the ten states used to describe contraceptive-pregnancy histories subsequent to entering the Taichung experiment.

All women entering the Taichung experiment accepted an IUD, i.e., in terms of the definitions in Table 6.9.1, they entered state E_6. Consider, then, a women who accepts an IUD at $t=0$ when she is age x. Given this event, let $A_{6j}(x,t)$ be the conditional probability that this IUD segment ended in an expulsion, $E_7 (j=7)$, a removal, $E_8 (j=8)$, or an *in situ* pregnancy, $E_9 (j=9)$, some time during the time interval $(0,t]$, $t>0$. As part of implementing the model on a computer, the functions $A_{6j}(x,t)$, $6\leq j\leq 9$, were estimated on an age-time lattice using a multiple decrement life table procedure.

When attempting to analyze and model a large and complicated data set, such as that for the Taichung experiment, a useful approach to follow is that of constructing a model based on simple components or modules. One simple module of the model was that for the first IUD segment described in the preceding paragraph. In order to illustrate ideas and simplify the discussion, only those modules, concerning events leading up to the first pregnancy after entering the experiment, will be described in mathematical detail in this section.

Toward this end, consider those women whose first IUD segment ended in either an expulsion, E_7, or a removal, E_8. Both these states were viewed as instantaneous. That is, as soon as a women experienced either an expulsion or removal, she would immediately enter another state in the set $A = [E_1, E_2, E_3, E_4, E_5, E_6, E_9]$. Her next segment could thus be the null state, E_3, the acceptance of a pill, E_4, some other (conventional) method of contraception, E_5, the reacceptance of an IUD, E_6, or an entrance into the pregnant state, E_9. Alternatively, she could enter one of the absorbing states, E_1 or E_2, by becoming menopausal, E_1, or by accepting a contraceptive sterilization, E_2.

Another simple module of the Taichung model consisted of probabilities for transitions from the states E_7 and E_8 into a state in the set A. For $i = 7$, 8, let $a_{ij}(x)$ be the conditional probability of the transition $E_i \rightarrow E_j \in A$, given that the first IUD segment ended in state E_i when a women was age x. From now on, those women whose first IUD segment ended in either an expulsion or a removal will be symbolized by the branches B_7 and B_8.

Following the termination of the initial IUD segment, women in each of the branches B_7 and B_8 may have experienced one or more additional IUD segments prior to a pregnancy. Consider a woman in branch B_k, $k = 7$, 8, who begins a subsequent IUD segment at $t = 0$ when she is age x. Given this event, let $A_{6j}(k, x, t)$ be the conditional probability that this IUD segment terminates in state E_j, $j = 6$, 7, 8, sometime during the time interval $(0, t]$, $t > 0$. Like the functions in the module for the initial IUD segment, these functions were estimated on an age-time lattice, using data pooled over all IUD segments within each branch B_k, $k = 7$, 8. If these later IUD segments terminated in either state E_7 or E_8 when a woman was age x, it was assumed that she then passed into a state in the set A according to the probabilities $a_{7j}(x)$ and $a_{8j}(x)$ for $E_j \in A$.

The most complex module in that part of the model, dealing with events prior to the first pregnancy in the branches B_7 and B_8, was that describing transitions among and out of the states in the set $S = [E_3, E_4, E_5]$, a subset of A. Chosen for this module was the AASM-process with $\mathscr{I}_1 = [E_1, E_2, E_6, E_9]$ as the set of absorbing states and $\mathscr{I}_2 = S$ as the set of transient states. With each transient state $E_i \in \mathscr{I}_2$, there was associated a set of one-step transition probabilities with the following interpretation. Suppose the initial IUD segment ended in state E_k, $k = 7$, 8. At some later time, taken as $t = 0$, a woman entered state $E_i \in \mathscr{I}_2$ when she was age x. Given these events, $A_{ij}(k, x, t)$ is the conditional probability of a transition into the state $E_j \in \mathscr{I}_1 \cup \mathscr{I}_2$ some time during the time interval $(0, t]$, $t > 0$. For each fixed i and k, the functions $A_{ij}(k, x, t)$, $E_j \in \mathscr{I}_1 \cup \mathscr{I}_2$, were estimated on an age-time lattice, using a multiple decrement life table procedure on pooled data prior to the first pregnancy on all entrances into the state E_i, $i = 3$, 4, 5, within each branch B_k, $k = 7$, 8. Of all the modules described so far, these AASM-processes involve the strongest assumptions. A test for the validity of these assumptions will be described in a subsequent section.

Having described the modules of the Taichung model leading up to the first pregnancy following entrance into the experiment, it is now possible to consider a probabilistic description of waiting times to the first pregnancy. In describing the data subsequent to the first pregnancy, it seemed advisable to classify pregnancies according to whether the last IUD segment prior to the pregnancy ended in an *in situ* pregnancy, an expulsion, or a removal. For, an *in situ* pregnancy was much more likely to be terminated by an induced abortion than others. Accordingly, pregnancies will be classified by $k=6$, 7, or 8. For a woman who entered the Taichung experiment at $t=0$ when she was of age x, let $F_{69}(k,x,t)$ be the conditional probability that the pregnant state E_9 is entered for the first time during $(0,t]$, $t>0$, and the last IUD segment prior to the pregnancy was of type $k=6$, 7, or 8.

Needed to calculate the functions $F_{69}(k,x,t)$ are certain other functions $H_{j9}(k,x,t)$ with the following interpretation. Suppose the first IUD segment ends at $t=0$ in state E_j, $j=7$, 8, when a woman is age x. Given this event, let $H_{j9}(k,x,t)$ be the conditional probability that the pregnant state E_9 is reached for the first time during $(0,t]$, $t>0$, and the last IUD segment preceding the pregnancy was of type $k=6$, 7, or 8. Then, by taking into account the three ways the initial IUD segment may end, a renewal argument yields the formula

$$F_{ij}(k,x,t)=\delta_{6k}A_{69}(x,t)+\sum_{v=7}^{8}\sum_{s=0}^{t}a_{6v}(x,s)H_{v9}(k,x+s,t-s),\qquad(6.9.1)$$

where $k=6$, 7, or 8, δ_{6k} is the Kronecker delta, and $a_{6v}(x,s)$ are the densities of the functions $A_{6v}(x,t)$, $v=7$, 8. It will be helpful to recall that the functions $A_{6v}(x,t)$, $v=6$, 7, 8, were estimated from data on the initial IUD segment.

From (6.9.1) it can be seen that the problem of calculating the function $F_{69}(k,x,t)$ reduces to that of calculating the H-functions on the right. These functions, in turn, depend on several other functions. One of these functions are absorption probabilities determined by an equation of type (6.2.12). Consider the AASM-module in branch B_k with $\mathcal{I}_1=[E_1,\,E_2,\,E_6,\,E_9]$ as the set of absorbing states and $\mathcal{I}_2=[E_3,\,E_4,\,E_5]$ as the set of transient states. Given that a woman in branch B_k, $k=7$, 8, enters transient state $E_i\in\mathcal{I}_2$ at $t=0$ when she is age x, let $P_{ij}(k,x,t)$ be the conditional probability of entering state $E_j\in\mathcal{I}_1$ some time during $(0,t]$, $t>0$. The matrix of functions $(P_{ij}(k,x,t)$, $E_i\in\mathcal{I}_2$ and $E_j\in\mathcal{I}_1)$, was calculated by applying an age-dependent version of Littman's algorithm in the matrix case.

Like the absorption probabilities just described, the H-functions on the right in (6.9.1) also satisfy a system of renewal type equations. To deduce this system of equations, it was necessary to define and calculate other functions in addition to the absorption probabilities just described. Again suppose the first IUD segment ended in state E_j, $j=7$, 8, at $t=0$ when a woman was age x. Given this event, let $F_{jk}(x,t)$ be the conditional probability that state E_k, $k=6$, 7, 8, or 9, was reached for the first time during $(0,t]$, $t>0$. Furthermore, given this event, let $F_{j9}(6,x,t)$ be the conditional probability that state E_9 was reached for the first time during $(0,t]$, $t>0$, and the last IUD segment prior to the pregnancy was of type 6, i.e., the segment ended in an *in situ* pregnancy.

Then, for $j=7$ or 8 and $k=6$ or 9, it follows that

$$F_{jk}(x,t)=a_{jk}(x)+\sum_{v=3}^{5}a_{jv}(x)P_{vk}(j,x,t).\qquad(6.9.2)$$

The a-functions on the right in (6.9.2) are from the simple modules governing immediate transitions out of the instantaneous states E_7 and E_8 in the branches B_7 and B_8. Observe that the P-functions on the right are the absorption probabilities for the AASM-modules described above. Equation (6.9.2) is thus also valid for $k=1$ and 2.

Similarly, for $j=7$ or 8 and $k=7$ or 8, it follows that

$$F_{jk}(x,t)=a_{j6}(x)A_{6k}(j,x,t)+\sum_{s=0}^{t}f_{j6}(x,s)A_{6k}(j,x+s,t-s), \qquad (6.9.3)$$

where $f_{j6}(x,t)$ is the density of the function $F_{j6}(x,t)$ in (6.9.2). Finally,

$$F_{j9}(6,x,t)=a_{j6}(x)A_{69}(j,x,t)+\sum_{s=0}^{t}f_{jk}(x,s)A_{69}(j,x+s,t-s), \qquad (6.9.4)$$

where the small f's are again densities. Note that the A-functions on the right in (6.9.3) and (6.9.4) are those for IUD segments in the branches B_7 and B_8 following the termination of the initial IUD.

The functions in the preceding paragraphs were sufficient to deduce renewal type equations for the H-functions on the right in Eq. (6.9.1). In deducing these equations, it was assumed that the states E_7 and E_8 were regenerative points of the process. Under this assumption, a renewal type argument may be invoked to show that the functions $H_{j9}(6,x,t)$, $j=7$ or 8, satisfy the system

$$H_{j9}(6,x,t)=F_{j9}(6,x,t)+\sum_{s=0}^{t}\sum_{v=7}^{8}f_{jv}(x,s)H_{v9}(6,x+s,t-s) \qquad (6.9.5)$$

of renewal equations. An analogous argument leads to the conclusion that the functions $H_{j9}(k,x,t)$ satisfy the system

$$H_{j9}(k,x,t)=\delta_{jk}F_{j9}(x,t)+\sum_{s=0}^{t}\sum_{v=7}^{8}f_{jv}(x,s)H_{v9}(k,x+s,t-s), \qquad (6.9.6)$$

for each fixed $k=7$ or 8 and $j=7$ and 8. Hence, (6.9.6) represents two sets of 2×2 systems of renewal equations with two unknowns, the H-functions for a given value of k.

Together, (6.9.5) and (6.9.6) comprise three sets of renewal equations. Each of these sets of equations may be solved by first applying Littman's algorithm to the 2×2 matrix of functions

$$\begin{bmatrix} f_{77}(x,t) & f_{78}(x,t) \\ f_{87}(x,t) & f_{88}(x,t) \end{bmatrix}. \qquad (6.9.7)$$

To compute the H-functions, the output of Littman's algorithm is then convolved with the inhomogeneous terms in each set of equations. It is important to note that in solving the renewal equations represented by (6.9.5) and (6.9.6) only one application of Littman's algorithm is required, since the coefficient functions of the H's are the same in all sets of equations.

This completes the description of the algorithms needed to compute the functions $F_{69}(k,x,t)$, $k=6$, 7, or 8, describing waiting times to the first pregnancy following entrance into the Taichung experiment. Another strong assumption underlying the calculation of these functions was that the states E_7

and E_8 were regenerative points of the process. A validity check of this assumption will also be given in the next section.

The next module of the model was concerned with the resolution of the first pregnancy. Let $A_{9j}(k, x, t)$, $j = 10$, 11, 12 and $k = 6$, 7, 8, be a set of functions with the following interpretation. Suppose that at $t = 0$, a woman experiences a pregnancy when she is of age x and her last IUD segment prior to the pregnancy ended in an *in situ* pregnancy ($k = 6$), an expulsion ($k = 7$), or a removal ($k = 8$). Given this event, let $A_{9j}(k, x, t)$ be the conditional probability that a pregnancy ends in a live birth ($j = 10$), induced abortion ($j = 11$), or some other type of pregnancy outcome ($j = 12$) some time during $(0, t]$, $t > 0$. When estimating these functions by a multiple decrement life table procedure, data for the cases $k = 7$ and $k = 8$ were pooled; but, data for the case $k = 6$ were processed separately, because a higher probability of induced abortion was associated with *in situ* pregnancies.

Following the resolution of a pregnancy, a woman passed into one of the anovulatory states E_{10}, E_{11}, or E_{12}, depending on the type of pregnancy outcome. After a sojourn time in one of these states of temporary sterility, the next state visited was a member of the set A. Whether an IUD was reaccepted depended on how the last IUD segment prior to the pregnancy ended. Consequently, the next module of the system consisted of the set of functions $A_{ij}(k, x, t)$, $i = 10$, 11, 12 and $k = 6$, 7, 8, with the following interpretation. Suppose that at $t = 0$ a woman, whose last IUD segment prior to the pregnancy was of type $k = 6$, 7, 8, enters state E_i, $i = 10$, 11, 12, when she is age x. Given these events, let $A_{ij}(k, x, t)$ be the conditional probability of a transition to state $E_j \in A$ some time during $(0, t]$, $t > 0$. A multiple decrement life table procedure was again used to estimate these functions for each combination of the indices i and k. A weakness in the formulation of this module was that the possibility of an overlap between contraception and anovulatory sterile periods was not taken into account.

After a state in the set A was entered, the structure of the model describing waiting times to the second pregnancy was similar to that portion of the model prior to the first pregnancy, except that path dependencies induced by the termination of IUD segments were taken into account. It was these path dependencies that made the model non-Markovian with respect to the set of states under consideration. The structure of the process after the second pregnancy was assumed to be the same as that governing waiting times between the first and second pregnancies. Littman (1977), Littman and Mode (1977), Mode (1978), and Mode and Soyka (1980) may be consulted for further mathematical details on linking the modules just described to build a model of the Taichung experiment.

6.10 Estimates of Transition Functions Associated with First IUD Segment in Taichung Model

Presented in this section are estimates of the transition functions associated with the first IUD segment in the Taichung model. Tables 6.10.1 and 6.10.2

contain estimates of the functions $A_{6j}(x,t)$, $j=7$, 8, 9, for Taichung limiters at selected t-values, according to the age groups: <30, 30–34, 35–39, and ≥ 40. Within each age group, the actuarial method for handling withdrawals, as described in Mode et al. (1977), was used to obtain estimates of these functions on a monthly time scale. Contained in Tables 6.10.3 and 6.10.4 are estimates of the transition probabilities $a_{ij}(x)$, governing transitions into the set $A=[E_1, E_2, E_3, E_4, E_5, E_6, E_9]$ following an expulsion or removal of the initial IUD.

According to the estimates in Tables 6.10.1 and 6.10.2, the most likely cause for the termination of the initial IUD segment in all age groups was removal. Within six years (72 months) following the acceptance of an IUD, the estimated percentages of women having the IUD removed in the age groups <30, 30–34, 35–39, and ≥ 40 were about 43, 40, 39, and 49, respectively. Moreover, at the end of six years, the percentage of women in these age groups experiencing either an expulsion or an *in situ* pregnancy were approximately equal. For example, within the age group <30 in Table 6.10.1, the estimated percentages of women experiencing expulsions and pregnancies within six years were 19.6 and 19.8, respectively. One of the most striking pieces of evidence for age-dependence in Tables 6.10.1 and 6.10.2 was that the fractions becoming pregnant decreased as the ages of the women increased. Thus, the percentages of women experiencing an *in situ* pregnancy within six years following the acceptance of the initial IUD were estimated as 19.8, 16.8, 10.7, and 6.6 for the age groups <30, 30–34, 35–39, and ≥ 40.

Table 6.10.3 contains estimates of the probabilities $a_{7j}(x)$, $E_j \in A$, governing transitions into the set A, following an expulsion of the initial IUD, for the four age groups under consideration. From the estimates in Table 6.10.3, it can be seen that a substantial majority of the women, who terminated the first

Table 6.10.1. Estimated probabilities of terminating initial IUD segment at selected months by cause of termination for Taichung limiters – age groups under 30 and 30–34

Month	Cause of termination			
	expulsion	removal	pregnancy	all causes
	Ages under 30 (1,865 cases)			
6	0.101	0.036	0.047	0.278
12	0.146	0.195	0.080	0.421
24	0.179	0.286	0.143	0.608
48	0.192	0.386	0.185	0.763
72	0.196	0.430	0.198	0.824
96	0.198	0.454	0.201	0.853
	Age group 30–34 (1,728 cases)			
6	0.058	0.090	0.037	0.185
12	0.084	0.146	0.068	0.298
24	0.101	0.227	0.118	0.446
48	0.119	0.332	0.151	0.602
72	0.128	0.391	0.168	0.687
96	0.130	0.435	0.169	0.734

Table 6.10.2. Estimated probabilities of terminating initial IUD segment at selected months by cause of termination for Taichung limiters – age groups 35–39 and 40 or over

Month	Cause of termination			
	expulsion	removal	pregnancy	all causes
	Age group 35–39 (1,107 cases)			
6	0.044	0.085	0.026	0.155
12	0.066	0.129	0.048	0.243
24	0.084	0.196	0.073	0.353
48	0.094	0.316	0.101	0.511
72	0.102	0.393	0.107	0.602
96	0.105	0.465	0.108	0.687
	Age group 40 or over (234 cases)			
6	0.056	0.099	0.009	0.164
12	0.073	0.155	0.026	0.254
24	0.082	0.211	0.035	0.328
48	0.091	0.371	0.043	0.505
72	0.095	0.492	0.066	0.653
96	0.095	0.611	0.066	0.772

Table 6.10.3. Estimated probabilities by age groups of direct transition to other states following an expulsion of initial IUD for Taichung limiters

State	Age group			
	<30 (310)[a]	30–34 (212)	35–39 (146)	≥40 (41)
E_1	0.000	0.000	0.000	0.000
E_2	0.003	0.000	0.007	0.000
E_3	0.377	0.453	0.521	0.583
E_4	0.013	0.005	0.034	0.000
E_5	0.052	0.038	0.075	0.122
E_6	0.452	0.405	0.322	0.293
E_9	0.103	0.099	0.041	0.000

[a] Number of cases where expulsion was observed but not followed immediately by a withdrawal.

Table 6.10.4. Estimated probabilities by age groups of direct transition to other states following removal of initial IUD for Taichung limiters

State	Age group			
	<30 (527)[a]	30–34 (623)	35–39 (520)	≥40 (367)
E_1	0.002	0.005	0.004	0.101
E_2	0.032	0.047	0.046	0.030
E_3	0.577	0.557	0.583	0.662
E_4	0.066	0.085	0.069	0.060
E_5	0.188	0.186	0.192	0.101
E_6	0.046	0.064	0.058	0.038
E_9	0.089	0.056	0.048	0.008

[a] Number of cases where removal was observed but not followed immediately by a withdrawal.

IUD segment with an expulsion, elected either to reaccept an IUD (entered state E_6) or chose no method of contraception (entered state E_3). As the ages of women increased, the fraction of women reaccepting an IUD decreased; while, the fraction entering the null state E_3 increased. See the rows in Table 6.10.3 corresponding to the states E_3 and E_6.

Table 6.10.4 contains estimates of the probabilities $a_{8j}(x)$, $E_j \in A$, following the removal of the initial IUD. As can be seen from Table 6.10.4, less than six percent of the women in all age groups reaccepted an IUD following a removal. In fact, the majority of these women, see the row in Table 6.10.4 corresponding to the null state E_3, elected not to use any method of contraception. A substantial fraction of women did, however, elect to use some conventional method of contraception as can be seen by viewing the estimated probabilities in the row corresponding to state E_5 in Table 6.10.4. Oral contraceptives were not widely used in the Taichung experiment; but, it is interesting to observe from Tables 6.10.3 and 6.10.4, see rows corresponding to state E_4, that a higher fraction of women accepted a pill following a removal than following an expulsion of the initial IUD.

In the next section, the model developed in Sect. 6.9 will be used to describe and summarize data on the Taichung limiters subsequent to the termination of the initial IUD segment and up to the first pregnancy. Attention will also be given to validating the model of Sect. 6.9.

6.11 Validation of Taichung Model

After choosing the set of states defined in Table 6.9.1 and deriving the equations in Sect. 6.9, it is natural to ask: in what sense is the proposed model a valid representation of the Taichung data? There seem to be no completely satisfactory answers to this question, particularly for models as complex as that described in Sect. 6.9. One approach to obtaining answers to the question was to check whether the proposed model was consistent with the data. For example, suppose it is possible to estimate a distribution function of the waiting times to some event, using a procedure depending on the assumptions underlying a model. Furthermore, suppose it is also possible to estimate this distribution function, using a procedure not depending on these assumptions. If the two distribution functions agree sufficiently well, the model could be judged valid, at least to the extent that the assumptions do not lead to contradictions with the data. In other words, the assumptions underlying the model have not been falsified. It was this approach that was followed in all attempts to validate the model of the Taichung data proposed in Sect. 6.9.

Consider, for example, the distribution functions $F_{jk}(x, t)$, $j = 7$ or 8 and $k = 1, 2, 6$, or 9, in (6.9.2), governing waiting times to first entrance into some state in the set $\mathcal{I}_1 = [E_1, E_2, E_6, E_9]$ after experiencing an expulsion (E_7) or a removal (E_8) of the initial IUD. Underlying the derivation of Eq. (6.9.2) was the assumption that times of first entrance into these states, within the branches

B_7 and B_8, were governed by AASM-processes with $\mathscr{I}_1$ as the set of absorbing states and $\mathscr{I}_2 = [E_3, E_4, E_5]$ as the set of transient states. These distribution functions were estimated under the model assumptions used in the derivation of Eq. (6.9.2). Also used in the estimation of these functions was a multiple decrement life table procedure, treating the states in $\mathscr{I}_1$ as competing risks. In this procedure, intermediate transitions among the states in the set $\mathscr{I}_2$ were ignored.

Presented in Table 6.11.1 are pairs of values of the distribution functions $F_{jk}(x, t)$ at $t = 36$ months for the four age groups, following the expulsion of the initial IUD. In each pair, the upper value represented a theoretical one based on model assumptions; the calculation of the lower value in each pair did not depend on model assumptions. The corresponding calculations following the removal of the initial IUD are displayed in Table 6.11.2. As can be seen from Tables 6.11.1 and 6.11.2, the two sets of calculations agree at thirty-six months for all age groups. Good agreement between the two estimates of distribution functions was also obtained at other t-values not reported here, indicating that the assumption of underlying AASM-process governing for entrance into a state in the set $\mathscr{I}_1$, following the termination of the initial IUD segment by either an expulsion or removal, did not lead to contradictions with the data.

Of substantive interest is the observation that the probability of reaccepting an IUD after an expulsion was much higher in all age groups than that following the removal of the initial IUD, as can be seen by comparing the IUD-columns in Tables 6.11.1 and 6.11.2. Within the Taichung experiment, the cost of not reaccepting an IUD after removal of the initial one was a higher probability of becoming pregnant; compare the pregnant-columns in Tables 6.11.1 and 6.11.2.

Another aspect of validating the model prior to the first pregnancy was to consider the H-functions on the right in Eq. (6.9.1). Covered in this portion of the model were IUD segments subsequent to the first but prior to the first pregnancy. Under the assumption that the states E_7 and E_8 are regenerative points of the process, the H-functions satisfy 2×2 systems of renewal type equations represented by (6.9.5) and (6.9.6). The H-functions were estimated by solving these three sets of equations, based on computer input from the Taichung data as described in the derivation of Eqs. (6.9.5) and (6.9.6). Each distribution function was also estimated by a life table procedure, according to a classification determined by the type of last IUD segment prior to the pregnancy. In this estimation procedure, only first visits to the pregnant state E_9 were considered following an expulsion or removal of the initial IUD.

Table 6.11.3 contains pairs of estimates of the H-functions at $t = 6$, 12, and 24 months, calculated according to the two estimation procedures, in that branch of the model representing those women who experienced an expulsion of the initial IUD. Only values for the age groups <30 and 30–34 are presented in Table 6.11.3. Corresponding pairs of values of the H-functions in that branch of the model, representing those women who experienced a removal of the initial IUD, are presented in Table 6.11.4. For example, in the upper panel of Table 6.11.3, columns two, three, and four contain estimates of the function $H_{79}(k, x, t)$, according to the classifications $k = 6$ (*in situ* pregnancy), $k=7$ (expulsion), and $k=8$ (removal), at the values $t=6$, 12, and 24 months and ages $x \le 30$. The numbers in the fifth column are the sums of those in columns two, three, and four.

From an inspection of the values in Tables 6.11.3 and 6.11.4, it can be seen that all pairs agree well. Thus, the assumption that E_7 and E_8 are regenerative points of the process, in that part of the model describing events subsequent to the first IUD segment but prior to the first pregnancy, also did not lead to contradictions between the data and the model. In contrast to estimates in Tables 6.11.1 and 6.11.2, the fractions of women eventually becoming pregnant following the expulsion or removal of the initial IUD are more nearly equal at each time point; compare values in the fifth columns of Tables 6.11.1 through 6.11.4.

At first sight, the close fit between the model and the data, as shown in Tables 6.11.1 through 6.11.4, would seem encouraging. Such close fits might, on the other hand, lead an investigator to suspect they could be an artifact associated with using the same data in the two procedures of calculation. Programming errors might also contribute to the close fits. Careful checks of the computer programs, which were written in FORTRAN, seemed to rule out any significant programming errors. One approach to allaying suspicions associated with the two procedures of calculation would be that of applying them to other large data sets. If these procedures did not always lead to good fits between a proposed model and the data, the results in Tables 6.11.1 through 6.11.4 would seem more credible.

A research exercise of this type was reported by Pickens and Mode (1981), using data on maternity histories from the 1973 United States National Survey of Family Growth. Illustrated in this paper are examples in which semi-Markovian models did not fit the data as well as that proposed for the Taichung experiment. Results of this type suggest that the close fits between the Taichung model and the data were not artifactual. A possible explanation of the lack of fit between the U.S. data and the proposed models was the question of data quality. Because the U.S. data were based on a single interview, rather than on several follows-ups as in the Taichung experiment, they are thought to be a less faithful representation of reality. Certain critical events in contraceptive pregnancy histories, such as removals and expulsions of IUDs, may not have been recorded in the U.S. data. Without information on critical events, affecting subsequent events in a set of histories, it would not have been possible to construct a model that faithfully represented the Taichung data.

A basic assumption entering into the derivation of Eqs. (6.9.5) and (6.9.6) for the H-functions was that the outcome of the first pregnancy was dependent upon the type of last IUD segment prior to the pregnancy. Presented in Table 6.11.5 are estimates of the probabilities of each type of pregnancy outcome classified by age group and type of termination of last IUD segment prior to the pregnancy. By viewing the estimates for Taichung limiters in Table 6.11.5, it is apparent that pregnancies, with an IUD *in situ*, were much more likely to end in either an induced or spontaneous abortion (E_6) than those for the other classifications E_7 and E_8. In the age group <30, for example, the percentages of *in situ* pregnancies ending with induced abortions and other types of outcomes, mostly spontaneous abortions, were about 49 and 19; see row in upper panel corresponding to E_6. As the ages of women increase in Table 6.11.5, so do the estimated probabilities of *in situ* pregnancies ending in induced abortions. When comparing the estimates of pregnancy outcome probabilities for Taichung limiters in Tables 6.5.4 and 6.11.5, it can be seen that the estimates of the probability of induced abortions following *in situ* pregnancies in Table 6.11.5 are higher than the general rates for induced abortions reported in the lower panel of Table 6.5.4. The evidence presented in Table 6.11.5 was crucial in making the decision to include the type of IUD segment prior to the first pregnancy as a factor in the quest for a valid model of the Taichung data.

The tables presented in Sect. 6.10 and this section were abstracted from the Ph.D. thesis, Littman (1977). More extensive tables and further details on the analysis of the Taichung data may be found in this thesis.

Table 6.11.1. Fractions in each of the absorbing states at thirty-six months following expulsion of initial IUD – empirical and theoretical calculations for Taichung limiters

Age group	States			
	menopausal	sterilized	IUD	pregnant
<30	0.004[a]	0.007	0.633	0.316
	0.004[b]	0.007	0.631	0.309
30–34	0.004	0.000	0.616	0.311
	0.000	0.000	0.614	0.310
35–39	0.000	0.034	0.577	0.261
	0.000	0.038	0.579	0.265
≥40	0.052	0.000	0.413	0.221
	0.059	0.000	0.426	0.229

[a] Theoretical calculations based on assumptions underlying Taichung model.
[b] Empirical calculations not dependent on model assumptions.

Table 6.11.2. Fractions in each of the absorbing states at thirty-six months following removal of initial IUD – empirical and theoretical calculations for Taichung limiters

Age group	States			
	menopausal	sterilized	IUD	pregnant
<30	0.004[a]	0.056	0.175	0.601
	0.004[b]	0.058	0.172	0.597
30–34	0.010	0.070	0.201	0.472
	0.010	0.072	0.203	0.474
35–39	0.035	0.080	0.193	0.340
	0.032	0.072	0.201	0.340
≥40	0.231	0.053	0.118	0.117
	0.218	0.045	0.122	0.129

[a] Theoretical calculations based on assumptions underlying Taichung model.
[b] Empirical calculations not dependent on model assumptions.

Table 6.11.3. Distribution functions from expulsion to pregnancy by type of last IUD termination prior to pregnancy

Month	Type of IUD termination			
	in situ pregnancy	expulsion	removal	all pregnancies
	Age group <30			
6	0.038[a]	0.272	0.009	0.319
	0.030[b]	0.280	0.017	0.327
12	0.077	0.360	0.023	0.460
	0.078	0.358	0.027	0.463
24	0.100	0.412	0.046	0.558
	0.106	0.414	0.062	0.582
	Age group 30–34			
6	0.024	0.231	0.003	0.258
	0.033	0.222	0.005	0.260
12	0.042	0.297	0.009	0.348
	0.062	0.280	0.010	0.352
24	0.068	0.353	0.021	0.442
	0.082	0.333	0.029	0.444

[a] Theoretical calculations based on assumptions underlying Taichung model.
[b] Empirical calculations not dependent on model assumptions.

Table 6.11.4. Distribution functions from removal to pregnancy by type of last IUD termination prior to pregnancy

Month	Type of IUD termination			
	in situ pregnancy	expulsion	removal	all pregnancies
	Age group <30			
6	0.006[a]	0.000	0.363	0.369
	0.004[b]	0.000	0.363	0.367
12	0.009	0.001	0.468	0.478
	0.008	0.000	0.482	0.490
24	0.015	0.001	0.578	0.594
	0.010	0.000	0.597	0.607
	Age group 30–34			
6	0.001	0.001	0.252	0.254
	0.003	0.000	0.268	0.271
12	0.007	0.002	0.346	0.355
	0.008	0.002	0.366	0.376
24	0.014	0.003	0.437	0.454
	0.012	0.003	0.456	0.471

[a] Theoretical calculations based on assumptions underlying Taichung model.
[b] Empirical calculations not dependent on model assumptions.

Table 6.11.5. Fractions of each pregnancy outcome by age and last IUD termination prior to first pregnancy for Taichung limiters

Type of IUD termination	Type of pregnancy outcome			Sample size
	live birth	induced abortion	other	
Age group <30				
E_6 [a]	0.315	0.492	0.193	291
E_7 [b]	0.560	0.364	0.076	119
E_8 [c]	0.595	0.329	0.076	316
Age group 30–34				
E_6	0.229	0.617	0.154	334
E_7	0.471	0.447	0.082	85
E_8	0.499	0.435	0.066	366
Age group 35–39				
E_6	0.198	0.634	0.168	202
E_7	0.490	0.431	0.079	51
E_8	0.348	0.573	0.079	227
Age group ≥40				
E_6	0.136	0.743	0.121	44
E_7	0.285	0.551	0.164	19
E_8	0.257	0.621	0.122	75

[a] Pregnancy with IUD *in situ.*
[b] Expulsion.
[c] Removal.

6.12 State and Fertility Profiles for Taichung Limiters

As the Taichung experiment progressed, at each point in time certain fractions of women in the experiment occupied various states of the model as defined in Table 6.9.1. Mathematically, these fractions may be expressed in terms of current state probabilities. Suppose a woman entered the Taichung experiment at $t=0$ when she was age x. Given this event, let $P_{6j}(x, t)$ be the probability she is in state E_j at time $t>0$. These current state probabilities may be positive at each $t>0$ for all states except the instantaneous states E_7 and E_8 for which they are identically zero. Another function of interest was $\mu_{6,10}(x, t)$, the mean number of live births experienced during the time interval $(0, t]$ for a woman of age x when entering the Taichung experiment at $t=0$. Presented in this section are selected computer generated graphs of the current state probabilities and the mean function $\mu_{6,10}(x, t)$.

It should be pointed out that the functions $P_{6j}(x, t)$ and $\mu_{6,10}(x, t)$ apply to women in marriages undissolved by death, separation, or other causes for a ten year period following entrance into the Taichung study. Although each graph begins at an arbitrary time origin labeled $t=0$, in reality women entered the Taichung study over an approximately three year period. Thus, in a sense, the graphs are an abstraction of the data in that the flow of time is not represented absolutely. Algorithms for calculating the functions $P_{6j}(x, t)$, $1 \le j \le 12$, and $\mu_{6,10}(x, t)$ have been discussed in Mode and Soyka (1980).

Contained in Fig. 6.12.1 are plots of the function $P_{6j}(x, t)$ for the states $E_1(j=1)$ and $E_3(j=3)$ for women aged 25–29 when entering the Taichung study. In absolute terms, the fractions of women occupying either the menopausal (E_1) or null (E_3) states never exceeded 0.08. Within this upper bound, however, the profile of the null state, representing women unprotected by contraception, exhibited an interesting pattern. At $t=0$, the fraction of women in the null state E_3 was zero, corresponding to all women accepting an IUD when entering the Taichung experiment. Associated with a sharp decline in IUD usage, as depicted in Fig. 6.12.2, was a steep rise in the null-curve, which reached a maximum of about 0.08 at $t=1.25$ years. Thereafter, the curve declined to a low of about 0.06 at $t=7.25$ and then seemed to level off at this fraction. The steady rise in the menopausal curve in Fig. 6.12.1 seems to be congruent with expectations, given women aged 25–29 when entering the Taichung study.

Figure 6.12.2 contains profiles of those states representing women protected by some method of contraception. Presented in this figure are plots, as functions of t, of the current state probabilities $P_{6j}(x, t)$ for the states E_j, $j=2, 3, 4, 5$, and 6, representing women (couples) currently sterilized (E_2), or currently using a pill (E_4), a conventional method of contraception (E_5), or an

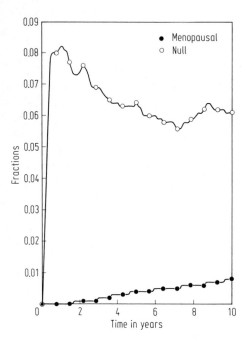

Fig. 6.12.1. Fractions of menopausal and unprotected women aged 25–29 when entering Taichung experiment

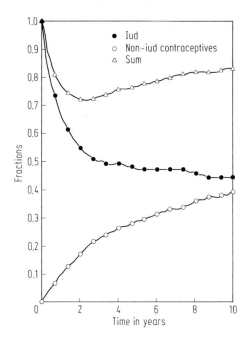

Fig. 6.12.2. Fractions of women protected by contraception aged 25–29 when entering Taichung experiment

IUD (E_6). In order to make the number of simultaneous plots manageable, the current state probabilities for the states E_2, E_4, and E_5 have been combined into a sum. Attention was again focused on women aged 25–29 when entering the Taichung experiment.

At $t=0$, the IUD curve in Fig. 6.12.2 is one, representing the fact that all women entered the Taichung study by accepting an IUD. As time increased, the IUD curve decreased rather sharply during the first four years but then decreased slowly and leveled off at about 0.44 in the tenth year. Among the reasons underlying this leveling off was the fact that many women were reaccepting IUDs.

Associated with a decrease in the IUD curve in Fig. 6.12.2 was a rise in the curve for the fraction of women (couples) who had accepted a sterilization or were using either a pill or a conventional method of contraception, i.e., non-IUD contraceptives. Sterilizations contributed most to the rise in this curve. For example, at ten years after the start of the study, the fraction of the cohort in one of these states was 0.41, with the states E_2, E_4, and E_5 contributing the approximate fractions 0.28, 0.04, and 0.09, respectively.

The upper curve in Fig. 6.12.2 is the sum of the two lower ones and represents the fraction of women in the Taichung study protected by some method of contraception as a function of time. That a substantial majority of the women were protected by some method of contraception is illustrated by the fact that this curve reached about 0.82 at ten years. A question that naturally arises is: what impact did this high level of contraceptive protection have on the fractions of women in states associated with pregnancy?

Presented in Fig. 6.12.3 are the profiles for the states associated with pregnancy for women aged 25–29 when entering the Taichung experiment. The functions plotted in this figure are $P_{6j}(x, t)$ for the pregnant state $E_9(j=9)$, and the temporary sterile state $E_{10}(j=10)$, $E_{11}(j=11)$, and $E_{12}(j=12)$ following a live birth, induced abortion, or some other type of pregnancy outcome. As can be seen from Fig. 6.12.3, the sum of the fractions of the cohort in the states E_9, E_{10}, E_{11}, and E_{12} never exceeded a value of about 0.22, the value of the sum a little beyond two years. Within this bound, the state profiles in Fig. 6.12.3 also exhibit interesting patterns.

As expected, associated with the steep rise in the null-curve in Fig. 6.12.3, representing the fraction of women in the null state E_3 unprotected by contraception, was a delayed rise in the fraction of women in the pregnant state E_9 as depicted by the pregnant-curve in Fig. 6.12.3. This curve reached a maximum of about 0.11 between $t=1.0$ and $t=1.25$ years but then declined.

A graphical representation of the ways unwanted pregnancies were resolved among Taichung limiters is presented in Fig. 6.12.3 for the curves representing sterile states following a live birth, induced abortion, or some other type of pregnancy outcome. Among the graphs representing current state probabilities for these states, that for E_{10}, the live birth-curve, dominates the others, indicating that a majority of unwanted pregnancies experienced by women, aged 25–29 when entering the study, were terminated with live births. The steep rise in the induced abortion-curve, during the first year following entrance into the study, and its subsequent leveling off at about 0.02 indicates that a substantial number of pregnancies were terminated by induced abortions. Compared to states E_{10} and E_{11}, the fractions of women in the state E_{11}, as represented by the other-curve in Fig. 6.12.3, were low throughout the time interval covered in the figure.

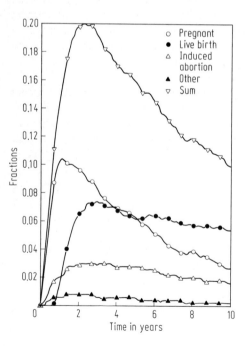

Fig. 6.12.3. Fractions of women in states associated with pregnancy aged 25–29 when entering Taichung experiment

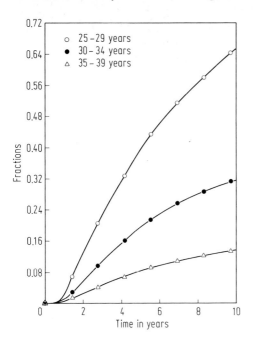

Fig. 6.12.4. Mean cumulative fertility of women aged 25–29, 30–34, and 35–39 when entering Taichung experiment

The forms of the state profiles for women aged 30–34 and 35–39 when entering the Taichung study were similar, in many ways, to those in Fig. 6.12.1, 6.12.2, and 6.12.3 for the age group 25–29. Among the features differentiating the state profiles for these older age groups from those of the younger was a greater incidence of induced abortions in resolving unwanted pregnancies. No further space will be used to present the state profiles for women aged 30–34 and 35–39 when entering the Taichung study.

Presented in Fig. 6.12.4 are graphs of the mean function $\mu_{6,10}(x,t)$ for women aged 25–29, 30–34, and 35–39 when entering the Taichung experiment. These graphs provide visual links between the state profiles presented in Figs. 6.12.1 to 6.12.3 (at least for the age group 25–29), and live births experienced by Taichung limiters. At 2.5 years following entry into the study, the mean number of live births experienced by women in the age groups 25–29, 30–34, and 35–39 was estimated at 0.18, 0.09, and 0.04, respectively. The corresponding estimates for ten years following entrance into the study were about 0.66, 0.32, and 0.14. Since the maximum time period an individual could be observed in the study was about nine years, these latter estimates should correspond fairly closely to the average fertility actually experienced by women who entered the study. The actual fertility experienced by these women is not completely known, however, because they either withdrew from the study or the observations were cut off prior to completing their fertility histories. Littman and Mode (1977) may be consulted for an attempt to estimate the actual fertility experienced by Taichung limiters.

6.13 Implications of the Taichung Experiment
for Evaluating Family Planning Programs

A feature differentiating the evaluation of the Taichung experiment from that of other family planning programs was that data on the subjects of study were collected over an extended period of time. Collecting longitudinal data in evaluating family planning programs seems to be the exception rather than the rule; for, even to this day attention is frequently focused on single segments of contraceptive use or short time periods in attempts to evaluate the demographic impact of family planning programs. Discussed in this section are three implications of the Taichung experiment that should be useful in evaluating other family planning programs. For the point of view of analyzing longitudinal data by linking AASM-processes in a kind of time series, the current state probabilities have proved to be very useful in presenting the vivid graphical evidence on which these implications rest.

The three implications may be stated as follows:

1. The act of accepting an IUD with the intention of averting pregnancies did not prevent subsequent pregnancies for many women.

2. By interacting with the program, women were, nevertheless, learning how to prevent unwanted pregnancies.

3. Short-term transient phenomena can easily mask the positive effects of program intervention, effects that simply require several years to become manifest. In the case of the Taichung experiment, a program evaluation based on the initial IUD segment or even one to three years of data would have been too pessimistic relative to longer term performance.

In the early 1960s, IUDs were viewed with great optimism in some circles as a method of preventing pregnancies on a long term basis. That this optimism was not completely justified is vividly demonstrated by the steep rise in the pregnant-curve in Fig. 6.12.3 to about 10% at two years following entrance into the Taichung study. This steep rise in the pregnant-curve justifies the first implication.

If women in the Taichung study had not been learning to prevent pregnancies, a leveling off of the pregnancy-curve after its initial rise would have been expected. Any subsequent decline in the curve would have been associated with women becoming physiologically sterile as they aged. From an inspection of the pregnancy-curve in Fig. 6.12.3, it can be seen that this leveling off did not occur for women aged 25–29 when entering the Taichung experiment. The shape of the pregnancy-curve was essentially the same for all age groups in the experiment, indicating that the onset of permanent sterility with advancing age was not the decisive factor underlying the rise and decline in the curve. The second implication, therefore, seems justified.

The third implication seems justified in view of the shape of the pregnancy-curve in Fig. 6.12.3. For, clearly, if observations had been cut off after two or three years, the decline in the pregnancy-curve would not have been observed.

Whether to observe single or multiple segments of contraceptive use in evaluating the impact of contraceptives on fertility is a methodological question of continuing interest. Pickens and Soyka (1980) and Pickens and Mode (1981) have also provided examples illustrating that confining attention to a single segment of contraceptive use can be misleading. Semi-Markovian processes also formed the basis for analyzing the Taiwanese and U.S. data used in these papers.

6.14 On Measuring the Fertility Impact of Family Planning Programs

A major reason for the existence of family planning programs, sponsored by either a private or governmental organization, is to assist women (couples) in averting unintended births. Family planning programs often compete with other programs for limited funds, leading administrators to seek advice in evaluating them. A common question asked in evaluating such programs is: how many unintended births did the program avert? Any answer to this question is bound to be speculative; for, an estimate of the number of births averted will always depend on what an analyst assumes the fertility histories of women would have been if they had not had access to the family planning methods offered by the program.

Reported in this section are the results of a computer simulation experiment in which the model, described in previous sections, was used to yield estimates of births averted by the Taichung experiment. Due to the limited amount of information available to most analysts, estimates of the fertility of women after accepting a method of contraception are also frequently based on a set of assumptions that are difficult to validate. A distinct advantage of using the Taichung model in a simulation experiment is that it captures numerical information describing the actual use of contraception and induced abortion to avert unintended births and is thus nearly free of arbitrary assumptions. A disadvantage of the Taichung model is that it is of limited applicability, because the model was centered around IUD usage rather than oral contraceptives that are now more widely used throughout the world.

A computer run was made in which the input for parities 0, 1, and 2 was the same as that described for runs III and IV in Table 6.6.2. At the end of an anovulatory period following the third live birth, it was assumed that all women in the simulated cohort accepted an IUD to avert unintended births. Thereafter, their fertility histories followed those of the women in the Taichung experiment. In other words, through the third live birth, it was assumed that the maternity histories of women followed those based on the analysis of KAP-IV data. After attaining the desired number of three live births, it was assumed that maternity histories followed those of the Taichung experiment in order to avert unintended births.

Besides estimates of births averted, fractions of women in a cohort using contraceptives dispersed by the program may also be used in evaluating program performance. In the above simulation exercise, as well as in the real world, not all women in a cohort would start using an IUD at the same time, because maternity histories are not synchronized. It is, therefore, of interest to study fractions of women protected by contraception as a function of duration of marriage, in order to provide insights as to how the non-synchronization of maternity histories affects estimates of women protected by contraception.

Depicted in Fig. 6.14.1 are the fractions of women protected by contraception as a function of duration of marriage for women marrying at age 20. All graphs in Fig. 6.14.1 were based on estimates for the current state probabilities. As expected from the computer input to the simulation run, the IUD-curve did not begin to rise from zero until about five years following marriage, the time some women in the cohort had experienced three live births. During the next five years, the IUD-curve rose rapidly and reached a level of about 0.43 at ten years following marriage. At that point, the fraction of women (couples) who had either been sterilized or were using a pill or conventional method of contraception was about 0.18. Altogether, at ten years following marriage, the fraction of women protected by methods of contraception offered by the program was about 0.61. At 15 years following marriage at age 20, the fraction of women using an IUD was about 0.40; but, the fraction of women in one of the states E_2, E_4, or E_5 had risen to about 0.29. Contrary to what might be expected, only about 69% of the women were protected at that time by some method of contraception offered by the program. Since all women in the simulated cohort were destined to enter the program, this estimate of percent protected after 15 years of marriage seems lower than might be expected. When evaluating family planning programs, it would, therefore, seem advisable to take into account such an apparent underestimate of percent protected due to non-synchronization of maternity histories in a cohort.

Table 6.14.1 contains estimates of births averted by the Taichung experiment when used as a simulation model. All calculations in the simulation run under consideration were carried out on a monthly time scale and yielded an

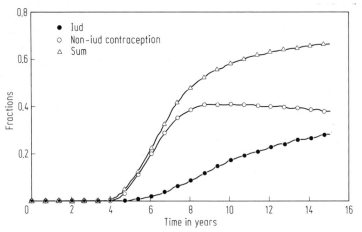

Fig. 6.14.1. Fractions of women protected by contraception as function of duration of marriage for women marrying at age 20

estimate of 3.390 births per woman marrying at age 20 and remaining married to age 50. In making comparisons with computer runs III and IV, as described in Table 6.6.2, it thus become necessary to adjust all calculations to a monthly time scale.

Recall that when calculations were carried out on a monthly time scale, computer run IV yielded an estimate of 4.251 live births per woman in a cohort directly comparable to that in the simulation run, see Table 6.8.3. Unfortunately, calculations in computer run III were carried out on a quarter year rather than a monthly time scale. These calculations yielded an estimate of 5.163 births per woman marrying at age 20 and remaining married to age 50; see the row in the upper panel of Table 6.7.2 corresponding to the symbol E_4. To adjust for downward bias in this estimate due to the quarter-year time scale, this estimate was multiplied by 1.0567, the ratio of 4.251 to 4.023 for the monthly and quarter-year time scale estimates in Table 6.8.3, to yield the estimate of 5.456 for run III in Table 6.14.1. Before making comparisons among runs III and IV and the simulation run, it should be recalled that the input to runs III and IV differed only in respect to the high levels of induced abortion used by Taichung limiters after the third live birth.

According to the estimates presented in Table 6.14.1, when compared to computer run III, in which essentially no induced abortion was used to avert births, a cohort with maternity histories like those in the Taichung simulation run would experience 2.066 fewer live births on the average. The difference between the means for computer run IV and the Taichung simulation run was 0.861. The difference between these two estimates of births averted was 1.205. When compared with 0.861, this difference suggests that induced abortions were a more potent factor than contraception in averting unintended births among women in the Taichung experiment. In the light of the calculations presented in Table 6.14.1, the Taichung experiment seems to have been an effective family planning program. At the same time, however, it pointed to a need for improving contraceptive technology in order to avert high levels of induced abortions.

With respect to assessing the quality of estimates in Table 6.14.1, it needs to be pointed out that calculations in runs III and IV were based on the renewal matrix algorithms described in Sect. 6.3; while, those in the Taichung simulation run were based on the modular procedure for parities 0, 1, and 2 as

Table 6.14.1. Estimates of births averted by Taichung medical IUD experiment for women marrying at age 20

Mean number of live births[a]			Mean births averted	
run III (1)	run IV (2)	Taichung simulation (3)	(1)–(3)	(2)–(3)
5.456[b]	4.251	3.390	2.066	0.861

[a] It was assumed women remained married to age 50.
[b] Adjusted upward from quarter-year time scale to monthly time scale by ratio and proportion.

described in Mode and Soyka (1977). Biases due to non-associativity of age-dependent convolutions, resulting from different aggregations of the age and time scale, may thus be present in the calculations for the Taichung simulation run. The estimate of 3.390 births per woman for this run does, however, seem reasonable. Moreover, it is believed that any biases present in the calculations are not sufficiently large to change any of the above conclusions regarding births averted. All calculations presented in Sect. 6.12 and this section were based on FORTRAN programs written by M.G. Soyka.

6.15 Conclusions and Further Reading

From the discussion in the preceding sections of this chapter, it is clear that the development and implementation of simulation models closely linked to data analyses commands a high price in terms of time and labor devoted to computer programming. It should also be emphasized that the labors of many people were involved in gathering and processing the data into a usable form. Special computer tapes were also made in order to decrease computer costs in making the many explorations of the data that were an integral and necessary part of developing the models considered in this chapter.

Although data analyses are a necessary part of developing fruitful linkages between theory and practice, special properties of data sets, such as emphasis on IUD's in the Taichung experiment, can inhibit developing more comprehensive views of population dynamics, because of the extensive time and labor required to analyze each large data set. It is also very difficult to make large two-dimensional numerical arrays, like those used in the age-dependent models considered in this chapter, user-friendly. In this connection, having age and time as covariates in parameterized models of competing risks could greatly expedite the computer generation of two-dimensional arrays with the same aggregation of the age and time scales. Numerical problems, due to non-associativity of age-dependent convolutions, would thus be eliminated.

Even with a capability for generating large two-dimensional arrays, the problem of having access to a computer with sufficient storage capacity to handle them remains. Computer costs arising from the manipulation of large arrays may also be a problem. Problems of this nature also arise in physics, where the accurate numerical processing of three-dimensional arrays is crucial to understanding certain physical phenomena. Like many simplified mathematical models, MATHIST, the user-friendly simulation system described in Sect. 5.5, which requires only one-dimensional arrays as computer input, is a useful half-way house between complicated real-world human populations and our desires to understand them. In fact, only after it became clear that it would be difficult to make age-dependent models user-friendly was the decision made to implement the simple models underlying MATHIST. Further experimentation with MATHIST may also provide helpful insights, regarding the degree to which age-dependent models are crucial to understanding the dynamics of human populations.

Measuring the impact of family planning programs on fertility has been a methodological concern for at least a decade. Among the computer systems that have been widely used in supplying solutions to the problem are PROJ-TARG, a system developed by Bogue et al. (1973), and CONVERSE, a system developed by Nortman et al. (1977). Unlike the simulation systems developed in this book, the designs of PROJTARG and CONVERSE were based on perceived algebraic relationships rather than on unifying principles with roots in theories of stochastic processes. Their underlying designs can perhaps be best understood by reading many lines of FORTRAN, a task that is both disagreeable and difficult.

Briefly, in CONVERSE only single segments of contraceptive use are taken into account. As has been observed in the Taichung experiment, focusing attention on single segments can lead to an overly pessimistic view of the program's capabilities for averting unintended births. Following the acceptance of a method of contraception, it is assumed in CONVERSE that, among those who do not discontinue the method immediately, the continuation of contraception is governed by an exponential distribution, with a parameter depending on the method. A factor that may lead to difficulties in interpreting the output from CONVERSE is that negative age-specific birth rates may arise when they are adjusted for contraception; rates that in theory should be nonnegative. Negative age-specific birth rates in the computer output suggest that the algebraic relations determining the mathematical design of CONVERSE are not internally consistent. An advantage of the simulation systems developed in this book is that their mathematical designs are internally consistent, because they are based on the unifying principles of renewal theory. When designs are based on such principles, problems with negative birth rates, which result from taking contraception into account, never arise.

Differences in the mathematical designs under systems can also lead to difficulties in applying them to measure the impact of a family planning program on fertility. Khoo (1979), for example, has used both PROJTARG and CONVERSE in measuring the impact of the Thai Family Planning Program on fertility. Even though the computer input to the two systems was the same, except for differences required by the designs, quite different estimates of births averted by the program during the period 1965–1975 were obtained. In general, estimates of births averted according to PROJTARG were uniformly higher than those for CONVERSE throughout the period. In 1975, the estimate of the number of births averted according to PROJTARG was 501,000; while that for CONVERSE was 398,900, a difference of 102,000. A drawback common to the designs of both PROJTARG and CONVERSE in estimated births averted by a family planning program is that no distinctions are made between spacers and limiters. Rapid changes in hardware that have occurred since PROJTARG and CONVERSE were developed also limit their use. Despite difficulties in their designs and other limitations, the pioneering efforts represented by PROJTARG and CONVERSE have not only stimulated further research but have also proved to be useful in many situations.

There is an extensive literature on evaluating family planning programs. Books on the subject include those of Chandrasekaran and Hermalin (1975) and that of Cutright and Jaffe (1977). Potter (1974) has also contributed a small monograph on measuring the impact of contraception on fertility. Watson and Laphan (1975) provided a world review of family planning programs and Freedman and Berelson (1976) have discussed the record of family planning programs. Other papers directly related to measuring the fertility impact of family planning programs include those of Mode and Littman (1976) and Teachman et al. (1978).

Papers of methodological interest in evaluating family planning programs include those of Chiang and VanDenBerg (1982) and Poole (1973). Unlike the terminating renewal processes considered in this book, the analysis of Poole was based on nonterminating renewal processes in which parity progression ratios were ignored. A paper of interest for further study is that of Shactman and Hogue (1976). In this paper, a time variable was included in the state space of a Markov chain. By studying this paper along with the stochastic processes discussed in this book, further insights for developing guidelines for choosing state spaces may be obtained. Finally, it should be noted that the stochastic processes discussed in this chapter have connections with multidimensional mathematical demography, see Land and Rogers (1982) for a collection of papers on the subject.

Problems and Miscellaneous Complements

1. Discussed in this problem is a method for finding numerical solutions to matrix-valued, age-dependent, renewal type equations of the form displayed in (6.2.13).

(a) Let $\mathbf{q}(x, t)$ be the $r_2 \times r_2$ matrix-valued function defined in connection with Eq. (6.2.13). Let $\mathbf{q}^{(0)}(x, t)$ be a $r_2 \times r_2$ matrix-valued function such that $\mathbf{q}^{(0)}(x, 0) = \mathbf{I}$ and $\mathbf{q}^{(0)}(x, t) = \mathbf{0}$ when $t \geq 1$ for every $x \geq 0$. In this discussion, $\mathbf{I}$ and $\mathbf{0}$ are the $r_2 \times r_2$ identity and zero matrix, respectively. Show that

$$\sum_{s=0}^{t} \mathbf{q}^{(0)}(x, s)\mathbf{q}(x+s, t-s) = \mathbf{q}(x, t)$$

and

$$\sum_{s=0}^{t} \mathbf{q}(x, s)\mathbf{q}^{(0)}(x+s, t-s) = \mathbf{q}(x, t)$$

for all points $\langle x, t \rangle$. Thus, one may conclude that $\mathbf{q}^{(0)}(x, t)$ is the identity function with respect to the operation of age-dependent convolutions for matrix-valued functions.

(b) Suppose $\mathbf{q}(x, 0) = \mathbf{0}$, a $r_2 \times r_2$ zero matrix, for all $x \geq 0$. The $r_2 \times r_2$ matrix-valued function $\mathbf{m}(x, t)$ is said to be an inverse of the function $\mathbf{q}^{(0)}(x, t) - \mathbf{q}(x, t)$

on a finite lattice of $\langle x, t \rangle$-points, if

$$\sum_{s=0}^{t} \mathbf{m}(x, s)(\mathbf{q}^{(0)}(x+s, t-s) - \mathbf{q}(x+s, t-s)) = \mathbf{q}^{(0)}(x, t)$$

at all points $\langle x, t \rangle$ on the lattice. Use an argument analogous to that used in proving Theorem 4.8.1 to show that, for every $x \geq 0$, $\mathbf{m}(x, 0) = \mathbf{I}$, and

$$\mathbf{m}(x, t) = \sum_{s=0}^{t-1} \mathbf{m}(x, s) \mathbf{q}(x+s, t-s)$$

for $t \geq 1$. This last equation is a recursive in $t \geq 1$ for every $x \geq 0$ and can easily be implemented on a computer. It can also be shown that $\mathbf{m}(x, t)$ is a right-inverse of the function $\mathbf{q}^{(0)}(x, t) - \mathbf{q}(x, t)$.

(c) If $*$ represents the operation of age-dependent convolution and the argument $\langle x, t \rangle$ is dropped, then Eq. (6.2.13) may be represented in the form

$$(\mathbf{q}^{(0)} - \mathbf{q}) * \mathbf{P} = \mathbf{D}.$$

Equation (6.2.14) results when the operation $*$ is applied to both sides of this equation. Observe that the associativity of $*$, $A*(B*C) = (A*B)*C$, makes (6.2.14) valid.

2. Outlined in this problem is a justification for the interpretation given the function in (6.2.23) and the formulas in (6.3.30) and (6.3.31).

(a) Let $\mathbf{q}(x, t)$ be the $r_2 \times r_2$ matrix-valued function defined in connection with Eq. (6.2.13); and, let $\mathbf{q}^{(0)}(x, t)$ be the identity defined in 1(a). For $n \geq 1$, define a sequence $\mathbf{q}^{(n)}(x, t)$ of matrix-valued functions by

$$\mathbf{q}^{(n)}(x, t) = \sum_{s=0}^{t} \mathbf{q}^{(n-1)}(x, s) \mathbf{q}(x+s, t-s);$$

and, let

$$\mathbf{Q}^{(n)}(x, t) = \sum_{s=0}^{t} \mathbf{q}^{(n)}(x, s).$$

Then, the renewal matrix $\mathbf{M}(x. t)$ for an absorbing age-dependent semi-Markov process may be represented in the form

$$\mathbf{M}(x, t) = \sum_{n=0}^{\infty} \mathbf{Q}^{(n)}(x, t).$$

At every point $\langle x, t \rangle$, only finitely many terms in the series are non-zero matrices.

(b) Let $X(n, j; t)$ be a random variable such that $X(n, j; t) = 1$ if transient state j is entered at the n-th step, $n \geq 0$, sometime during the time interval $(0, t]$, $t > 0$, and let $X(n, j; t) = 0$ otherwise. Then,

$$X(j; t) = \sum_{n=0}^{\infty} X(n, j; t)$$

is a random variable giving the number of visits to state j during the time interval $(0, t]$. Suppose state i is entered at age x and let $\langle i, x \rangle$ symbolize this event. Then.

$$E[X(n, j; t)|\langle i, x\rangle] = Q_{ij}^{(n)}(x, t),$$

the ij-th element in the matrix $\mathbf{Q}^{(n)}(x, t)$. Therefore,

$$E[X(j; t)|\langle i, x\rangle] = \sum_{n=0}^{\infty} E[X(n, j; t)|\langle i, x\rangle] = M_{ij}(x, t).$$

When this last equation is expressed in the notation of the stochastic process discussed in Sect. 6.3, the interpretation given the function in (6.3.23) is justified.

(c) Let $Z(t)$ be the random function defined in the discussion preceding (6.2.11) and define a random function $Y(j; t)$ such that $Y(j; t) = 1$ if $Z(t) = j$ and let $Y(j; t) = 0$ otherwise. Then,

$$E[Y(j; t)|\langle i, x\rangle] = P_{ij}(x, t),$$

the current state probability defined in (6.2.11). The random function

$$W(j; t) = \sum_{s=1}^{t} Y(j; s)$$

gives the number of time units spent in state j during the time interval $(0, t]$, $t > 0$. By taking conditional expectations, it follows that

$$E[W(j; t)|\langle i, x\rangle] = \sum_{s=1}^{t} P_{ij}(x, s).$$

Equation (6.3.31) is thus justified, after appropriate changes in notation.

3. Formulas (6.4.1) and (6.4.4) were used in implementing the age-dependent model of maternity histories developed in Sect. 6.3. Outlined in this problem are the arguments used in their derivations.

(a) Consider the model discussed in Sect. 3.2 with $m = 2$ competing risks and suppose it is determined by two independent latent random variables Y_1 and Y_2 with distribution functions F_1 and F_2. Let f_1 and f_2 be their densities. Use the arguments of Sec. 3.3 to show that

$$P[X \leq x, C = 1] = G_1(x) = \int_0^x (1 - F_2(s)) f_1(s) ds, \qquad x \geq 0.$$

Arguments of the type used in deriving the above equation may be used to justify the formulas in (6.4.1).

(b) When the latent random variables $Y_1, Y_2, \ldots, Y_m$ described in Sect. 3.3 are independent with distribution functions $F_1, F_2, \ldots, F_m$, then the latent risk function μ_j of the distribution function F_j is determined by Eq. (3.2.10). If it is recalled that $S(x) = 1 - G(x)$ for $x \geq 0$, then formula (6.4.4) follows immediately from Eq. (3.2.10).

References

Barrett, J.C. (1969): A Monte Carlo Simulation of Human Reproduction. *Genus* 25:1–22

Berman, S.M. (1963): A Note on Extreme Values, Competing Risks and Semi-Markov Processes. *Annals of Mathematical Statistics* 34:1104–1106

Bogue, D.J., Edmonds, S. and Bogue, E.J. (1973): *An Empirical Model for Demographic Evaluation of Impact of Contraception and Marital Status on Birth Rates.* Community and Family Study Center. University of Chicago

Chandrasekaran, C. and Hermalin, A.I. (Editors) (1975): *Measuring the Effect of Family Planning Programs on Fertility.* Ordina Editions, 4830 Dolhain, Belgium

Chiang, C.L. and VanDenBerg, B.J. (1982): A Fertility Table for the Analysis of Human Reproduction. *Mathematical Biosciences* 62:237–251

Cutright, P. and Jaffe, F.S. (1977): *Impact of Family Planning Programs on Fertility – The U.S. Experience.* Praeger Publishers, New York, Washington, London

Freedman, R. and Berelson, B. (1976): The Record of Family Planning Programs. *Studies in Family Planning* 7:1–40

Freedman, R., Coombs, L.C., Chang, M.C. and Sun, T.H. (1974): Trends in Fertility, Family Size Preferences, and Practice of Family Planning – Taiwan, 1965–1973. *Studies in Family Planning* 5:270–288

Freedman, R. and Takeshita, J.Y. (1969): *Family Planning in Taiwan.* Princeton University Press, Princeton, N.J.

Hoem, J.M. (1972): Inhomogeneous semi-Markov Processes, Select Actuarial Tables, and Duration-Dependence in Demography. In *Population Dynamics* (T.N.E. Greville, ed.) pages 251–296. New York, Academic Press

Kemeny, J.G. and Snell, J.L. (1976): *Finite Markov Chains.* Springer-Verlag, New York, Heidelberg, Berlin

Khoo, S.E. (1979): Measuring the Thai Family Planning Program's Impact on Fertility Rates – A Comparison of Computer Models. *Studies in Family Planning* 10: 137–145

Land, K.C. and Rogers, A. (Editors) (1982): *Multidimensional Mathematical Demography.* Academic Press, New York

Leridon, H. (1977): *Human Fertility – The Basic Components.* University of Chicago Press, Chicago, Illinois

Littman, G.S. (1977): Development and Validation of a Non-Markovian Stochastic Model for the Taichung Medical IUD Experiment. Ph.D. Thesis, Drexel University Library, Philadelphia, Pa. 19104

Littman, G.S. and Mode, C.J. (1977): A non-Markovian Stochastic Model for the Taichung Medical IUD Experiment. *Mathematical Biosciences* 34:279–302

Mode, C.J. (1976): On the Calculation of the Probability of Current Family-Planning Status in a Cohort of Women. *Mathematical Biosciences* 31:105–120

Mode, C.J. (1977): Some Techniques in Building and Validating Stochastic Models of Human Reproduction. *Bulletin of Mathematical Biology* 39:693–704

Mode, C.J. (1978): Algorithms for a Stochastic Model of Human Fertility Under Contraception. *Mathematical Biosciences* 38:217–246

Mode, C.J. (1982): Increment-Decrement Life Tables and Semi-Markovian Processes From a Sample Path Perspective. In *Multidimensional Mathematical Demography* (K.C. Land and A. Rogers, editors) pages 535–565. New York, Academic Press

Mode, C.J., Avery, R.C., Littman, G.S. and Potter, R.G. (1977): Methodological Issues Underlying Multiple Decrement Life Table Analysis. *Demography* 14:87–96

Mode, C.J. and Littman, G.S. (1976): Implications of Fertility Patterns in Korea – A Study by Computer Simulation. *International Journal of Biomedical Computing* 7:289–305

Mode, C.J. and Soyka, M.G. (1977): A Stochastic Model of Human Fertility Under Natural Fecundability – Algorithms and Numerical Examples. *Mathematical Biosciences* 35:187–209

Mode, C.J. and Soyka, M.G. (1980): Linking Semi-Markov Processes in Time Series – An Approach to Longitudinal Data Analysis. *Mathematical Biosciences* 51:141–164

Moots, B.L. (1975): Factors Affecting the Impact of IUDs on Fertility Control. Pages 71–86 in F. Hefnawi and S.J. Segal (eds.). *Analysis of Intrauterine Contraception.* North-Holland, Amsterdam

Nortman, D.L., Potter, R.G., Kirmeyer, S.W. and Bongaarts, J. (1977): *Model Relations Between Birth Rates and Birth Control Practices*. Rolumes I and II. The Population Council, New York, N.Y.

Pickens, G.T. and Mode, C.J. (1981): Sequence of Events Following Adoption of Contraception – An Exploratory Analysis of 1973 Fertility History Data. *Social Biology* 28:111–125

Pickens, G.T. and Soyka, M.G. (1980): Effects of Changed Contraceptive Patterns on Fertility in Taiwan – Applications of a Non-Markovian Stochastic Model. *International Journal of Biomedical Computing* 11:1–19

Pittenger, D.B. (1973): An Exponential Model of Female Sterility. *Demography* 10:113–124

Poole, W.K. (1973): Fertility Measures Based on Birth Interval Data. *Theoretical Population Biology* 4:357–387

Potter, R.G. (1974): *Some Techniques for Measuring the Impact of Contraception*. Asian Population Studies Series No. 18, United Nations, New York

Potter, R.G., Ford, K. and Moots, B. (1975): Competition Between Spontaneous and Induced Abortion. *Demography* 12: 129–141

Potter, R.G., Mode, C.J., Soyka, M.G. and Avery, R.C. (1979): Measuring Potential Fertility Through Null Segments – An Exploratory Analysis. *Social Biology* 26:314–328

Shactman. R.H. and Hogue, C.J. (1976): Markov Chain Model for Events Following Induced Abortion. *Operations Research* 24:916–932

Soyka, M.G. (1983): *Computer-Aided Applications of non-Markovian Macro-Simulation Models to the Study of Human Populations*. Ph.D. Thesis, Drexel University Library, Philadelphia, Pa. 19104

Teachman, J.D., Hogan, D.P. and Bogue, D.J. (1978): A Components Method for Measuring the Impact of a Family Planning Program on Birth Rates. *Demography* 15:113–129

Tuan, C.H. (1958): Reproduction Histories of Chinese Women in Rural Taiwan. *Population Studies* 12:40–50

Watson, W.B. and Laphan, R.J. (1975): Family Planning Programs – World Review 1974. *Studies in Family Planning* 6:205–322

Chapter 7. Population Projection Methodology Based on Stochastic Population Processes

7.1 Introduction

Models of maternity histories in female birth cohorts, although interesting in their own right, are only one facet of the dynamics of human populations. In order to gain insights into the implication of these models from the standpoint of the evolutionary development of human populations, they need to be linked, primarily through cohort fertility density functions, to a methodology for making population projections. The primary purpose of this chapter is to provide such a linkage to a population projection methodology based on a class of stochastic processes in discrete time with either time homogeneous or time inhomogeneous laws of evolution.

A focal point of attention in the class of stochastic processes to be considered in this chapter is the evolutionary development of a birth cohort. If individuals in a one-sex population are accounted for in terms of lines of descent stemming from the initial individuals in the population at some epoch, then the population process may be represented as a generalized age-dependent branching process in discrete time, with either time homogeneous or inhomogeneous laws of evolution. In their present state of development, renewal type equations are the principal tool used in the analysis of such branching processes. If, on the other hand, individuals in a one-sex population are accounted for only in terms of birth cohorts, then the population process may be represented in terms of random sums of random functions.

Both representations of the population process are useful. For processes with time homogeneous laws of evolution, the branching process formulation, together with renewal theory, yields many formulas from stable population theory as asymptotic results, as well as formulas usually not associated with this theory, in a simple and rigorous way. But, when computational algorithms for the covariance functions, describing stochastic variability among realizations of the population process, are desired, then a representation based on random sums of random functions provides a tractable viewpoint from which the desired algorithms may be derived. A recursive algorithm of the Leslie matrix type for computing the mean age structure of the population may be derived from either representation of the population process. Unlike many approaches to making population projections, the Leslie matrix is not a point of departure in the analysis but arises as a computing algorithm based on the assumptions underlying the process, assumptions that will be spelled out in more detail as the chapter unfolds.

A unifying framework of sufficient generality to encompass not only formulas from stable population theory, frequently attributable to such pioneers
as A.J. Lotka and R.A. Fisher, but also the Leslie matrix presents acute expository problems. Rather than following the less formal expository style of the
preceding chapters, a decision was made to make the presentation in a more
mathematical style, a style that expedited the presentation of a variety of results in a rather compact form. As will become apparent to the reader from the
study of the numerical examples presented in Sects. 7.17 and 7.18, much remains to be done in exploring the numerical implications of the algorithms
presented in this chapter.

7.2 Basic Functions Underlying a Branching Process

Ages vary continuously among individuals of a population. But, in the computer implementation of stochastic models describing the evolution of human
populations, it will be useful to group individuals into age classes. To fix ideas,
let the time unit be a year and consider the age intervals $[x, x+1)$ – closed on
the left and open on the right – for $x = 0, 1, 2, \ldots$. Infants belong to the age
class $[0, 1)$, one-year olds to the age class $[1, 2)$, and so on. As is the custom in
western societies, ages of individuals will be reckoned according to their last
birthday. From now on ages will be symbolized by values in the set $A = [x
= 0, 1, 2, \ldots, r]$ of integers, where r is some maximum age.

Just as ages of individuals in a population may be grouped, so we can
imagine observing a population at discrete points in time. Let $T = [t : t = 0, \pm 1,
\pm 2, \ldots]$ be the set of time points under consideration. When a population
evolves according to time inhomogeneous laws of evolution, it will be useful to
index each cohort according to its epoch $t \in T$ of birth. A similar convention
was used in connection with the study of continuous time models of mortality
and fertility in Sects. 2.7 and 4.9. In the idealized discrete time branching process considered in this chapter, however, it will be supposed that ages of individuals in a population increase in lock-step as t increases. Thus, if an individual is of age x at epoch t, then he is of age $x+1$ at epoch $t+1$.

For any given sex, survivorship in the population will be governed by period survival probabilities $p(t, x)$ defined for all $t \in T$ and $x \in A$. These probabilities will be interpreted as follows. Suppose an individual is age $x-1$, $x \geq 1$, at
epoch $t-1$. Then, $p(t, x)$ is the conditional probability he survives to age x at
epoch t. Infants credited to the population at epoch t, will be represented by
the age $x = 0$. By definition, $p(t, 0) = 1$ for all $t \in T$. The conditional probability
that a live individual of age $x-1$ at epoch $t-1$ is dead by epoch t is $q(t, x) =
1 - p(t, x)$.

From the perspective of survivorship across age groups at epoch t, the period survival function at epoch t is

$$S_p(t, x) = \prod_{v=0}^{x} p(t, v), \tag{7.2.1}$$

for $x \geq 0$. It is this survival function that is estimated at selected ages in period life tables. Next consider a cohort of births credited to the population at epoch t; and, for every member of the cohort, let $X(t)$ be a random variable representing age at death. Then, the survival function for a cohort born at epoch t is defined by

$$P[X(t) > x] = S_c(t, x) = \prod_{v=0}^{x} p(t+v, v),$$ (7.2.2)

for $x \geq 0$. The p.d.f. of the random variable $X(t)$ is

$$P[X(t) = x] = g_c(t, x) = S_c(t, x-1)q(t+x, x),$$ (7.2.3)

for $1 \leq x \leq r$. Suggestions for the computer implementation of formulas (7.2.1), (7.2.2), and (7.2.3) are given in Problem 1. It is also shown in Problem 2 that (7.2.2) implies (7.2.3).

Conditional versions of the functions in (7.2.2) and (7.2.3) will also be needed for subsequent developments. Let the functions $S_c(t, x, y)$ and $g_c(t, x, y)$ be defined by

$$S_c(t, x, y) = P[X(t) > x + y \mid X(t) > x]$$

and (7.2.4)

$$g_c(t, x, y) = P[X(t) = x + y \mid X(t) > x].$$

Then,

$$S_c(t, x, y) = \frac{S_c(t, x+y)}{S_c(t, x)},$$ (7.2.5)

for $y \geq 0$, and

$$g_c(t, x, y) = \frac{g_c(t, x+y)}{S_c(t, x)}$$ (7.2.6)

for $y \geq 1$ and $g_c(t, x, 0) = 0$. In (7.2.5) and (7.2.6) $x \geq 0$, $x \in A$.

Births credited to the population at any epoch $t \in T$ will be defined in terms of random functions of the type defined by (4.3.1) and (4.3.2). For any member of a cohort born at epoch $t \in T$, let $N(t, x)$ be the number of births experienced by age x, i.e., during the time interval $(t, t+x]$. Then for every $x \geq 1$, the mean function $\varphi_c(t, x)$ is defined by

$$\varphi_c(t, x) = E[N(t, x) - N(t, x-1)].$$ (7.2.7)

The function $\varphi_c(t, x)$ is interpreted as the mean number of births credited to the population at epoch $t+x$ by a woman age x who was born at epoch t.

In terms of a density of the type appearing in (4.3.7), the function in (7.2.7) would be computed by approximating the integral

$$\varphi_c(t, x) = \int_{x-1}^{x} S_c(t, y)b_c(t, y)dy$$ (7.2.8)

for $1 \leq x \leq r$. It is through formula (7.2.8) that the output of models of maternity histories, as described in chapters IV and V, would enter as computer input to the population projection model. In the computer implementation of such models, all calculations are carried out on a monthly time scale. Hence, the integral in (7.2.8) would be approximated as a discrete sum of twelve terms. In these approximations, the cohort survival function $S_c(t, y)$ could also be evaluated on a monthly time scale by using either a monotone interpolation procedure or a parametric model. From now on it will be supposed that for every t, $\varphi_c(t, x) = 0$ for all x except for those x such that $15 \leq x \leq 50$, the child-bearing ages.

In subsequent developments, the function $\varphi_c(t, x)$ in (7.2.7) will also need to be considered from the perspective of a woman alive at age x who was born at epoch t. Let $\varphi_c(t, x, y)$ be the mean number of births credited to the population at epoch $t + x + y$ by a woman of age $x + y$ who was born at epoch t. Then,

$$\varphi_c(t, x, y) = \frac{\varphi_c(t, x + y)}{S_c(t, x)} \tag{7.2.9}$$

for $y \geq 1$, $15 \leq x + y \leq 50$ and $\varphi_c(t, x, y) = 0$ otherwise.

Although population projections are often made with respect to one sex, the functions in (7.2.8) and (7.2.9) apply to offspring of both sexes. Let $p(q)$ be the probability an offspring is male (female) with $p + q = 1$. Then, if the functions in (7.2.8) and (7.2.9) are to serve as input to a population projection with respect to the female population, they must be multiplied by q. A good estimate of q is $\hat{q} = \frac{100}{205}$.

7.3 Basic Random Functions and Their Means

When a population is considered at an epoch t, there will be some number of individuals belonging to each age class. Let $n(t, x)$ be the number of individuals of age x at epoch t. In order to develop a model for the evolution of a human population with respect to one sex within the framework of branching processes with time inhomogeneous laws of evolution, it will be useful to consider lines of descent emanating from each individual in the population at epoch t. Only female members of the initial population at epoch t and their female descendants born after epoch t will be considered in these lines of descent. The basic random functions to be considered for any individual of age x at epoch t may be defined as follows:

Definition 7.3.1:
$Z_B(t, x, y, s)$ – the number of live births to females of age y in the line at epoch $t + s$;
$Z_D(t, x, y, s)$ – the number of deaths to females of age y in the line at epoch $t + s$;
$Z(t, x, y, s)$ – the number of live individuals of age y in the line at epoch $t + s$.

Before proceding, it will be useful to describe the domains of the random functions given in Definition 7.3.1. All random functions are defined for $t \in T$, $x \in A$, and $s \geq 1$; whenever $s < 1$, all random functions will equal zero. For every fixed t, x, and s, $Z_B(t, x, y, s) = 0$ unless $15 \leq y \leq 50$. Furthermore, for every t, x, and s, the random functions $Z_D(t, x, y, s)$ and $Z(t, x, y, s)$ are defined only for y such that $1 \leq y \leq r$. As implied by (7.2.3), deaths are credited to the population only at the ages $1 \leq y \leq r$.

In any line of descent, births credited to the population at any epoch will be accounted for through the random function $Z_B(t, x, y, s)$. It should also be noted that only births after epoch t are counted in the random function $Z_B(t, x, y, s)$ so that the initial individual is not included in $Z_B(t, x, y, s)$. If, however, the initial female dies at age $y > x$, then she is counted in $Z_D(t, x, y, s)$; if she is alive at age $y > x$, then she is counted in $Z(t, x, y, s)$. Finally, because births cannot occur beyond the childbearing ages, $Z_B(t, x, y, s) = 0$ for all t, y, and s whenever $51 \leq x \leq r$.

As will be demonstrated subsequently, under certain assumptions, the random functions in Def. 7.3.1 satisfy stochastic renewal equations. Contained in these renewal equations are other random functions associated with each individual of age x in the initial population at epoch t. These random functions may be defined as follows:

Definition 7.3.2:

$Z_B^{(0)}(t, x, y, s)$ – the number of births to the initial individual at age $y > x$ credited to the population at epoch $t + s$;

$Z_D^{(0)}(t, x, y, s)$ – a random indicator function such that $Z_D^{(0)}(t, x, y, s) = 1$ if the initial individual dies at age $y > x$ at epoch $t + s$ and $Z_D(t, x, y, s) = 0$ otherwise;

$Z^{(0)}(t, x, y, s)$ – a random indicator function such that $Z^{(0)}(t, x, y, s) = 1$ if the initial individual is alive at age $y > x$ at epoch $t + s$ and $Z(t, x, y, s) = 0$ otherwise.

Certain conditional expectations of the random functions in Def. 7.3.2 will be needed subsequently. Let $\delta_0(t)$ be an indicator function defined for $t \in T$ such that $\delta_0(t) = 1$ if $t = 0$ and $\delta_0(t) = 0$ if $t \neq 0$. If an individual is age x at epoch t, then she was born at epoch $t - x$. Hence, a little reflection leads to the conclusion that

$$E[Z_B^{(0)}(t, x, y, s) | X(t - x) > x] = q \varphi_c(t - x, x, y - x) \delta_0(x + s - y), \qquad (7.3.1)$$

where q is the probability an infant is female. Similarly,

$$E[Z_D^{(0)}(t, x, y, s) | X(t - x) > x] = g_c(t - x, x, y - x) \delta_0(x + s - y), \qquad (7.3.2)$$

and

$$E[Z^{(0)}(t, x, y, s) | X(t - x) > x] = S_c(t - x, x, y - x) \delta_0(x + s - y). \qquad (7.3.3)$$

In what follows, it will be helpful to define the expectations in (7.3.1), (7.3.2), and (7.3.3) as zero whenever $y \leq x$.

A basic assumption underlying the branching process under consideration is that, with the birth of each individual, a line of descent begins independent of all others. For an individual, born at epoch $t+i$, $i\geq 1$, random functions of the form $Z_B(t+i,0,y,s)$, $Z_D(t+i,0,y,s)$, and $Z(t+i,0,y,s)$, $s\geq 1$, are in force. If there is some number $\Delta(i)$ of individuals born at epoch $t+i$, then the random function associated with each individual is independent of all others. For example, if live individuals of age y are under consideration, let $Z_j(t+i,0,y,s)$, $1\leq y\leq\Delta(i)$, be these random functions. The random functions $Z_{Bj}(t+i,0,y,s)$ and $Z_{Dj}(t+i,0,y,s)$, $1\leq j\leq\Delta(i)$, for births and deaths to individual of age y are defined similarly. For each fixed j, these random functions are governed by the same probabilistic laws as the random functions $Z_B(t+i,0,y,s)$, $Z_D(t+i,0,y,s)$, and $Z(t+i,0,y,s)$.

If an individual is of age x at epoch $t+x$, then

$$\Delta(x,i)=N(t-x,x+i)-N(t-x,x+i-1),\tag{7.3.4}$$

for $i\geq 1$, is the random number of births contributed to the population at epoch $t+i$. Only female infants will be counted in (7.3.4). By using the property that a branching process starts anew with the birth of each individual, it can be seen that the following stochastic renewal equations are valid:

$$Z_B(t,x,y,s)=Z_B^{(0)}(t,x,y,s)+\sum_{i=1}^{r}\sum_{j=1}^{\Delta(x,i)}Z_{Bj}(t+i,0,y,s-i);\tag{7.3.5}$$

$$Z_D(t,x,y,s)=Z_D^{(0)}(t,x,y,s)+\sum_{i=1}^{r}\sum_{j=1}^{\Delta(x,i)}Z_{Dj}(t+i,0,y,s-i);\tag{7.3.6}$$

and

$$Z(t,x,y,s)=Z^{(0)}(t,x,y,s)+\sum_{i=1}^{r}\sum_{j=1}^{\Delta(x,i)}Z_j(t+i,0,y,s-i).\tag{7.3.7}$$

Many equations of interest may be derived from the stochastic renewal equations in (7.3.5), (7.3.6), and (7.3.7). Of particular interest in this section are the mean functions:

$$M_B(t,x,y,s)=E[Z_B(t,x,y,s)],\tag{7.3.8}$$

$$M_D(t,x,y,s)=E[Z_D(t,x,y,s)],\tag{7.3.9}$$

and

$$M(t,x,y,s)=E[Z(t,x,y,s)].\tag{7.3.10}$$

For the time being, it will be assumed that these mean functions are finite at all points $\langle t,x,y,s\rangle$; subsequently, it will be shown that this is indeed the case.

Theorem 7.3.1: When the assumptions underlying the branching process under consideration are in force, the mean functions in (7.3.8), (7.3.9), and (7.3.10) satisfy the equations:

$$M_B(t,x,y,s)=q\varphi_c(t-x,x,y-x)\delta_0(x+s-y)$$
$$+q\sum_{i=1}^{r}\varphi_c(t-x,x,i)M_B(t+i,0,y,s-i);\tag{7.3.11}$$

$$M_D(t, x, y, s) = g_c(t-x, x, y-x)\delta_0(x+s-y)$$

$$+q \sum_{i=1}^{r} \varphi_c(t-x, x, i)M_D(t+i, 0, y, s-i); \qquad (7.3.12)$$

and

$$M(t, x, y, s) = S_c(t-x, x, y-x)\delta_0(x+s-y)$$

$$+q \sum_{i=1}^{r} \varphi_c(t-x, x, i)M(t+i, 0, y, s-i). \qquad (7.3.13)$$

Proof: To establish the validity of these equations, it suffices to prove the validity of Eq. (7.3.13). For every i and j

$$E[Z_j(t+i, 0, y, s-i)] = M(t+i, 0, y, s-i). \qquad (7.3.14)$$

Because the summands in the second term on the right in (7.3.7) are independent of $X(t-x)$ and $\Delta(x, i)$, the conditional expectation of this term given $X(t-x)$ and $\Delta(x, i)$ is

$$\sum_{i=1}^{r} \Delta(x, i)M(t+i, 0, y, s-i). \qquad (7.3.15)$$

Equation (7.3.13) now follows by using (7.3.15) and taking expectations in (7.3.7). ☐

It is important to observe that Eqs. (7.3.11), (7.3.12), and (7.3.13) become age-dependent ordinary renewal equations for the functions $M_B(t, 0, y, s)$, $M_D(t, 0, y, s)$, and $M(t, 0, y, s)$ when $x=0$. A more precise discussion of (7.3.4) and why the factor q enters into Eqs. (7.3.11), (7.3.12), and (7.3.13) is given in Problem 3.

7.4 Explicit Formulas for Mean Functions

As just noted, when $x=0$, Eqs. (7.3.11), (7.3.12), and (7.3.13) become renewal equations of the type

$$a(t, s) = b(t, s) + \sum_{v=0}^{s} c(t, v)a(t+v, s-v), \qquad s \geq 0, \qquad (7.4.1)$$

where the functions $b(t, s)$ and $c(t, s)$ are given and the function $a(t, s)$ is to be calculated. When $c(t, 0)=0$ for all t, then there is a renewal density function $m(t, s)$ such that $m(t, 0)=1$ and

$$m(t, s) = \sum_{v=0}^{s-1} m(t, v)c(t+v, s-v) \qquad (7.4.2)$$

for $s \geq 1$ and every $t \in T$. Having determined the renewal density function recursively, the solution of equation (7.4.1) is given by

$$a(t, s) = \sum_{v=0}^{s} m(t, s)b(t+v, s-v), \qquad (7.4.3)$$

for any t and $s \geq 0$. A matrix version of Eqs. (7.4.1), (7.4.2), and (7.4.3) has been discussed in Problem 1 of Chap. 6.

To solve Eqs. (7.3.11), (7.3.12), and (7.3.13) for the case $x=0$, let $c(t, s) = q\varphi_c(t, 0, s) = q\varphi_c(t, s)$ and then calculate the renewal density $m_c(t, s)$ according to formula (7.4.2). It will be convenient to let $m_c(t, s) = 0$, if $s < 0$, in what follows. Then, by choosing $b(t, s)$ as the first term on the right in (7.3.11), (7.3.12), and (7.3.14), explicit formulas for $M_B(t, 0, y, s)$, $M_D(t, 0, y, s)$, and $M(t, 0, y, s)$ may be derived, using (7.4.3). The resulting formulas are summarized in the next theorem.

Theorem 7.4.1: For the case $x=0$, the solutions of Eqs. (7.3.11), (7.3.12), and (7.3.13) are

$$M_B(t, 0, y, s) = q m_c(t, s-y)\varphi_c(t+s-y, 0, y), \tag{7.4.4}$$

$$M_D(t, 0, y, s) = m_c(t, s-y)g_c(t+s-y, 0, y), \tag{7.4.5}$$

and

$$M(t, 0, y, s) = m_c(t, s-y)S_c(t+s-y, 0, y), \tag{7.4.6}$$

for $s \geq y$. When $s < y$, all these mean functions vanish.

Proof: It suffices to prove (7.4.6). By applying formula (7.4.3), it follows from (7.4.7) that

$$M(t, 0, y, s) = \sum_{v=0}^{s} m_c(t, v)S_c(t+v, 0, y)\delta_0(s-y-v). \tag{7.4.7}$$

But, all summands on the right are zero unless $v = s-y$. Hence, (7.4.6) is valid. ☐

Explicit formulas for the means $M_B(t, x, y, s)$, $M_D(t, x, y, s)$, and $M(t, x, y, s)$, for any $x \geq 0$, may be written down by substituting (7.4.4), (7.4.5), and (7.4.6) on the right in (7.3.11), (7.3.12), and (7.3.13). These formulas take a simple form, if a function $\xi(t, x, y, s)$ is defined by

$$\xi(t, x, y, s) = q \sum_{i=1}^{r} \varphi_c(t-x, x, i)m_c(t+i, s-y-i), \tag{7.4.8}$$

for all $x \geq 0$.

Theorem 7.4.2: For any $x \geq 0$,

$$M_B(t, x, y, s)$$
$$= q\varphi_c(t-x, x, y-x)\delta_0(x+s-y) + q\varphi_c(t+s-y, 0, y)\xi(t, x, y, s), \tag{7.4.9}$$

$$M_D(t, x, y, s)$$
$$= g_c(t-x, x, y-x)\delta_0(x+s-y) + g_c(t+s-y, 0, y)\xi(t, x, y, s), \tag{7.4.10}$$

and

$$M(t, x, y, s)$$
$$= S_c(t-x, x, y-x)\delta_0(x+s-y) + S_c(t+s-y, 0, y)\xi(t, x, y, s). \tag{7.4.11}$$

Proof: By using formula (7.4.6), it can be shown that

$$q \sum_{i=1}^{r} \varphi_c(t-x, x, i) M(t+i, 0, y, s-i) = S_c(t+s-y, 0, y) \xi(t, x, y, s). \qquad (7.4.12)$$

Equation (7.4.11) now follows by substituting this result into (7.3.13). Equations (7.4.9) and (7.4.10) may be proved similarly. □

Because the function $\varphi_c(t, s)$ vanishes for s outside the childbearing ages for every t, the renewal density $m_c(t, s)$ has a special form; Problem 4 may be consulted for details.

7.5 Leslie Matrix Type Recursive Formulas for Mean Functions

In principle, the formulas in Theorem 7.4.2 provide a means for calculating the mean functions $M_B(t, x, y, s)$, $M_D(t, x, y, s)$, and $M(t, x, y, s)$ for any combination of x, y, and s for a fixed t. Problems in implementing these formulas could arise, however, because the space needed to store the rather large three dimensional arrays could go beyond the available storage capacity of a computing facility. It is thus important to search for ways of reducing the dimensionality of arrays that need to be manipulated in a computer.

One approach to reducing this dimensionality is to define the mean functions $v(t, y, s)$, $v_B(t, y, s)$, and $v_D(t, y, s)$ by

$$v(t, y, s) = \sum_{x=0}^{r} n(t, x) M(t, x, y, s), \qquad (7.5.1)$$

$$v_B(t, y, s) = \sum_{x=0}^{r} n(t, x) M_B(t, x, y, s), \qquad (7.5.2)$$

and

$$v_D(t, y, s) = \sum_{x=0}^{r} n(t, x) M_D(t, x, y, s). \qquad (7.5.3)$$

According to formula (7.5.1), if there are $n(t, x)$ females of age $x = 0, 1, 2, \ldots, r$ in the population at epoch t, then $v(t, y, s)$ is the mean number of live females of age y in the population at epoch $t+s$, $s \geq 1$. Similarly, $v_B(t, y, s)$ and $v_D(t, y, s)$ are the mean number of births and deaths, respectively, credited to females of age y in the population at epoch $t+s$.

From Theorem 7.4.2, it follows that each of the functions in (7.5.1), (7.5.2), and (7.5.3) may be represented as the sums

$$v(t, y, s) = v^{(0)}(t, y, s) + v^{(1)}(t, y, s), \qquad (7.5.4)$$

$$v_B(t, y, s) = v_B^{(0)}(t, y, s) + v_B^{(1)}(t, y, s), \qquad (7.5.5$$

and

$$v_D(t, y, s) = v_D^{(0)}(t, y, s) + v_D^{(1)}(t, y, s). \tag{7.5.6}$$

Upon inspecting Eqs. (7.4.9), (7.4.10), and (7.4.11), it can be seen that the $v^{(0)}$-functions are given by

$$v^{(0)}(t, y, s) = \sum_{x=0}^{r} n(t, x) S_c(t-x, x, y-x) \delta_0(x+s-y), \tag{7.5.7}$$

$$v_B^{(0)}(t, y, s) = q \sum_{x=0}^{r} n(t, x) \varphi_c(t-x, x, y-x) \delta_0(x+s-y), \tag{7.5.8}$$

and

$$v_D^{(0)}(t, y, s) = \sum_{x=0}^{r} n(t, x) g_c(t-x, x, y-x) \delta_0(x+s-y). \tag{7.5.9}$$

Furthermore, the $v^{(1)}$-functions are given by

$$v^{(1)}(t, y, s) = S_c(t+s-y, 0, y) \eta(t, y, s), \tag{7.5.10}$$
$$v_B^{(1)}(t, y, s) = q \varphi_c(t+s-y, 0, y) \eta(t, y, s), \tag{7.5.11}$$

and

$$v_D^{(1)}(t, y, s) = g_c(t+s-y, 0, y) \eta(t, y, s). \tag{7.5.12}$$

The function $\eta(t, y, s)$ is defined by

$$\eta(t, y, s) = \sum_{x=0}^{r} n(t, x) \xi(t, x, y, s). \tag{7.5.13}$$

The mean functions in (7.5.1), (7.5.2), and (7.5.3) have the remarkable property that they may be calculated recursively.

Theorem 7.5.1: For every $y=0, 1, 2, \ldots, r$ and $s \geq 0$, the functions $v(t, y, s)$, $v_B(t, y, s)$, and $v_D(t, y, s)$ are finite and satisfy the following recursive system.
  For $s=0$,

$$v(t, y, 0) = n(t, y), \tag{7.5.14}$$

for $y=0, 1, 2, \ldots, r$.
  If $y \geq 1$ and $s \geq 1$,

$$v(t, y, s) = p(t+s, y) v(t, y-1, s-1), \tag{7.5.15}$$
$$v_B(t, y, s) = q \varphi_c(t+s-y, y-1, 1) v(t, y-1, s-1), \tag{7.5.16}$$

and

$$v_D(t, y, s) = q(t+s, y) v(t, y-1, s-1). \tag{7.5.17}$$

  For every $s \geq 1$,

$$v(t, 0, s) = \sum_{y=15}^{50} v_B(t, y, s). \tag{7.5.18}$$

Proof: If the recursive relations (7.5.15), (7.5.16), and (7.5.17) are valid, then, because $n(t, y)$ is a finite integer for every y, the functions $v(t, y, s)$, $v_B(t, y, s)$, and $v_D(t, y, s)$ must be finite for every y and $s \geq 1$. Next observe that if equation (7.5.16) is valid, then (7.5.18) is also a recursive relationship. Thus, all that remains to be done in proving the theorem is to establish the validity of Eqs. (7.5.15), (7.5.16), and (7.5.17).

To prove the validity of Eq. (7.5.15), observe that for a fixed y and s all summands on the right in (7.5.7) are zero unless $x = y - s$. Therefore,

$$v^{(0)}(t, y, s) = n(t, y - s)S_c(t + s - y, y - s, s).$$ (7.5.19)

But,

$$v^{(0)}(t, y - 1, s - 1) = n(t, y - s)S_c(t + s - y, y - s, s - 1),$$ (7.5.20)

and, from (7.2.2) and (7.2.5), it follows that

$$S_c(t + s - y, y - s, s) = S_c(t + s - y, y - s, s - 1)p(t + s, y).$$ (7.5.21)

Therefore,

$$v^{(0)}(t, y, s) = p(t + s, y)v^{(0)}(t, y - 1, s - 1).$$ (7.5.22)

Turning next to Eq. (7.5.10), observe that $S_c(t + s - y, 0, y) = S_c(t + s - y, y)$ and

$$S_c(t + s - y, y) = S_c(t + s - y, y - 1)p(t + s, y).$$ (7.5.23)

From (7.4.8) and (7.5.13), it can be seen that $\eta(t, y - 1, s - 1) = \eta(t, y, s)$. Therefore, Eq. (7.5.10) may be written in the form

$$v^{(1)}(t, y, s) = p(t + s, y)v^{(1)}(t, y - 1, s - 1).$$ (7.5.24)

Adding Eqs. (7.5.22) and (7.5.24) yields Eq. (7.5.15).

To prove the validity of Eq. (7.5.16), observe that the equation

$$\varphi_c(t - x, x, y - x) = \varphi_c(t - x, y - 1, 1)S_c(t - x, x, y - x - 1),$$ (7.5.25)

is valid for every x and y. By using this equation, it can be shown that Eq. (7.5.8) reduces to

$$v_B^{(0)}(t, y, s) = q\varphi_c(t + s - y, y - 1, 1)v^{(0)}(t, y - 1, s - 1).$$ (7.5.26)

Similarly, by using the equation

$$\varphi_c(t + s - y, 0, y) = \varphi_c(t + s - y, y)$$
$$= \varphi_c(t + s - y, y - 1, 1)S_c(t + s - y, y - 1),$$ (7.5.27)

it can be shown that Eq. (7.5.11) may be written in the form

$$v_B^{(1)}(t, y, s) = q\varphi_c(t + s - y, y - 1, 1)v^{(1)}(t, y - 1, s - 1).$$ (7.5.28)

Adding Eqs. (7.5.26) and (7.5.28) yields Eq. (7.5.16).

Arguments similar to those used above may be used to establish the validity of the equations

$$v_D^{(0)}(t, y, s) = g_c(t + s - y, y - 1, 1)v^{(0)}(t, y - 1, s - 1)$$ (7.5.29)

and

$$v_D^{(1)}(t, y, s) = g_c(t+s-y, y-1, 1) v^{(1)}(t, y-1, s-1). \tag{7.5.30}$$

But, from (7.2.3) and (7.2.6) it follows that

$$g_c(t+s-y, y-1, 1) = q(t+s, y). \tag{7.5.31}$$

When Eqs. (7.5.29) and (7.5.30) are added, Eq. (7.5.17) follows. □

Equations (7.5.14), (7.5.15), and (7.5.18) are actually equivalent to projecting a population according to a Leslie matrix scheme. For every t and s define a $(r+1) \times (r+1)$ matrix $\mathbf{L}(t, s)$ with elements $L_{ij}(t, s)$. All these elements are zero except for $L_{0j}(t, s) = q \varphi_c(t+s-j, j-1, 1)$, $15 \leq j \leq 50$, and $L_{j, j-1}(t, s) = p(t+j, j)$, $1 \leq j \leq r$. Then, if $v(t, s)$ is a $(r+1) \times 1$ column vector with elements $v(t, y, s)$, $0 \leq y \leq r$, it follows that

$$v(t, s) = \mathbf{L}(t, s) v(t, s-1), \tag{7.5.32}$$

for $s \geq 1$.

Instead of starting with a branching process formulation, Eq. (7.5.32) could also be a point of departure in developing a methodology for making population projections. It seems likely that the elements $L_{j, j-1}(t, s) = p(t+j, j)$, as derived from a branching process formulation, would also be chosen by an investigator on a purely intuitive basis. Choosing the elements $L_{0j}(t, s)$ would, however, seem more problematic. Among the advantages of the branching process formulation is that the elements $L_{0j}(t, s)$, $15 \leq j \leq 50$, can be derived in a precise way. From (7.2.9), it can be seen that these elements have the form

$$L_{0j}(t, s) = \frac{q \varphi_c(t+s-j, j)}{S_c(t+s-j, j-1)}, \tag{7.5.33}$$

for $15 \leq j \leq 50$. Although cohort survival functions are not usually calculated in making population projections, it is the cohort survival function that appears in the denominator of (7.5.33). A useful approximation to (7.5.33) is discussed in Problem 5.

7.6 A Brief Review of Literature

Rather than as a model *per se* for the development of a one-sex population, a Leslie type matrix appears in the results of the previous section as an algorithm for the recursive calculation of sums of mean functions in a branching process with time inhomogeneous laws of evolution. There is an extensive literature on both the Leslie matrix and branching process; consequently, a brief review of this literature seems appropriate.

Smith and Keyfitz (1977) have given excerpts from papers dealing with attempts at forecasting and predicting future population along with an account of the theory those attempts stimulated. A common phenomenon in the evolution of science is that one or more investigators may anticipate the results of others. Evidently, this phenomenon occurred in the use of the so-called Leslie matrix in population mathematics. Although other investigators had done arithmetic and had used formalism essentially equivalent to that implied by the Leslie matrix (see Smith and Keyfitz (1977) for details), these techniques became wide spread only after the publication of the papers, Leslie (1945, 1948), which contained the formalism now widely used.

Among the first to use stochastic models in making population projections was Pollard (1966), who used concepts taken from the theory of multitype Galton-Watson branching processes. The Leslie matrix type structure used by Pollard was sufficiently simple to permit computer implementation. A review of the earlier literature on multitype Galton-Watson branching processes was given by Harris (1963); the book, Pollard (1973), may also be consulted for details on mathematical models used in the study of human populations.

The branching processes used in this chapter, however, have their roots in an independent line of development, starting with the general age-dependent branching processes in continuous time formulated in the papers Crump and Mode (1968, 1969) and Jagers (1969). Further developments of this theory may be found in the book, Jagers (1975); a theory of general multitype age-dependent branching processes in continuous time was developed in the book Mode (1971). The stochastic structures considered in the foregoing work were preferred to the type suggested by Pollard, because they were sufficiently rich to accommodate many factors known to affect the evolution of human populations. Indeed, the developments in Chaps. IV, V, and VI of this book may be viewed as an attempt to supply concrete examples, capable of computer implementation, of the general reproductive processes that were set down as part of the theory of generalized age-dependent branching processes in continuous time developed previously.

Continuous time theories have much aesthetic appeal but often prove to be impractical when attempts are made to implement them on a computer. The papers, Mode (1974a, 1976), represent initial attempts at formulating and developing age-dependent branching processes in discrete time with a view towards implementing these models on a computer and applying them to problems in demography. Contained in the papers Mode (1974b) and Mode and Littman (1974) are further attempts at theory and the results of computer experiments. Additional refinements of generalized age-dependent branching processes in discrete time and their relation to problems in demography were set forth in the paper, Mode and Busby (1981). Throughout this work, an attempt was made to preserve the formalism of generalized branching processes in continuous time.

In their initial stages of development, models of population growth based on generalized branching processes in discrete time called for the manipulation of three-dimensional arrays in a computer, a task that could prove to be troublesome. In earlier work, G.S. Littman had used crude approximations to sums of mean functions that were solutions to renewal type equations. Sums were taken to reduce three-dimensional arrays to two dimensions. These approximations were similar to calculations based on a Leslie matrix. Littman's calculations had the property that they would give results predicted by limit theorems from the theory of branching processes with time homogeneous laws of evolution. But, it remained to be discovered whether Littman's approximations were, in fact, exact. Motivated by further computer experiments, G.T. Pickens was able to establish Theorem 7.5.1 for branching processes with time homogeneous laws of evolution, a result that was first reported by Pickens et al. (1983). Theorem 7.5.1 for branching processes with time inhomogeneous laws of evolution is new.

Another line of development in stochastic models of population growth is represented by the papers, Cohen (1976, 1977a, 1977b). In these papers, the elements of Leslie matrices are treated as realizations of a Markov process or perhaps some stochastic process with stationary laws of evolution. Properties of products of stochastic matrices play an essential role in this development. It is of interest to note that if the elements of a Leslie matrix implied by the results of Theorem 7.5.1 were viewed as random, then it would seem to make sense to interpret them as conditional expectations given some "environmental" conditions. Another paper of interest in the vast literature on Leslie matrices is that of Sykes (1969). In subsequent developments of this chapter, however, matrix theory will play no essential role.

7.7 Stochastic Variability in Population Structure as a Gaussian Process

Described and analyzed in this section are integer-valued random functions $Y(t, y, s)$, $Y_B(t, y, s)$, and $Y_D(t, y, s)$ whose means are displayed in (7.5.1), (7.5.2), and (7.5.3). As before, suppose there are $n(t, x) \geq 1$ individuals of age x in the population at epoch t; and, for $1 \leq k \leq n(t\,x)$, let $Z_k(t, x, y, s)$, $Z_{Bk}(t, x, y, s)$, and $Z_{Dk}(t, x, y, s)$ be independent and identically distributed random functions with the same distributions as the random functions $Z(t, x, y, s)$, $Z_B(t, x, y, s)$, and $Z_D(t, x, y, s)$ defined in Def. 7.3.1. Then, for $s \geq 1$, the random functions under consideration have the form:

$$Y(t, y, s) = \sum_{x=0}^{r} \sum_{k=1}^{n(t, x)} Z_k(t, x, y, s), \tag{7.7.1}$$

$1 \leq y \leq r$;

$$Y_B(t, y, s) = \sum_{x=0}^{r} \sum_{k=1}^{n(t, x)} Z_{Bk}(t, x, y, s), \tag{7.7.2}$$

$15 \leq y \leq r$; and

$$Y_D(t, y, s) = \sum_{x=0}^{r} \sum_{k=1}^{n(t, x)} Z_{Dk}(t, x, y, s), \tag{7.7.3}$$

for $1 \le y \le r$. The number of births entering the population at epoch $t+s$ is given by the random function

$$Y(t, 0, s) = \sum_{y=15}^{50} Y_B(t, y, s). \qquad (7.7.4)$$

At any epoch $t+s$, $s \ge 1$, the population structure is given by the values of the random functions in (7.7.1), (7.7.2), (7.7.3), and (7.7.4). The principal purpose of this section is to give a characterization of the variability in this structure not only at any epoch $t+s$ but also among epochs $t+s_k$, $1 \le k \le n$, for any n, where $1 \le s_1 \le s_2 \le \ldots \le s_n$.

As a first step towards characterizing this variability, the covariance functions of the random functions in (7.7.1), (7.7.2), and (7.7.3) will be considered. From now on, it will be assumed that all these covariance functions are finite valued. Because the summands on the right in (7.7.1) are independent, it follows that the covariance function of the random function $Y(t, y, s)$ for live individuals is given by

$$\text{cov}\,[Y(t, y_1, s_1), Y(t, y_2, s_2)]$$
$$= \sum_{x=0}^{r} n(t, x)\,\text{cov}\,[Z(t, x, y_1, s_1), Z(t, x, y_2, s_2)], \qquad (7.7.5)$$

where $y_1, y_2 = 1, 2, \ldots, r$ and $1 \le s_1 \le s_2$. A similar argument may be used to show that the covariance function of the random function $Y_B(t, y, s)$ for births is given by

$$\text{cov}\,[Y_B(t, y_1, s_1), Y_B(t, y_2, s_2)]$$
$$= \sum_{x=0}^{r} n(t, x)\,\text{cov}\,[Z_B(t, x, y_1, s_1), Z_B(t, x, y_2, s_2)]. \qquad (7.7.6)$$

In this case, $y_1, y_2 = 15, 16, \ldots, 50$ and $1 \le s_1 \le s_2$. Finally, the covariance function of the random function $Y_D(t, y, s)$ for deaths is given by

$$\text{cov}\,[Y_D(t, y_1, s_1), Y_D(t, y_2, s_2)]$$
$$= \sum_{x=0}^{r} n(t, x)\,\text{cov}\,[Z_D(t, x, y_1, s_1), Z_D(t, x, y_2, s_2)]. \qquad (7.7.7)$$

In (7.7.7), $y_1, y_2 = 1, 2, \ldots, r$ and $1 \le s_1 \le s_2$.

Also needed in characterizing the variability in population structure are the cross covariances among the random functions in (7.7.1), (7.7.2), and (7.7.3). For $1 \le s_1 \le s_2$, these functions are determined by the following formulas:

$$\text{cov}\,[Y(t, y_1, s_1), Y_B(t, y_2, s_2)]$$
$$= \sum_{x=0}^{r} n(t, x)\,\text{cov}\,[Z(t, x, y_1, s_1), Z_B(t, x, y_2, s_2)]; \qquad (7.7.8)$$

$$\text{cov}\,[Y(t, y_1, s_1), Y_D(t, y_2, s_2)]$$
$$= \sum_{x=0}^{r} n(t, x)\,\text{cov}\,[Z(t, x, y_1, s_1), Z_D(t, x, y_2, s_2)]; \qquad (7.7.9)$$

and

$$\operatorname{cov}\left[Y_B(t, y_1, s_1), Y_D(t, y_2, s_2)\right]$$

$$= \sum_{x=0}^{r} n(t, x) \operatorname{cov}\left[Z_B(t, y_1, s_1), Z_D(t, x, y_2, s_2)\right]. \tag{7.7.10}$$

In (7.7.8), (7.7.9), and (7.7.10) the ranges of the variables y_1 and y_2 are defined as in (7.7.5), (7.7.6), and (7.7.7).

As a final step in defining the required covariance functions, it will be necessary to consider covariance associated with the random function $Y(t, 0, s)$ in (7.7.4), representing births. For $1 \le s_1 \le s_2$, these covariance functions are:

$$\operatorname{cov}\left[Y(t, 0, s_1), Y(t, y_2, s_2)\right] = \sum_{y_1=15}^{50} \operatorname{cov}\left[Y_B(t, y_1, s_1), Y(t, y_2, s_2)\right] \tag{7.7.11}$$

for $0 \le y_2 \le r$;

$$\operatorname{cov}\left[Y(t, 0, s_1), Y_B(t, y_2, s_2)\right] = \sum_{y_1=15}^{50} \operatorname{cov}\left[Y_B(t, y_1, s_1), Y_B(t, y_2, s_2)\right]; \tag{7.7.12}$$

for $15 \le y_2 \le 50$; and

$$\operatorname{cov}\left[Y(t, 0, s_1), Y_D(t, y_2, s_2)\right] = \sum_{y_1=15}^{50} \operatorname{cov}\left[Y_B(t, y_1, s_1), Y_D(t, y_2, s_2)\right], \tag{7.7.13}$$

for $1 \le y_2 \le r$.

At any epoch $t+s$, $s \ge 1$, the total number of living members of the population is given by the random function

$$Y(t, s) = \sum_{y=0}^{r} Y(t, y, s). \tag{7.7.14}$$

The mean of this random function is

$$v(t, s) = \sum_{y=0}^{r} v(t, y, s); \tag{7.7.15}$$

and its covariance function is

$$\operatorname{cov}\left[Y(t, s_1), Y(t, s_2)\right]$$

$$= \sum_{y_1=0}^{r} \sum_{y_2=0}^{r} \operatorname{cov}\left[Y(t, y_1, s_1), Y(t, y_2, s_2)\right], \tag{7.7.16}$$

for $1 \le s_1 \le s_2$.

Having defined the necessary covariance functions, a result characterizing stochastic variability in population structure may be stated.

Theorem 7.7.1: Suppose $n(t, x)$ is large for every $x = 0, 1, 2, \ldots, r$ and some t.

(i) Then, the variability in population structure is approximately that of a Gaussian process. The mean functions of this process are those given in Theorem 7.5.1; the covariance functions of this process are those listed in (7.7.5) through (7.7.13).

(ii) The variability in population size (7.7.14) is also approximately that of a

Gaussian process. The mean and covariance functions of this process are those in (7.7.15) and (7.7.16).

Proof: To prove assertion (i), let $\text{var}[Y(t, y, s)] = \text{cov}[Y(t, y, s), Y(t, y, s)]$ be the variance function of the random function $Y(t, y, s)$ in (7.7.1). Then, let $\varepsilon(t, y, s)$ be the standardized random variable

$$\varepsilon(t, y, s) = (Y(t, y, s) - v(t, y, s))/(\text{var}[Y(t, y, s)])^{1/2}. \qquad (7.7.17)$$

Next, consider the random function $X(t, x, y, s)$ and the constant function $b(t, x, y, s)$ defined by

$$(n(t, x)\, \text{var}[Z(t, x, y, s)])^{1/2} X(t, x, y, s)$$

$$= \sum_{k=1}^{n(t, x)} (Z_k(t, x, y, s) - M(t, x, y, s)); \qquad (7.7.18)$$

and

$$b(t, x, y, s) = (n(t, x)\, \text{var}[Z(t, x, y, s)])^{1/2}/(\text{var}[Y(t, x, y, s)])^{1/2}. \qquad (7.7.19)$$

Given these definitions, the random function in (7.7.17) may be represented in the form

$$\varepsilon(t, y, s) = \sum_{x=0}^{r} b(t, x, y, s) X(t, x, y, s). \qquad (7.7.20)$$

But, for every fixed x, y, and s, it follows from the central limit theorem for independent and identically distributed random variables with a finite variance, that the random variable $X(t, x, y, s)$ converges in distribution to a normal random variable $\zeta(t, x, y, s)$ with mean zero and variance one. Therefore, when $n(t, x)$ is large for all x, the distribution of the random variable in (7.7.20) is approximately that of the random variable

$$\sum_{x=0}^{r} b(t, x, y, s)\zeta(t, x, y, s), \qquad (7.7.21)$$

where the ζ's are normally and independently distributed with a common mean of zero and variance one. The random variable in (7.7.21) clearly has mean zero and variance one; moreover, being a linear combination of normal independent random variables, it is also normally distributed.

Thus, when all $n(t, x)$ are large, the random variable in (7.7.20) has an approximate normal distribution with mean zero and variance one for every y and s. Hence, for every y and s,

$$Y(t, y, s) = v(t, y, s) + (\text{var}[Y(t, y, s)])^{1/2} \varepsilon(t, y, s) \qquad (7.7.22)$$

is approximately normally distributed with mean $v(t, y, s)$ and $\text{var}[Y(t, y, s)]$.

A similar argument may be used to show that the random functions $Y_B(t, y, s)$ and $Y_D(t, y, s)$ in (7.7.2) and (7.7.3) are approximately Gaussian for every y and s. Furthermore, a similar argument may be used to show that any linear combination of the random variables $Y(t, y_i, s_i)$, $Y_B(t, y_i, s_i)$, and $Y_D(t, y_i, s_i)$, $1 \le i \le n$, is also approximately Gaussian. In particular, the random function in (7.7.4) is approximately Gaussian. Since the covariances of these random functions are as stated, assertion (i) is proved.

Assertion (ii) also follows, because the random function in (7.7.14) is a linear combination of the random functions in (7.7.1) and (7.7.2). □

Keyfitz (1981, 1982) has discussed the limits of population forecasts and the philosophical question: Can Knowledge Improve Forecasts? In limits to population forecasts, Keyfitz was concerned with reliability of published estimates of future population; reliability of population projections was measured in terms of the past experience of about 1,000 population forecasts. Thus, the variability discussed was that among assumptions and techniques used by different forecasters, rather than the stochastic variability discussed here. It should be kept in mind that the stochastic variability discussed in this section refers to variability among realizations of a process, given a fixed set of functions specifying the underlying laws of evolution.

7.8 A Representation of Population Structure Based on Birth Cohorts

As was demonstrated in the proof of Theorem 7.7.1, when $n(t, x)$ is large for all ages $x = 0, 1, 2, \ldots, r$, then the population structure after epoch t evolves approximately as a Gaussian process. Since the evolution of population structure is approximately Gaussian, its stochastic variability may be characterized in terms of covariances functions. Although the representations of the random functions $Y(t, y, s)$, $Y_B(t, y, s)$, $Y_D(t, y, s)$, and $Y(t, 0, s)$, given in (7.7.1) through (7.7.4) in terms of lines of descent, were useful in establishing Theorem 7.7.1, they are less useful in deriving algorithms for computing covariance functions. Set forth in this section is an alternative representation of these random functions based on birth cohorts. This alternative representation not only provides deeper insights into the population process under consideration but also appears to be more useful for designing algorithms for generating realizations of the process by Monte Carlo methods.

Of central importance to the population process is the birth process represented by the random function $Y(t, 0, s)$ in (7.7.4). To make the notation more descriptive, the birth process will be represented by the random function $Y_B(t, s) = Y(t, 0, s)$ in this section. A degree of simplification may also be attained in computing the covariance function of the birth process, if the ages of females in the population are set forth explicitly in representation of the random function $Y_B(t, s)$.

Because it is assumed that childbearing in any cohort does not begin until age 15, only females in the initial population will contribute births during the first fifteen epochs following t. If a female is of age x at epoch t, then she is of age $x + s$ at epoch $t + s$, $1 \leq s \leq 15$. Moreover, only those females with ages x at t such that $15 \leq x + s \leq 50$ may contribute births to the population at epoch

$t+s$. Therefore, the random function $Y_B(t, s)$ takes the form

$$Y_B(t, s) = \sum_{x=15-s}^{50-s} Y_B(t, x+s, s), \tag{7.8.1}$$

for $1 \le s \le 15$.

Under the assumption 50 is the end of the childbearing ages, females in the population at epoch t may contribute births to the population at those epochs $t+s$ such that $16 \le s \le 50$. Those female births credited to the population at epochs $t+1, \ldots, t+s-1$ may also contribute births to the population at epochs $t+s$, $16 \le s \le 50$. Hence, the random function $Y_B(t, s)$ takes the form

$$Y_B(t, s) = \sum_{z=15}^{s-1} Y_B(t, z, s) + \sum_{x=0}^{50-s} Y_B(t, x+s, s), \tag{7.8.2}$$

for $16 \le s \le 50$.

Females in the population at epoch t will no longer contribute births to the population after epoch $t+50$. Therefore, if $s > 50$, then the random function $Y_B(t, s)$ has the form

$$Y_B(t, s) = \sum_{z=15}^{50} Y_B(t, z, s), \tag{7.8.3}$$

as indicated in (7.7.4).

The alternative representation described in this section is based on the observation that the summands on the right in (7.8.1), (7.8.2), and (7.8.3) may in turn be represented as sums of random functions. For any epoch t, define a random function $\varDelta(t, x, s)$ by $\varDelta(t, x, s) = N_f(t, x+s) - N_f(t, x+s-1)$, where $x \ge 0$ and $s \ge 1$. Let $\varDelta_k(t, x, s)$, $k \ge 1$, be independent random functions such that, for every k, $\varDelta_k(t, x, s)$ has the same probability laws as the random function $\varDelta(t, x, s)$. Then, the random functions $Y_B(t, x+s, s)$ appearing on the right in (7.8.1) and (7.8.2) may be represented in the form

$$Y_B(t, x+s, s) = \sum_{k=1}^{n(t, x)} \varDelta_k(t-x, x, s), \tag{7.8.4}$$

for $1 \le s \le 15$.

If an individual is of age z at epoch $t+s$, $s \ge 16$, then she was born at epoch $t+s-z$. The random function $Y_B(t, z, s)$ appearing on the right in (7.8.2) and (7.8.3) may be represented in the form

$$Y_B(t, z, s) = \sum_{k=1}^{Y_B(t, s-z)} \varDelta_k(t+s-z, 0, z), \tag{7.8.5}$$

for $s \ge 16$. Observe that the sum in (7.8.4) has a fixed number $n(t, x)$ of summands; while that in (7.8.5) has a random number $Y_B(t, s-z)$ of summands.

Representations analogous to those in (7.8.4) and (7.8.5) may also be set down for the random function $Y(t, y, s)$ for live individuals of age $y \ge 1$ at epoch $t+s$, $s \ge 1$. For any member of a cohort born at epoch t, define a random function $U(t, x, s)$ by $U(t, x, s) = 1$, if she survives to age $x \ge 0$ and is still alive at age $x+s$; and, let $U(t, x, s) = 0$ otherwise. Let $U_k(t, x, s)$, $k \ge 1$, be independent random functions such that, for every k, $U_k(t, x, s)$ has the same probability laws as the random function $U(t, x, s)$. Then for $1 \le s \le r$ and $y \ge s$, the random function $Y(t, y, s)$ may be represented in the form

$$Y(t, y, s) = \sum_{k=1}^{n(t, y-s)} U_k(t+s-y, y-s, s). \qquad (7.8.6)$$

But, for $1 \le y < s \le r$,

$$Y(t, y, s) = \sum_{k=1}^{Y_B(t, s-y)} U_k(t+s-y, 0, y). \qquad (7.8.7)$$

When $s > r$, Eq. (7.8.7) also holds for all ages $1 \le y \le r$.

To take deaths into account, let $V(t, x, s)$ be a random function such that $V(t, x, s) = 1$, if a member of a cohort born at epoch t is alive at age $x \ge 0$ but dies at age $x+s$; and, let $V(t, x, s) = 0$ otherwise. Let $V_k(t, x, s)$, $k \ge 1$, be independent random functions such that, for all k, $V_k(t, x, s)$ has the same probability laws as the random function $V(t, x, s)$. Then, for $1 \le s \le r$ and $y \ge s$,

$$Y_D(t, y, s) = \sum_{k=1}^{n(t, y-s)} V_k(t+s-y, y-s, s); \qquad (7.8.8)$$

and for $1 \le y < s \le r$,

$$Y_D(t, y, s) = \sum_{k=1}^{Y_B(t, s-y)} V_k(t+s-y, 0, y). \qquad (7.8.9)$$

Like equation (7.8.7), Eq. (7.8.9) holds for all ages $1 \le y \le r$ when $s > r$.

Problem 6 may be consulted for a discussion connecting the random functions $\Delta(t, x, s)$, $U(t, x, s)$, and $V(t, x, s)$ with those in Def. 7.3.2.

7.9 Covariance Functions for the Birth Process and Live Individuals

Given the representations of the random functions describing the evolution of the population structure presented in Sect. 7.8, formulas for the covariance functions may be derived. As will be discussed below, the computer implementation of some of these formulas is feasible. Only the covariance functions for births and live individuals will be considered in this section. Covariance functions for live individuals may be expressed in terms of the covariance function of the birth process; consequently, this function will be considered first.

The covariance function of the birth process may be expressed in terms of two covariances defined in Sect. 7.8. The first of these functions is

$$\text{cov}\left[Y_B(t, x+s_1, s_1), Y_B(t, x+s_2, s_2)\right]$$
$$= n(t, x) \, \text{cov}\left[\Delta(t-x, x, s_1), \Delta(t-x, x, s_2)\right], \qquad (7.9.1)$$

defined for $0 \leq x \leq 50$ and $1 \leq s_1 \leq s_2$. The second function is

$$\text{cov}\left[Y_B(t, z_1, s_1), Y_B(t, z_2, s_2)\right]$$
$$= \text{cov}\left[Y_B(t, s_1 - z_1), Y_B(t, s_2 - z_2)\right] q^2 \varphi_c(t + s_1 - z_1, z_1) \varphi_c(t + s_2 - z_2, z_2),$$

$$(7.9.2)$$

provided that $s_1 - z_1 \neq s_2 - z_2$. Observe that the expression on the right in (7.9.2) contains $\text{cov}\left[Y_B(t, s_1), Y_B(t, s_2)\right]$, the covariance function of the birth process. Note also that (7.9.2) is a special case of the general formula in part (c) of Problem 7. When $s_1 - z_1 = s_2 - z_2$, then the covariance in (7.9.2) becomes a variance so that the formula needs to be modified according to the general formula in part (a) of Problem 7. The details of this modification will be left to the reader.

Theorem 7.9.1: The covariance function of the birth process is determined as follows:

(i) If $1 \leq s_1 \leq s_2 \leq 15$, then

$$\text{cov}\left[Y_B(t, s_1), Y_B(t, s_2)\right]$$
$$= \sum_{x=15-s_1}^{50-s_2} \text{cov}\left[Y_B(t, x + s_1, s_1), Y_B(t, x + s_2, s_2)\right]. \qquad (7.9.3)$$

(ii) If $1 \leq s_1 \leq 15$, $16 \leq s_2 \leq 50$, and $50 - s_2 \geq 15 - s_1$, then formula (7.9.3) holds; if $50 - s_2 < 15 - s_1$ for s_1 and s_2 in this range, then $\text{cov}\left[Y_B(t, s_1), Y_B(t, s_2)\right] = 0$.

(iii) If $16 \leq s_1 \leq s_2 \leq 50$, then

$$\text{cov}\left[Y_B(t, s_1), Y_B(t, s_2)\right]$$
$$= \sum_{z_1=15}^{s_1-1} \sum_{z_2=15}^{s_2-1} \text{cov}\left[Y_B(t, z_1, s_1), Y_B(t, z_2, s_2)\right]$$
$$+ \sum_{x=0}^{50-s_2} \text{cov}\left[Y_B(t, x + s_1, s_1), Y_B(t, x + s_2, s_2)\right]. \qquad (7.9.4)$$

(iv) If $1 \leq s_1 \leq s_2$ and $s_2 \geq 51$, then

$$\text{cov}\left[Y_B(t, s_1), Y_B(t, s_2)\right]$$
$$= \sum_{z_1=15}^{50} \sum_{z_2=15}^{50} \text{cov}\left[Y_B(t, z_1, s_1), Y_B(t, z_2, s_2)\right]. \qquad (7.9.5)$$

Proof: Regarding assertion (i), formula (7.9.3) follows from the observation that in covariance of two sums $Y_B(t, s_1)$ and $Y_B(t, s_2)$ of the form given in (7.8.1), only those terms such that $15 - s_1 \leq x \leq 50 - s_2$ have nonzero covariances. An inspection of (7.8.1) and (7.8.2) and a similar argument may be used to prove assertions (ii) and (iii). Assertion (iv) follows directly from (7.8.3). Formula (7.9.1) is an immediate consequence of (7.8.4); (7.9.2) follows from (7.8.5) and an application of part (c) of Problem 7. ∎

A method for calculating the covariance function on the right in (7.9.1) will be developed in the next section. Given values of this covariance function for all $0 \leq x \leq 50$ and $1 \leq s_1 \leq s_2 \leq 50$ such that $15 \leq x + s_1 \leq 50$ and $15 \leq x + s_2 \leq 50$, assertions (i) and (ii) may be used in applying formula (7.9.3) to calculate $\text{cov}[Y_B(t, s_1), Y_B(t, s_2)]$, the covariance function of the birth process, for all pairs $\langle s_1, s_2 \rangle$ satisfying the conditions $1 \leq s_1 \leq 15$ and $1 \leq s_2 \leq 50$. Different values of x in the random function $Y_B(t, x + s, s)$ correspond to different birth cohorts so that these random functions have the property that they are uncorrelated for different values of x. This property has been exploited in substantially reducing the number of terms in the sum on the right in (7.9.3) as compared to that in (7.9.5). It should also be noted that if only the variance function of the birth process is required during the first 15 years of a projection, then formula (7.9.3) may be applied for $s_1 = s_2 = s$ and $1 \leq s \leq 15$.

The covariance functions in (7.9.1) and (7.9.2) must be used in applying formula (7.9.4) to calculate the covariance function of the birth process at all pairs $\langle s_1, s_2 \rangle$ such that $16 \leq s_1 \leq s_2 \leq 50$. Formulas (7.9.1), (7.9.2), and (7.9.4) are recursive in the sense that, for any pair $\langle s_1, s_2 \rangle$, values of the covariance function of the birth process at other pairs $\langle s_1^*, s_2^* \rangle$ such that $s_1^* < s_1$ and $s_2^* < s_2$ will be required to calculate $\text{cov}[Y_B(t, s_1), Y_B(t, s_2)]$. Formulas (7.9.1), (7.9.2), and (7.9.5) are also recursive in this sense. Note that this recursive property follows from the condition that the age structure of the population is known at epoch t.

Having derived formulas for calculating the covariance function of the birth process, formulas for the covariances among live individuals at any epoch $t + s$, $s \geq 1$, may be set down. As can be seen from formulas (7.7.14) and (7.7.16) these covariances arise in the variance of the random function $Y(t, s)$, representing population size at any epoch $t + s$.

Theorem 7.9.2: At any epoch $t + s$, $s \geq 1$, and ages y_1 and y_2 such that $1 \leq y_1 \leq r$ and $1 \leq y_2 \leq r$, the covariance function $\text{cov}[Y(t, y_1, s), Y(t, y_2, s)]$ is determined as follows:

(i) If $y_1 = y_2 = y \geq s$, then

$$\text{var}[Y(t, y, s)] = n(t, y - s) S_c(t + s - y, y - s, s)(1 - S_c(t + s - y, y - s, s)). \quad (7.9.6)$$

If $y_1 > y_2 \geq s$, then

$$\text{cov}[Y(t, y_1, s), Y(t, y_2, s)] = 0. \quad (7.9.7)$$

(ii) If $1 \leq y < s$, then

$$\text{var}[Y(t, y, s)] = v(t, 0, s - y) S_c(t + s - y, y)(1 - S_c(t + s - y, y))$$
$$+ S_c^2(t + s - y, y) \text{var}[Y_B(t, s - y)]. \quad (7.9.8)$$

If $1 \leq y_1 < y_2 < s$, then

$$\text{cov}[Y(t, y_1, s), Y(t, y_2, s)]$$
$$= \text{cov}[Y_B(t, s - y_1), Y_B(t, s - y_2)] S_c(t + s - y_1, y_1) S_c(t + s - y_2, y_2). \quad (7.9.9)$$

Proof: Formula (7.9.6) follows directly from (7.8.6) by noting that

$$\text{var}[U_k(t + s - y, y - s, s)]$$
$$= S_c(t + s - y, y - s, s)(1 - S_c(t + s - y, y - s, s)), \quad (7.9.10)$$

for every k. Equation (7.9.7) also follows from (7.8.6) by noting that $Y(t, y_1. s)$ and $Y(t, y_2, s)$ are uncorrelated when $y_1 \neq y_2$. This proves assertion (i).

Formula (7.9.8) follows from an application of part (a) of Problem 7 to (7.8.9); formula (7.9.9) follows from an application of part (c) of Problem 7 to (7.8.9) for different values of y. Assertion (ii) is thus proved. ▯

Also needed in the calculation of the variance of $Y(t, s)$, total population size at epoch $t + s$, are covariances among births $Y_B(t, s) = Y(t, 0, s)$ entering the population and live individuals $Y(t, y, s)$ of ages $1 \leq y \leq r$ at epoch $t + s$. Contained in the next theorem are formulas for these desired covariance functions.

Theorem 7.9.3: (i) If $1 \leq s \leq 15$, then

$$\text{cov}\,[Y_B(t, s), Y(t, y, s)]$$
$$= n(t, y - s)\,q\,S_c(t - y + s, y - s, s)[b_c^*(t - y + s, y) - \varphi_c(t - y + s, y - s, s)] \quad (7.9.11)$$

for $15 \leq y \leq 50$; if $1 \leq y < 15$ or $50 < y \leq r$, then

$$\text{cov}\,[Y_B(t, s), Y(t, y, s)] = 0. \quad (7.9.12)$$

(ii) If $16 \leq s \leq 50$, then Eq. (7.9.12) holds if $y \geq s$; if $1 \leq y < s$, then

$$\text{cov}\,[Y_B(t, s), Y(t, y, s)]$$
$$= \sum_{z=15}^{s-1} \text{cov}\,[Y_B(t, z, s), Y(t, y, s)]. \quad (7.9.13)$$

(iii) If $s \geq 51$ and $y \geq s$, then Eq. (7.9.12) holds; if $1 \leq y < s$, then

$$\text{cov}\,[Y_B(t, s), Y(t, y, s)] = \sum_{z=15}^{50} \text{cov}\,[Y_B(t, z, s), Y(t, y, s)]. \quad (7.9.14)$$

(iv) The covariances on the right in (7.9.13) and (7.9.14) are determined as follows:
If $y \neq z$, then

$$\text{cov}\,[Y_B(t, z, s), Y(t, y, s)]$$
$$= \text{cov}\,[Y_B(t, s - z), Y_B(t, s - y)]\,q\,\varphi_c(t + s - z, z)S_c(t + s - y, y). \quad (7.9.15\,a)$$

If $y = z$, then

$$\text{cov}\,[Y_B(t, y, s), Y(t, y, s)]$$
$$= v_B(t, s - y)\,\text{cov}\,[\Delta(t + s - y, 0, y), U(t + s - y, 0, y)]$$
$$+ \text{var}\,[Y_B(t, s - y)]\,q\,\varphi_c(t + s - y, y)S_c(t + s - y, y). \quad (7.9.15\,b)$$

Proof: To prove assertion (i), observe that (7.8.4) and (7.8.6) imply that

$$\text{cov}\,[Y_B(t, y, s), Y(t, y, s)]$$
$$= n(t, x)\,\text{cov}\,[\Delta(t - x, x, s), U(t - x, x, s)] \quad (7.9.16)$$

for all y such that $y = x + s$. In Problem 8, it is shown that the covariance on the right in (7.9.16) is that given in (7.9.11) for $15 \leq y \leq 50$. If either $1 \leq y < 15$ or $50 < y \leq r$, then $\text{cov}\,[Y_B(t, y, s), Y(t, y, s)] = 0$, because different birth cohorts are involved and the number of summands in (7.8.4) and (7.8.6) is constant. Hence, in view of (7.8.1), assertion (i) follows.

If $16 \leq s \leq 50$, births entering the population are represented by (7.8.2). When Eq. (7.8.2) is in force, an argument similar to that used in proving assertion (i) may be used to prove assertion (ii). When Eq. (7.8.3) is in force, an argument similar to that used in proving assertion (i) may also be used to establish assertion (iii).

When $y \neq z$, an application of Problem 7 may be used to establish (7.9.15a). If $y = z$, an application of part (d) of Problem 7 may be used to establish (7.9.15b). This proves assertion (iv). □

A question that naturally arises in exploring the feasibility of implementing the formulas in Theorem 7.9.1 on a computer is: how many values of the covariance function of the birth process need to be computed and stored in a computer, if a population projection is to be made for $n \geq 1$ epochs? The number in question is the number of pairs $\langle s_1, s_2 \rangle$ such that $1 \leq s_1 \leq s_2 \leq n$. Let $N_B(n)$ be this number. Then, a little reflection leads to the conclusion that

$$N_B(n) = \frac{n(n+1)}{2}. \tag{7.9.17}$$

Presented in Table 7.9.1 are values of $N_B(n)$ for chosen values of n. Most population projections are made for periods ranging from 15 to 50 years into the future. From an inspection of Table 7.9.1, it would appear that the calculation of values of the covariance function of the birth process for projection of up to 50 years would appear to be feasible, given the facilities of many computing laboratories. Whether such calculations would indeed be feasible, depends on how efficiently the covariance function on the right in (7.9.1) can be computed. In population processes with time homogeneous laws of evolution, the covariance function in (7.9.1) does not depend on the epoch t and would thus need to be computed for only one birth cohort. When a population process is governed by time inhomogeneous laws of evolution, however, the covariance function on the right in (7.9.1) may have to be calculated for many birth cohorts, which could make the computer implementation of the formulas in Theorem 7.9.1 problematic.

Table 7.9.1. Number of values of covariance function of birth process as a function of epochs in population projection

Epochs n	Number of values $N_B(n)$
15	120
25	325
35	630
45	1,035
50	1,275
100	5,050

When $s_1 = s_2 = s$, then formula (7.7.16) reduces to var $[Y(t, s)]$, the variance in total population size $Y(t, s)$ at epoch $t + s$, $s \geq 1$. Altogether the sum on the right in (7.7.16) contains $(r + 1)^2$ terms; but, because of the symmetry of the covariance function, only $(r + 1)(r + 2)/2$ of these terms are distinct. If $r = 100$, then 5,151 distinct covariances would have to be computed in order to compute the variance of total population size at any epoch $t + s$. Since 5,151 is a rather large number, it is important to know whether any covariances may be zero.

As can be seen from (7.9.7) and (7.9.12), most of these covariances are zero for the first 15 epochs, $1 \leq s \leq 15$, of a projection. The only nonzero terms in the sum on the right in (7.7.16) for the variance of $Y(t, s)$ would be the variances in (7.9.6) and (7.9.8) and the covariances in (7.9.9). When the projection goes beyond 15 epochs, $s > 15$, more nonzero terms would appear in this sum as can be seen by inspecting (7.9.6), (7.9.8), (7.9.9), (7.9.13), and (7.9.14). Even though the number of nonzero terms in var $[Y(t, s)]$ increases as s increases, it appears feasible to calculate var $[Y(t, s)]$ up to 50 epochs, $1 \leq s \leq 50$, particularly, for population processes with time homogeneous laws of evolution. It is of interest to note that if $s > r$, then all 5,151 terms in var $[Y(t, y)]$ would have to be calculated, an operation that may not be feasible.

Formula (7.9.6) provides insights into the magnitude of var $[Y(t, y, s)]$, the variance of the number of individuals of age $y \geq s$ in the population at epoch $t + s$, $1 \leq s \leq r$. For according to (7.9.6), the inequality

$$\text{var}\,[Y(t, y, s)] \leq n(t, y - s) \tag{7.9.18}$$

is valid for $y \geq s$ and $1 \leq s \leq r$. If $S_c(t + s - y, y - s, s)$ is near either zero or one, then var $[Y(t, y, s)]$ will be substantially smaller than $n(t, y - s)$. This observation suggests that during the first 15 epochs of a population projection, the variance in total population size will be small relative to

$$n(t) = \sum_{x=0}^{r} n(t, x), \tag{7.9.19}$$

the total population size at epoch t. It is more difficult to foresee the relative magnitudes of the variances and covariances in (7.9.8), (7.9.9), (7.9.13), and (7.9.14). But, from an inspection of these formulas, it seems clear that the most interesting aspect of assessing stochastic variation in making population projections will be that of calculating the covariance function of the birth process.

It is clear that formulas for all the covariances in Sect. 7.7 could be derived, using the techniques developed in this section. The task of developing all these formulas will, however, not be undertaken, because many of these formulas would not seem to be of great substantive interest.

7.10 Product Moments of the Actual and Potential Birth Processes

A function that is fundamental to the calculation of the covariance functions discussed in Sect. 7.9 is a covariance function of the form

$$
\begin{aligned}
&\operatorname{cov}\left[\varDelta(t, x, s_1), \varDelta(t, x, s_2)\right]\\
&= E\left[\varDelta(t, x, s_1)\varDelta(t, x, s_2)| X(t)>x\right]\\
&\quad - q^2 \varphi_c(t, x, s_1)\varphi_c(t, x, s_2),
\end{aligned}
\tag{7.10.1}
$$

appearing on the right in (7.9.1). As a first step in deriving formulas for calculating the product moment function in (7.10.1), it will be helpful to recall that

$$
\varDelta(t, x, s) = N_f(t, x+s) - N_f(t, x+s-1).
\tag{7.10.2}
$$

Let $\varLambda_f(t, x, s_1, s_2)$ be a product moment function defined by

$$
\varLambda_f(t, x, s_1, s_2) = E\left[N_f(t, x+s_1)N_f(t, x+s_2)| X(t)>x\right].
\tag{7.10.3}
$$

Then, the product moment on the right in (7.10.1) may be expressed in the form

$$
\begin{aligned}
&E\left[\varDelta(t, x, s_1)\varDelta(t, x, s_2)| X(t)>x\right]\\
&= \varLambda_f(t, x, s_1, s_2) - \varLambda_f(t, x, s_1, s_2 - 1)\\
&\quad - \varLambda_f(t, x, s_1 - 1, s_2) + \varLambda_f(t, x, s_1 - 1, s_2 - 1).
\end{aligned}
\tag{7.10.4}
$$

Thus, once the product moment function in (7.10.3) is determined, the covariance function in (7.10.1) may be computed. The purpose of this section is to set down formulas for calculating the product moment function $\varLambda_f(t, x, s_1, s_2)$.

The function defined in (7.10.3) is, in turn, expressible in terms of two other product moment functions. One of these functions involves the actual birth process $N(t, s)$ for a cohort of females born at epoch t. It is defined by

$$
\varLambda_N(t, x, s_1, s_2) = E\left[N(t, x+s_1)N(t, x+s_2)| X(t)>x\right].
\tag{7.10.5}
$$

A second function, involving the potential birth process, is defined by

$$
\varLambda_K(t, x, s_1, s_2) = E\left[K(t, x+s_1)K(t, x+s_2)\right].
\tag{7.10.6}
$$

In (7.10.3) through (7.10.6), x, s_1, and s_2 are integers such that $15 \le x+s_v \le 50$ for $v = 1, 2$; t is any epoch such that $t = 0, \pm1, \pm2, \dots$.

Discussed in Problem 3 is a relationship connecting the random functions $N_f(t, s)$ and $N(t, s)$; also discussed in this problem is an equation connecting the product moments $E[N_f(t, s)N_f(t, s+\tau)]$, for $\tau > 0$, and the moments in (7.10.5). Let the mean function $m(t, x, s)$ be defined by

$$
m(t, x, s) = E\left[N(t, x+s)| X(t)>x\right].
\tag{7.10.7}
$$

Then, by using an argument similar to that used in deriving the result in Problem 3(c), it can be shown that

$$\Lambda_f(t, x, s_1, s_2) = q(1-q)m(t, x, s_1) + q^2 \Lambda_N(t, x, s_1, s_2),\tag{7.10.8}$$

for $s_1 \leq s_2$.

The next result concerns a formula connecting the functions $\Lambda_N(t, x, s_1, s_2)$ and $\Lambda_K(t, x, s_1, s_2)$. All calculations involving the potential birth process are done on a monthly time scale, a scale that is much finer than the yearly time scale of a population projection. To simplify the argumentation, a relationship connecting the functions $\Lambda_N(t, x, s_1, s_2)$ and $\Lambda_K(t, x, s_1, s_2)$ will be derived in continuous time. It should be remembered, however, that in the computer implementation of this relationship, all integrals are calculated as sums of integrands evaluated on a monthly time scale. In what follows, $g_c(t, x, s)$ is the conditional p.d.f. associated with the survival function $S_c(t, x, s)$.

Theorem 7.10.1: If $0 \leq s_1 \leq s_2$, then the formula connecting the functions $\Lambda_N(t, x, s_1, s_2)$ and $\Lambda_K(t, x, s_1, s_2)$ is given by

$$\Lambda_N(t, x, s_1, s_2) = \int_0^{s_1} g_c(t, x, v) \Lambda_K(t, x, v, v) dv$$

$$+ \int_{s_1}^{s_2} g_c(t, x, v) \Lambda_K(t, x, s_1, v) dv$$

$$+ S_c(t, x, s_2) \Lambda_K(t, x, s_1, s_2).\tag{7.10.9}$$

Proof: As indicated in formula (7.10.9), the derivation is conditional on $X(t) > x$, the event that any given member of a cohort born at epoch t survives to age x.

If $X(t) = x + v$ and $0 < v \leq s_1$, then the equations connecting the actual and potential birth processes are given by

$$N(t, x+s_1) = K(t, x+v)$$

and $\hspace{11cm}$ (7.10.10)

$$N(t, x+s_2) = K(t, x+v).$$

Therefore,

$$N(t, x+s_1)N(t, x+s_2) = K^2(t, x+v)\tag{7.10.11}$$

for all $0 < v \leq s_1$. The first term on the right in (7.10.9) follows by taking expectations in (7.10.11) and then integrating on v.

The other two terms on the right in (7.10.9) may be derived by a similar argument. If $X(t) = x + v$ and $s_1 < v \leq s_2$, then the equations

$$N(t, x+s_1) = K(t, x+s_1)$$

and $\hspace{11cm}$ (7.10.12)

$$N(t, x+s_2) = K(t, x+v)$$

are valid and lead to the second term on the right in (7.10.9).

Finally, if $X(t) > x + s_2$, then the equations

$$N(t, x + s_1) = K(t, x + s_1)$$

and   (7.10.13)

$$N(t, x + s_2) = K(t, x + s_2)$$

are valid and may be used to derive the third term on the right in (7.10.9). ∎

To derive a formula for the product moment function $\Lambda_K(t, x, s_1, s_2)$, it will be necessary to return to a consideration of assumptions underlying an evolutionary potential birth process of the type discussed in Sect. 4.9. For a cohort of females born at epoch t, let $T_0(t)$ be a random variable representing age at marriage; and, let $T_v(t)$, $v \geq 1$, be a monotone increasing sequence of random variables, representing the ages at which members of the cohort experience live births. As in Sect. 4.2, it will be assumed that the random variables $Z_v(t) = T_v(t) - T_{v-1}(t)$, $v \geq 1$, representing waiting times among live births, are independent with p.d.f.'s $h(t, v-1, s)$.

Of central importance in the derivation of a computing formula for $\Lambda_K(t, x, s_1, s_2)$ is the distribution function

$$F^{(n)}(t, s) = P[T_n(t) \leq s] \tag{7.10.14}$$

of the waiting times to the n-th live birth, $n \geq 1$, in a cohort of females born at epoch t. As in Sect. 4.2, let $g_0(t, x)$ be the p.d.f. of the waiting times to marriage in a cohort of females born at epoch t and let $H^{(n)}(t, s)$ be the distribution function of the waiting times to the n-th live birth following marriage. Then, under the assumption $T_0(t)$ is independent of the sequence $Z_v(t)$, $v \geq 1$, the distribution function in (7.10.14) is given by

$$F^{(n)}(t, s) = \int_0^s g_0(t, v) H^{(n)}(t, s - v) dv, \tag{7.10.15}$$

for $n \geq 1$. Another function entering into a formula for $\Lambda_K(t, x, s_1, s_2)$ is the distribution function

$$F_{ij}(t, s) = P[Z_i(t) + \ldots + Z_j(t) \leq s], \tag{7.10.16}$$

defined for $2 \leq i \leq j$. Let $h(t, i, j, v)$ be the convolution of the p.d.f.'s $h(t, v-1, v)$ for $i \leq v \leq j$. Then,

$$F_{ij}(t, s) = \int_0^s h(t, i, j, v) dv. \tag{7.10.17}$$

An assumption underlying MATHIST, the computer simulation model of maternity histories described in Chap. 5, was that each birth cohort of females could experience up to some maximum number of births by the end of the childbearing ages. Under this assumption, the potential birth process $K(t, s)$ for a cohort of females born at epoch t may be represented as a finite sum of indicator functions. Let $I_v(t, s) = 1$ if $T_v(t) \leq s$ and let $I_v(t, s) = 0$ if $T_v(t) > s$. Then, if b is the maximum number of births that may occur in any birth cohort of females,

$$K(t, s) = \sum_{v=1}^{b} I_v(t, s). \tag{7.10.18}$$

Since a formula for $\Lambda_K(t, x, s_1, s_2)$ may be derived from $\Lambda_K(t, 0, s_1, s_2)$ by translation, it suffices to derive an expression for $\Lambda_K(t, 0, s_1, s_2)$.

Theorem 7.10.2: Let $f^{(v)}(t, v)$ be the p.d.f. of the distribution function $F^{(v)}(t, v)$. For $s_1 \leq s_2$ let $J_v(t, s_1, s_2)$ be a function defined by

$$J_v(t, s_1, s_2) = v F^{(v)}(t, s_1) + \sum_{j=v+1}^{b} \int_0^{s_1} f^{(v)}(t, v) F_{v+1, j}(s_2 - v) dv \tag{7.10.19}$$

for $v = 1, 2, \ldots, b-1$; and, let

$$J_b(t, s_1, s_2) = b F^{(b)}(t, s_1). \tag{7.10.20}$$

Then,

$$\Lambda_K(t, 0, s_1, s_2) = \sum_{v=1}^{b} J_v(t, s_1, s_2). \tag{7.10.21}$$

Proof: From (7.10.18), it follows that

$$\Lambda_K(t, 0, s_1, s_2) = \sum_{i \geq j} E[I_i(t, s_1) I_j(t, s_2)]$$
$$+ \sum_{i < j} E[I_i(t, s_1) I_j(t, s_2)]. \tag{7.10.22}$$

But, if $i \geq j$ and $s_1 \leq s_2$, then $T_i(t) \leq s_1$ implies $T_j(t) \leq s_2$. Therefore,

$$E[I_i(t, s_1) I_j(t, s_2)] = P[T_i(t) \leq s_1, T_j(t) \leq s_2]$$
$$= P[T_i(t) \leq s_1] = F^{(i)}(t, s_1). \tag{7.10.23}$$

If $i < j$, then

$$E[I_i(t, s_1) I_j(t, s_2)] = \int_0^{s_1} f^{(i)}(t, v) F_{i+1, j}(t, s_2 - v) dv. \tag{7.10.24}$$

Let v be a fixed index such that $1 \leq v \leq b$. Then, if $i = v$, the first sum on the right in (7.10.22) contributes the term $v F^{(v)}(t, s_1)$; the second term on the right in (7.10.22) contributes the second term on the right in (7.10.19). Summing (7.10.19) and (7.10.20) over $v = 1, 2, \ldots, b$ shows that (7.10.22) is equivalent to (7.10.21). ☐

Having derived a formula for the product moment function $\Lambda_K(t, 0, s_1, s_2)$ for the potential birth process, the next step is to set down a computational procedure for the second order difference appearing on the right in (7.10.4). Let D_1 be a difference operator defined by

$$D_1 \Lambda_f(t, x, s_1, s_2) = \Lambda_f(t, x, s_1, s_2) - \Lambda_f(t, x, s_1 - 1, s_2). \tag{7.10.25}$$

The difference operator D_2 is defined similarly with respect to s; the second order difference operator D_{12} is defined iteratively by $D_{12} = D_1 D_2$. With these definitions, the product moment in (7.10.4) may be expressed in the form

$$E[\Delta(t, x, s_1)\Delta(t, x, s_2)|X(t)>x] = D_{12}\Lambda_f(t, x, s_1, s_2)$$
$$= D_{21}\Lambda_f(t, x, s_1, s_2). \tag{7.10.26}$$

By applying the second order difference operator to equation (7.10.8), it follows that

$$E[\Delta(t, x, s_1)\Delta(t, x, s_2)|X(t)>x] = q^2 D_{12}\Lambda_N(t, x, s_1, s_2). \tag{7.10.27}$$

Two cases, $s_1 < s_2$ and $s_1 = s_2 = s$, need to be considered in evaluating the function $D_{12}\Lambda_N(t, x, s_1, s_2)$. For the case, $s_1 < s_2$, an application of the second order difference operator D_{12} to Eq. (7.10.9) yields the formula

$$D_{12}\Lambda_N(t, x, s_1, s_2) = \int_{s_2-1}^{s_2} g_c(t, x, v)D_1\Lambda_K(t, x, s_1, v)dv$$
$$+ S_c(t, x, s_2)D_1\Lambda_K(t, x, s_1, s_2)$$
$$- S_c(t, x, s_2-1)D_1\Lambda_K(t, x, s_1, s_2-1). \tag{7.10.28}$$

The formula for $D_{12}\Lambda_N(t, x, s, s)$ is slightly more complicated. In order to express this function compactly, it will be helpful to define a function $D^*(t, x, s_1, s_2)$ by

$$D^*(t, x, s_1, s_2) = \Lambda_K(t, x, s_2, s_2) - 2\Lambda_K(t, x, s_1, s_2). \tag{7.10.29}$$

Then,

$$D_{12}\Lambda_N(t, x, s, s) = \int_{s-1}^{s} g_c(t, x, v)D^*(t, x, s, v)dv$$
$$+ S_c(t, x, s)D^*(t, x, s-1, s)$$
$$+ S_c(t, x, s-1)\Lambda_K(t, x, s-1, s-1). \tag{7.10.30}$$

The next step is to derive formulas for the differences $D_1\Lambda_K(t, x, s_1, s_2)$ and $D^*(t, x, s_1, s_2)$, appearing in (7.10.28) and (7.10.30). Because $\Lambda_K(t, x, s_1, s_2) = \Lambda_K(t, 0, x+s_1, x+s_2)$, it suffices to restrict attention to the function $\Lambda_K(t, 0, s_1, s_2)$. From (7.10.21), it follows that

$$D_1\Lambda_K(t, 0, s_1, s_2) = \sum_{v=1}^{b} D_1 J_v(t, s_1, s_2), \tag{7.10.31}$$

if $s_1 < s_2$. Moreover, by applying the first order difference operator to Eqs. (7.10.19) and (7.10.20), it follows that

$$D_1 J_v(t, s_1, s_2) = v \int_{s_1-1}^{s_2} f^{(v)}(t, u)du$$
$$+ \sum_{j=v+1}^{b} \int_{s_1-1}^{s_1} f^{(v)}(t, u)F_{v+1,j}(t, s_2-u)du. \tag{7.10.32}$$

Equation (7.10.32) is valid for all $1 \leq v \leq b$, provided that the second term on the right is set equal to zero when $v = b$. Observe that formulas (7.10.31) and (7.10.32) are sufficient to determine the second and third terms on the right in

(7.10.28); all that remains to be done in implementing formula (7.10.28) is to devise a scheme for computing the integral on the right in (7.10.28).

In devising a scheme for the computer implementation of formula (7.10.30), it will be helpful to take advantage of the special properties of the function $\Lambda_K(t, 0, s, s)$. When $s_1 = s_2 = s$, then the integrals on the right in (7.10.19) become ordinary convolutions. Hence, the J-functions in (7.10.21) take the form

$$J_v(t, s, s) = v F^{(v)}(t, s) + \sum_{j=v+1}^{b} F^{(j)}(t, s), \tag{7.10.33}$$

a form that is valid for all $1 \leq v \leq b$ if the second term on the right is set equal to zero when $v = b$. It is important to observe from (7.10.33) that the function

$$\Lambda_K(t, 0, s, s) = \sum_{v=1}^{b} J_v(t, s, s) \tag{7.10.34}$$

can be computed efficiently, using a FFT agorithm. Thus, the third term on the right in (7.10.30) can be computed efficiently. Formulas (7.10.19), (7.10.20), and (7.10.21) can also be used to calculate the function $\Lambda_K(t, 0, s-1, s)$ needed in evaluating the function $D^*(t, x, s-1, s)$ in the second term on the right in (7.10.30). The problem of calculating the integral on the right in (7.10.30), however, still remains.

In principle, the integrals on the right in (7.10.28) and (7.10.30) could be calculated with sufficient accuracy, if the integrands were known on a monthly time scale. It should also be noted that the functions $D_1 \Lambda_K(t, x, s_1, s_2)$ and $D^*(t, x, s_1, s_2)$ appearing in the integrands in (7.10.28) and (7.10.30) can, in principle, be calculated on a monthly time scale. But, the size of the arrays generated in calculating these functions on a monthly time scale are so large that the operation would not be feasible for many computers. Calculating the functions in (7.10.14) and (7.10.16) on a monthly time scale, using a FFT algorithm is, however, a feasible operation for most computers. These results could then be used to calculate the functions $D_1 \Lambda_K(t, x, s_1, s_2)$ and $D^*(t, x, s_1, s_2)$ for yearly values of $s_1 \leq s_2$. This operation is again feasible; for, the number of pairs of integers $\langle s_1, s_2 \rangle$ such that $15 \leq s_1 \leq s_2 \leq 50$ is only 666.

After calculating the functions $D_1 \Lambda_K(t, x, s_1, s_2)$ and $D^*(t, x, s_1, s_2)$ on a yearly time scale, a spline interpolation scheme could be used to interpolate them down to a monthly time scale as required for the calculation of the integrals on the right in (7.10.28) and (7.10.30). Similarly, a monotone spline interpolation procedure could be used to interpolate the cohort survival function $S_c(t, x, s)$ down to a monthly time scale, in order to get an estimate of the p.d.f. $g_c(t, x, s)$. Using this scheme, the integrals in (7.10.28) and (7.10.30) could be estimated.

From the scheme just outlined, it is clear that the second order difference $D_{12} \Lambda_N(t, x, s_1, s_2)$ can be calculated for all integers x and pairs of integers $\langle s_1, s_2 \rangle$ such that $15 \leq x + s_1 \leq x + s_2 \leq 50$. Even though this calculation is feasible for a single birth cohort, it may become problematic if it is to be repeated for as many as fifty birth cohorts, the number needed for a population projection of fifteen years, for a population process with time inhomogeneous laws of evolution. When a population process has time homogeneous laws of evolution, however, the calculations will have to be carried out for only one birth cohort. Therefore, the above scheme should be useful in estimating the level of stochastic variability in the time homogeneous case.

7.11 Product Moment Functions as Solutions of Renewal Equations

An alternative approach to setting down computational formulas for the covariance functions of the population process is that of deriving renewal equations in terms of the basic random functions defined in Sect. 7.3. Unlike those defined in Sect. 7.8, these random functions are defined in terms of lines of descent. Specifically, formulas for the product moments going into the covariances on the right in (7.7.5), (7.7.6), and (7.7.8) will be derived. These covariances are of particular interest, because they enter into the formula for the covariance function of total population size given in (7.7.16).

Let $Z_B(t, x, y, s)$ and $Z(t, x, y, s)$ be the random functions in Definition 7.3.1; and, let E_x stand for conditional expectation given that $X(t-x)>x$. Then, the three product moment functions to be considered in this section are defined by

$$\Gamma(t, y, z, s, \tau) = E_x[Z(t, x, y, s)Z(t, x, z, s+\tau)], \tag{7.11.1}$$

$$\Gamma_B(t, x, y, z, s, \tau) = E_x[Z_B(t, x, y, s)Z_B(t, x, z, s+\tau)], \tag{7.11.2}$$

and

$$\Gamma^*(t, x, y, z, s, \tau) = E_x[Z(t, x, y, s)Z_B(t, x, z, s+\tau)]. \tag{7.11.3}$$

In the above, τ is an integer such that $s+\tau\geq 1$ and $s\geq 1$. When a product moment function is defined as in (7.11.1), the covariance function on the right in (7.7.5) may be written in the form

$$\operatorname{cov}[Z(t, x, y_1, s_1), Z(t, x, y_2, s_2)]$$
$$= \Gamma(t, x, y_1, y_2, s_1, \tau) - M(t, x, y_1, s_1)M(t, x, y_2, s_2), \tag{7.11.4}$$

where $s_2 = s_1 + \tau$. Similar formulas may be written down for the covariance functions on the right in (7.7.6) and (7.7.8).

A more succinct representation of the stochastic renewal equations in (7.3.5) and (7.3.7) will be useful in deriving renewal equations for the product moment functions defined above. To this end, let

$$W_B(t+i, 0, y, s-1) = \sum_{j=1}^{\Delta(x, i)} Z_{Bj}(t+i, 0, y, s-i)$$

and

$$\tag{7.11.5}$$

$$W(t+i, 0, y, s-i) = \sum_{j=1}^{\Delta(x, i)} Z_j(t+i, 0, y, s-i).$$

Also define two random functions by

$$Z_B^{(1)}(t, x, y, s) = \sum_{i=1}^{r} W_B(t+i, 0, y, s-i)$$

and   (7.11.6)

$$Z^{(1)}(t, x, y, s) = \sum_{i=1}^{r} W(t+i, 0, y, s-i).$$

Then, Eqs. (7.3.5) and (7.3.7) may be represented in the compact forms

$$Z_B(t, x, y, s) = Z_B^{(0)}(t, x, y, s) + Z_B^{(1)}(t, x, y, s)$$

and   (7.11.7)

$$Z(t, x, y, s) = Z^{(0)}(t, x, y, s) + Z^{(1)}(t, x, y, s).$$

Theorem 7.11.1: There are functions $f_B(t, x, y, z, s, \tau)$, $f(t, x, y, z, s, \tau)$, and $g(t, x, y, z, s, \tau)$ such that the product moment functions in (7.11.1), (7.11.2), and (7.11.3) satisfy the renewal equations

$$\Gamma_B(t, x, y, z, s, \tau)$$

$$= f_B(t, x, y, z, s, \tau) + q \sum_{i=1}^{r} \varphi_c(t-x, x, i)\Gamma_B(t, 0, y, z, s-i, \tau);$$ (7.11.8)

$$\Gamma(t, x, y, z, s, \tau)$$

$$= f(t, x, y, z, s, \tau) + q \sum_{i=1}^{r} \varphi_c(t-x, x, i)\Gamma(t, 0, y, z, s-i, \tau);$$ (7.11.9)

and

$$\Gamma^*(t, x, y, z, s, \tau)$$

$$= g(t, x, y, z, s, \tau) + q \sum_{i=1}^{r} \varphi_c(t-x, x, i)\Gamma^*(t, 0, y, z, s-i, \tau).$$ (7.11.10)

Proof: To demonstrate the method of proof, it will suffice to establish the validity of equation (7.11.9). By using the second equation in (7.11.7), it can be seen that the expectation on the right in (7.11.1) has four terms. They are:

$$f^{(1)}(t, x, y, z, s, \tau) = E_x[Z^{(0)}(t, x, y, s)Z^{(0)}(t, x, y, s)];$$
$$f^{(2)}(t, x, y, z, s, \tau) = E_x[Z^{(1)}(t, x, y, s)Z^{(0)}(t, x, z, s+\tau)];$$ (7.11.11)
$$f^{(3)}(t, x, y, z, s, \tau) = E_x[Z^{(0)}(t, x, y, s)Z^{(1)}(t, x, z, s+\tau)];$$

and

$$E_x[Z^{(1)}(t, x, y, s)Z^{(1)}(t, x, z, s+\tau)].$$ (7.11.12)

The validity of Eq. (7.11.9) follows from an analysis of the expectation in (7.11.12). Let

$$\gamma(t-x, x, i, i) = E_x[\Delta(x, i)(\Delta(x, i)-1)]$$

and   (7.11.13)

$$\gamma(t-x, x, i, j) = E_x[\Delta(x, i)\Delta(x, j)]$$

for $i \neq j$. A function that arises in this analysis is

$$f^{(4)}(t, x, y, z, s, \tau)$$
$$= \sum_{i, j} \gamma(t - x, x, i, j) M(t + i, 0, y, s - i) M(t + j, 0, z, s + \tau - j). \qquad (7.11.14)$$

By taking the product of the random functions in (7.11.12) and using the representations in (7.11.5) and (7.11.6), it can be shown that the expectation in (7.11.12) is the sum of the function in (7.11.14) and the second term on the right in (7.11.9). Therefore, if the function $f(t, x, y, z, s, \tau)$ is defined by

$$f(t, x, y, z, s, \tau) = \sum_{\nu = 1}^{4} f^{(\nu)}(t, x, y, z, s, \tau), \qquad (7.11.15)$$

Eq. (7.11.9) follows. The validity of Eqs. (7.11.8) and (7.11.10) may be established using similar techniques. □

The functions $f_B(t, x, y, z, s, \tau)$ and $g(t, x, y, z, s + \tau)$ are completely analogous to the function defined in (7.11.15). Each of these functions has three terms similar to those in (7.11.11); the fourth terms are analogous to the function in (7.11.14). Since these functions will be used subsequently, they will be set down here. They are:

$$f_B^{(4)}(t, x, y, z, s, \tau)$$
$$= \sum_{i, j} \gamma(t - x, x, i, j) M_B(t + i, 0, y, s - i) M_B(t + j, 0, z, s + \tau - j) \qquad (7.11.16)$$

and

$$g^{(4)}(t, x, y, z, s, \tau)$$
$$= \sum_{i, j} \gamma(t - x, x, i, j) M(t + i, 0, y, s - i) M_B(t + j, 0, z, s + \tau - j). \qquad (7.11.17)$$

In principle, by calculating a renewal density function as described in Sect. 7.4, the renewal equations of Theorem 7.11.1 could be solved numerically for the product moment functions under consideration. The algorithms presented in Sects. 7.9 and 7.10, which were based on the alternative representation of the population structure presented in Sect. 7.8, are, however, more amenable to computer implementation. The principal value of the renewal equations in Theorem 7.11.1 is that they provide a means for deducing asymptotic formulas for the covariance functions in a population evolving according to time homogeneous laws of evolution.

In order to evaluate these functions numerically, it will be necessary to compute the functions in (7.11.13), which may be expressed in terms of the second order differences on (7.10.27). Thus,

$$\gamma(t - x, x, i, i) = q^2 D_{12} A_N(t - x, x, i, i) - q \varphi_c(t - x, x, i); \qquad (7.11.18)$$

and

$$\gamma(t - x, x, i, j) = q^2 D_{12} A_N(t - x, x, i, j) \qquad (7.11.19)$$

for $i \neq j$. In the time homogeneous case, these expressions do not depend on the argument $t - x$, representing the epoch of birth for a cohort.

7.12 Asymptotic Formulas for Mean and Covariance Functions in the Time Homogeneous Case

For population processes with time homogeneous laws of evolution, well known limit theorems from renewal theory may be used to deduce a rather large class of asymptotic formulas. Many of these formulas will be familiar to demographers and belong to what is widely known as stable population theory. Keyfitz (1968), Smith and Keyfitz (1977), and Coale (1972) may be consulted for details; the paper, Mode (1974a), also contains historical references. Various authors have deduced many of these formulas by intuitive arguments; but, in a branching process formulation, they automatically appear in limit theorems, suggesting that they are applicable only after a population governed by time homogeneous laws has undergone a long period of evolution.

When $x=0$ and the laws governing the evolution of a population are time homogeneous, the renewal equations in Theorems 7.3.1 and 7.11.1 become ordinary renewal equations of the general form

$$a(s)=b(s)+ \sum_{v=0}^{s} c(v)a(s-v),\qquad(7.12.1)$$

where $s\geq 0$, the nonnegative functions b and c are given, and the a-function is to be determined. Two results play a major role in this and subsequent sections. In one of these results, Laplace transforms of the form

$$B(\lambda)= \sum_{v=0}^{\infty} b(v) \exp[-v\lambda]$$

and $\qquad(7.12.2)$

$$C(\lambda)= \sum_{v=0}^{\infty} c(v) \exp[-v\lambda],$$

are of fundamental importance. In all the applications to be considered, $B(0)$ and $C(0)$ are finite so that the series in (7.12.2) converge for $\lambda\geq k$, a number that may be negative.

Contained in the following lemma are two key results from which many others follow.

Lemma 7.12.1: (i) Let λ_0 be a number such that

$$C(\lambda_0)=1\qquad(7.12.3)$$

and suppose the derivative $C'(\lambda_0)$ is finite. Then,

$$\lim_{s\uparrow\infty} \exp[-\lambda_0 s]a(s)=\frac{B(\lambda_0)}{-C'(\lambda_0)}.\qquad(7.12.4)$$

(ii) If the limit

$$\lim_{s\uparrow\infty} b(s)=b,\qquad(7.12.5)$$

exists and is finite and $C(0) < 1$, then

$$\lim_{s \uparrow \infty} a(s) = \frac{b}{1 - C(0)}. \tag{7.12.6}$$

Assertion (i) is merely a restatement of a result on page 331 of Feller (1968); a hint for proving assertion (ii) is given in Problem 9.

In order to apply assertion (i) in the above lemma to the equations in Theorem 7.3.1, it will be helpful to recast Eq. (7.3.11) in a form applicable to population processes with time homogeneous laws of evolution. For a time homogeneous process, the functions $M_B(t, x, y, s)$ and $\varphi_c(t, x, y, s)$ will be represented by $M_B(x, y, s)$ and $\varphi_c(x, y, s)$. In what follows, it will be assumed that similar changes in notation have been made in Eqs. (7.3.12) and (7.3.13). Furthermore, a cohort survival function $S_c(t, x)$ will be denoted by $S_c(x)$.

When the renewal equations of Theorem 7.3.1 are under consideration in the time homogeneous case, the second Laplace transform in (7.12.2) takes the form

$$\Phi_c(\lambda) = q \sum_{i=1}^{r} \varphi_c(0, i) \exp[-i\lambda]. \tag{7.12.7}$$

A number α such that $\Phi_c(\alpha) = 1$ will be called the Malthusian parameter for reasons that will be made clear subsequently. It also turns out that a famous function from population mathematics arises in studying the asymptotic behavior of solutions to the renewal equations in Theorem 7.3.1. This function, which is known as the Fisherian reproductive value for a woman aged x, is defined by

$$V(x) = \frac{q \exp[\alpha x]}{S_c(x)} \sum_{v=x}^{r} \varphi_c(v) \exp[-\alpha v]. \tag{7.12.8}$$

Many students of population mathematics probably first encountered this function in the writings of Sir Ronald Fisher, see for example Fisher (1958), page 27.

Studied in the next theorem is the asymptotic behavior of the mean functions in Theorem 7.3.1 in the time homogeneous case.

Theorem 7.12.1: For population processes with time homogeneous laws of evolution,

$$\lim_{s \uparrow \infty} \exp[-\alpha s] M_B(x, y, s) = d^{-1} V(x) q \varphi_c(y) \exp[-\alpha y]; \tag{7.12.9}$$

$$\lim_{s \uparrow \infty} \exp[-\alpha s] M_D(x, y, s) = d^{-1} V(x) g_c(y) \exp[-\alpha y]; \tag{7.12.10}$$

and

$$\lim_{s \uparrow \infty} \exp[-\alpha s] M(x, y, s) = d^{-1} V(x) S_c(y) \exp[-\alpha y], \tag{7.12.11}$$

where

$$d = q \sum_{i=1}^{r} i \exp[-i\alpha] \varphi_c(i).$$

Proof: It will suffice to establish the validity of Eq. (7.12.9). When $x=0$, Eq. (7.3.11) becomes an ordinary renewal equation so that assertion (i) of Lemma 7.12.1 applies. For this equation,

$$B(\alpha)=q\varphi_c(y)\exp[-\alpha y]. \tag{7.12.12}$$

Because $V(0)=1$, an application of (7.12.4) yields (7.12.9) for the case $x=0$.

Upon multiplying (7.3.11) by $\exp[-\alpha s]$, letting $s\uparrow\infty$, and using (7.12.9) for the case $x=0$, a little algebraic manipulation leads to the conclusion that (7.12.9) is valid for all $x\neq 0$. In these manipulations, it will be helpful to set $\varphi_c(x,0)=0$ and let the sum on the right in (7.3.11) run from 0 to r. ☐

An alternative way of stating (7.12.11) is to assert that there exists a function $R(x,y,s)$ such that

$$M(x,y,s)=\exp[\alpha s]d^{-1}V(x)S_c(y)\exp[-\alpha y]+R(x,y,s), \tag{7.12.13}$$

and

$$\lim_{s\uparrow\infty}\exp[-\alpha s]R(x,y,s)=0 \tag{7.12.14}$$

at all points $\langle x,y\rangle$. From (7.12.13), it can be seen that if $\alpha>0$ and s is large, then $M(x,y,s)$ increases geometrically with s. It thus makes sense to call α the Malthusian parameter. The parameter α is also known as Lotka's r, see Lotka (1956) for some historical details and a list of his papers on mathematical demography. From an inspection of equation (7.12.13), it also makes sense to call $V(x)$ the reproductive value of a women aged x; for, it is indicative of the mean number of female descendants a woman of age x at some initial epoch will contribute to the future population.

As might be expected, the results in Theorem 7.12.1 also have interesting implications regarding the asymptotic behavior of the product moment functions in Theorem 7.11.1 for population processes with time homogeneous laws of evolution. To lighten the notation, drop the variable t in the functions defined in (7.11.4), (7.11.16), and (7.11.18). Similarly, let the function $\gamma(t-x,x,i,j)$ be represented by $\gamma(x,i,j)$ and define a function $\gamma^*(x)$ by

$$\gamma^*(x)=\sum_{i,j}\gamma(x,i,j)\exp[-\alpha(i+j)]. \tag{7.12.15}$$

Then, for every fixed x,y,z, and τ, it follows from Theorem 7.12.1 that

$$\lim_{s\uparrow\infty}\exp[-\alpha(2s+\tau)]f^{(4)}(x,y,z,s,\tau)$$
$$=d^{-2}\gamma^*(x)S_c(y)S_c(z)\exp[-\alpha(y+z)]; \tag{7.12.16}$$
$$\lim_{s\uparrow\infty}\exp[-\alpha(2s+\tau)]f_B^{(4)}(x,y,z,s,\tau)$$
$$=d^{-2}\gamma^*(x)\varphi_c(y)\varphi_c(z)\exp[-\alpha(y+z)]; \tag{7.12.17}$$

and

$$\lim_{s\uparrow\infty}\exp[-\alpha(2s+\tau)]g^{(4)}(x,y,z,s,\tau)$$
$$=d^{-2}\gamma^*(x)S_c(y)\varphi_c(z)\exp[-\alpha(y+z)]. \tag{7.12.18}$$

Given the limit formulas just set down, it is now possible to derive asymptotic formulas for the product moment functions in Theorem 7.11.1. As an aid to representing these formulas in a compact way, it will be helpful to define some additional quantities and functions. Let $\varphi^* = \Phi_c(2\alpha)$, where $\Phi_c(\lambda)$ is the Laplace transform in (7.12.7). Let $V(x, \alpha) = V(x)$, the function defined in (7.12.8), and, define a function $V^*(x)$ by $V^*(x) = V(x, 2\alpha)$. Then for $\alpha > 0$ define a function $h^*(x)$ by

$$h^*(0) = (1 - \varphi^*)^{-1} d^{-2} \gamma^*(0), \tag{7.12.19}$$

if $x = 0$, and by

$$h^*(x) = d^{-2}(\gamma^*(x) + \gamma^*(0))((1 - \varphi^*)^{-1} V^*(x)), \tag{7.12.20}$$

if $x \neq 0$. With these definitions, a limit theorem for the product moment functions in Theorem 7.11.1 may be stated as follows.

Theorem 7.12.2: In a population process with time homogeneous laws of evolution suppose $\alpha > 0$. Then, for x, y, z, and τ fixed,

$$\lim_{s \uparrow \infty} \exp[-\alpha(2s + \tau)] \Gamma_B(x, y, z, s, \tau)$$
$$= h^*(x) \varphi_c(y) \varphi_c(z) \exp[-\alpha(y + z)]; \tag{7.12.21}$$
$$\lim_{s \uparrow \infty} \exp[-\alpha(2s + \tau)] \Gamma(x, y, z, s, \tau)$$
$$= h^*(x) S_c(y) S_c(z) \exp[-\alpha(y + z)]; \tag{7.12.22}$$

and

$$\lim_{s \uparrow \infty} \exp[-\alpha(2s + \tau)] \Gamma^*(x, y, z, s, \tau)$$
$$= h^*(x) S_c(y) \varphi_c(z) \exp[-\alpha(y + z)]. \tag{7.12.23}$$

Proof: To illustrate the method of proof, it will suffice to demonstrate the validity of (7.12.22). In the time homogeneous case, Eq. (7.11.9) becomes an ordinary renewal equation when $x = 0$. In Problem 10, it is shown that all the functions in (7.11.11) converge to zero as $s \uparrow \infty$. Hence, $\exp[\alpha(2s + \tau)] f(0, y, z, s, \tau)$ approaches the limit in (7.12.16) as $s \uparrow \infty$ for the case $x = 0$. Because $\alpha > 0$ implies $\varphi^* = \Phi_c(2\alpha) < 1$, an application of assertion (ii) in Lemma 7.12.1 establishes (7.12.22) for the case $x = 0$. Another application of (7.12.16) together with the validity of (7.12.22) for the case $x = 0$, may be used to establish the validity of (7.12.22) for the case $x \neq 0$. □

Asymptotic formulas for covariance functions of the type appearing in (7.11.4) may easily be derived as a corollary to Theorems 7.12.1 and 7.12.2.

Corollary 7.12.1: Suppose that $\alpha > 0$ is a population process with time homogeneous laws of evolution. Then, for x, y, z, and τ fixed,

$$\lim_{s \uparrow \infty} \exp[-\alpha(2s + \tau)] \operatorname{cov}[Z_B(x, y, s), Z_B(x, z, s + \tau)]$$
$$= f^*(x) \varphi_c(y) \varphi_c(z) \exp[-\alpha(y + z)]; \tag{7.12.24}$$
$$\lim_{s \uparrow \infty} \exp[-\alpha(2s + \tau)] \operatorname{cov}[Z(x, y, s), Z(x, z, s + \tau)]$$
$$= f^*(x) S_c(y) S_c(z) \exp[-\alpha(y + z)]; \tag{7.12.25}$$

and

$$\lim_{s \uparrow \infty} \exp\left[-\alpha(2s+\tau)\right] \mathrm{cov}\left[Z(x, y, s), Z_B(x, z, s+\tau)\right]$$
$$= f^*(x) S_c(y) \varphi_c(y) \exp\left[-\alpha(y+z)\right], \tag{7.12.26}$$

where

$$f^*(x) = h^*(x) - d^{-2} V^2(x). \tag{7.12.27}$$

If $\tau=0$ and $y=z$, then the covariance on the left in (7.12.25) becomes a variance so that the lefthand side is nonnegative for all x, y, and s. Therefore, $f^*(x) \geq 0$ for all x. It thus follows that the asymptotic expressions on the right in (7.12.24), (7.12.25), and (7.12.26) are nonnegative at all points $\langle x, y, z \rangle$.

Corollary 7.12.1 implies that in population processes with time homogeneous laws of evolution, stochastic fluctuations may be rather large at some epoch $t+s$ following an initial epoch t. For according to (7.12.24), when s is large, the covariance function of the birth process involves formulas of the type

$$\mathrm{cov}\left[Z_B(x, y, s), Z_B(x, z, s+\tau)\right]$$
$$\simeq \exp\left[\alpha(2s+\tau)\right] f^*(x) \varphi_c(y) \varphi_c(z) \exp\left[-\alpha(y+z)\right]. \tag{7.12.28}$$

Clearly, if $\alpha > 0$, this covariance function can be large when s is large, since the righthand side of (7.12.28) is nonnegative at all points $\langle x, y, z \rangle$. It is also clear that asymptotic formulas for all the covariances in Sect. 7.7 could be derived in the time homogeneous case, but, this task will not be undertaken here. According to (7.7.6), the asymptotic covariance function of the birth process would involve sums on x of the covariances on the right in (7.12.28). Under the assumption that 50 is the upper age of childbearing, $f^*(x)=0$ for all $x > 50$, indicating that the asymptotic formulas for the covariances would be applicable only to the descendants of individuals with ages $x \leq 50$ at epoch t. This observation suggests that formula (7.12.28) may not be a useful approximation unless s is beyond 50.

7.13 Period Demographic Indicators in Populations with Time Inhomogeneous Laws of Evolution

Period demographic indicators have been used extensively to describe the current state of human populations; in some cases, these indicators enter into the formulation and implementation of policies directed towards slowing population growth. Ross (1975) has given an account of ways in which various countries have used demographic indicators to measure the impact of family planning programs. A comprehensive discussion of problems encountered in measuring the effect of family programs on fertility may be found in papers contained in Chandrasekaran and Hermalin (1975). As indicated in this and other literature, success and/or failures of policies are sometimes judged in terms of changes in demographic indicators, such as the crude birth rate, over time. Using demographic indicators in this way gives rise to a need for a methodology for projecting them over time in terms of changes in the basic components underlying the demographic evolution of human populations.

Considered in this section are some widely used demographic indicators. For convenience of exposition, these indicators will be partitioned into three classes. These classes will be designated as distributional, age-specific rates, and scalar rates, summarizing some population property with respect to all ages or some specific age.

Among the distributional indicators of long-standing interest is the period age distribution. In a population process with time inhomogeneous laws of evolution, let $Y(t, y, s)$ and $Y(t, s)$ be the random functions defined in (7.7.1) and (7.7.14). Then, the age distribution at epoch $t+s$ is defined by the random function

$$\hat{a}(t, y, s) = \frac{Y(t, y, s)}{Y(t, s)}, \tag{7.13.1}$$

for $Y(t, s) \neq 0$, $1 \leq y \leq r$, and $s \geq 1$. In a real population, the value of the random function in (7.13.1) could, in principle, be determined by a complete census of the female population at epoch $t+s$; in Monte Carlo simulations, the value of this function could also be determined. But, in a macrosimulation system of the type under consideration, $\hat{a}(t, y, s)$ needs to be approximated by some function that may be easily calculated; moreover, some knowledge, regarding the appropriateness of the approximation, should be available. As will be shown below, the ratio of means, $a(t, y, s) = v(t, y, s)/v(t, s)$, will be a good approximation to $\hat{a}(t, y, s)$ in populations evolving from a large population at epoch t. Equations (7.5.1) and (7.7.15) may be consulted for definitions of means in the ratio $a(t, y, s)$.

Another period distribution of interest is the age of childbearing, a distribution that may be defined as follows. Let $Y_B(t, y, s)$ be the random function defined in (7.7.2), giving the number of females of age y in the population at epoch $t+s$; and, let $Y(t, 0, s) = Y_B(t, s)$ be the total number of births at epoch $t + s$, see (7.7.4). Then, the age distribution of childbearing at epoch $t+s$ is defined by

$$\hat{a}_B(t, y, s) = \frac{Y_B(t, y, s)}{Y_B(t, s)}, \tag{7.13.2}$$

for $15 \leq y \leq 50$ and $s \geq 1$. Like the age distribution, this period distribution may be approximated by the ratio of means $a_B(t, y, s) = v_B(t, y, s)/v_B(t, s)$, where $v_B(t, s) = v(t, 0, s)$; consult (7.5.2) and (7.5.18) for definitions of these means.

Another distributional period indicator of interest is the age distribution of death. Problem 11 may be consulted for a discussion of this indicator.

As indicated above, when the number of individuals $n(t, x)$ of age x in the population at epoch t is large for all $x = 0, 1, \ldots, r$, then the strong law of large numbers may be used to show that the random functions $\hat{a}(t, y, s)$ and $\hat{a}_B(t, y, s)$ are closely approximated by the ratios $a(t, y, s)$ and $a_B(t, y, s)$. In what follows, a symbol of the form $o(\cdot)$ will be any random function such that $o(\cdot) \to 0$ with probability one as $n(t, x) \uparrow \infty$ for all $x = 0, 1, \ldots, r$.

Theorem 7.13.1: Suppose $n(t, x)$ is large for all $x=0, 1, 2, ..., r$ and the laws governing the evolution of a population are time inhomogeneous. Then,

$$\hat{a}(t, y, s) = a(t, y, s) + o(t, y, s) \tag{7.13.3}$$

and

$$\hat{a}_B(t, y, s) = a_B(t, y, s) + o(t, y, s). \tag{7.13.4}$$

Proof: To illustrate the ideas underlying the proof of this theorem, it will suffice to sketch the proof of the validity of (7.13.3). Let $Z_k(t, x, y, s)$, $k \geq 1$, be the independent random functions going into the sum on the right in (7.7.1). Then, by the strong law of large numbers, there exists a random function $o(t, x, y, s)$ such that

$$\frac{1}{n(t, x)} \sum_{k=1}^{n(t, x)} Z_k(t, x, y, s) = M(t, x, y, s) + o(t, x, y, s). \tag{7.13.5}$$

Let

$$n(t) = \sum_{x=0}^{r} n(t, x),$$

the total population size at epoch t. Then, (7.5.1), (7.7.1), and (7.13.5) imply that

$$\frac{Y(t, y, s)}{n(t)} = \frac{v(t, y, s)}{n(t)} + o(t, y, s). \tag{7.13.6}$$

Similarly,

$$\frac{Y(t, s)}{n(t)} = \frac{v(t, s)}{n(t)} + o(t, y, s). \tag{7.13.7}$$

Equation (7.13.3) now follows by taking the ratio of (7.13.6) to (7.13.7) and doing some algebraic manipulation. □

A systematic investigation of the probabilistic behavior of the random functions on the right in (7.13.3) and (7.13.4) would be of some interest but it will not be undertaken here.

Two types of age-specific rates will be considered. A random function $\hat{R}_B(t, y, s)$ defined by

$$\hat{R}_B(t, y, s) = \frac{Y_B(t, y, s)}{Y(t, y, s)}, \tag{7.13.8}$$

for $15 \leq y \leq 50$ and $s \geq 1$, will be interpreted as the age-specific birth rate function. A second random function defined by

$$\hat{R}_D(t, y, s) = \frac{Y_D(t, y, s)}{Y(t, y-1, s-1)}, \tag{7.13.9}$$

for $1 \leq y \leq r$ and $s \geq 1$, will be interpreted as the age-specific death rate function. Like those defined above, these random functions may be approximated by the ratios, $R_B(t, y, s) = v_B(t, y, s)/v(t, y-1, s-1)$ and $R_D(t, y, s) = v_D(t, y, s)/v(t, y-1, s$

-1), of means. It is also essential to observe that these ratios may be expressed in terms of the basic functions governing the evolution of a population process. For, from Eqs. (7.5.15) and (7.5.16) in *Theorem* 7.5.1, it follows that

$$R_B(t, y, s) = \frac{q \varphi_c(t+s-y, y-1, 1)}{p(t+s, y)};$$

(7.13.10)

similarly Eq. (7.5.17) implies that

$$R_D(t, y, s) = q(t+s, y).$$

(7.13.11)

The definitions given in (7.13.8) and (7.13.9) are somewhat arbitrary and may not always match those used by investigators dealing with vital statistics. These definitions, as well as those given below, should, therefore, be viewed as operational ones for the computer macro-simulation system under consideration. In fact, the choices of definitions for demographic indicators used in this section were motivated in part by the desire to derive asymptotic formulas that were easily interpreted. For example, the function $R_D(t, y, s)$ in (7.13.11) is easy to interpret as a conditional period probability of death; a discussion of the function in (7.13.10) may be found in Problem 12. In the case of population processes with time homogeneous laws of evolution, ease of interpretation refers to the ability to derive some familiar formulas of stable population theory as asymptotic results. Some of these results will be illustrated in the next section.

Four scalar demographic indicators will be singled out for consideration. One such indicator is the rate of population growth at epoch $t+s$, which is defined by the random function

$$\hat{R}(t, s) = \frac{Y(t, s) - Y(t, s-1)}{Y(t, s-1)},$$

(7.13.12)

for $s \geq 1$. Two other period indicators are the crude birth and death rates at epoch $t+s$. These indicators are defined, respectively, by the random functions

$$\hat{R}_B(t, s) = \frac{Y_B(t, s)}{Y(t, s)},$$

(7.13.13)

and

$$\hat{R}_D(t, s) = \frac{Y_D(t, s)}{Y(t, s)},$$

(7.13.14)

where

$$Y_D(t, s) = \sum_{y=1}^{r} Y_D(t, y, s)$$

(7.13.15)

and $s \geq 1$. Finally, because some indicator of infant mortality is always of interest, the random function

$$\hat{D}_I(t, s) = \frac{Y_D(t, 1, s)}{Y_B(t, s)}, \quad s \geq 1,$$

(7.13.16)

will be used as this indicator. Observe that it is the ratio of the number of infant deaths credited to the population at epoch $t+s$ to the number of births entering the population at that epoch.

The random functions just defined may also be approximated by ratios of mean functions. For the rate of population growth at epoch $t+s$, this ratio is $R(t, s) = (v(t, s) - v(t, s-1))/v(t, s-1)$. The ratios for the crude birth and death rates are $R_B(t, s) = v_B(t, s)/v(t, s)$ and $R_D(t, s) = v_D(t, s)/v(t, s)$. For the case of infant mortality the ratio is $D_I(t, s) = v_D(t, 1, s)/v_B(t, s)$. The function $v_D(t, s)$ is the mean of the random function in (7.13.15).

Theorems completely analogous to *Theorem 7.13.1* could be written down for the period age-specific and scalar indicators just defined but the details will be omitted. It should be remembered, however, that in the computer implementation of the simulation system under consideration, time series of these indicators will be calculated as ratios of means. In populations evolving from large initial populations, these ratios will approximate the "true" random functions in the sense of *Theorem 7.13.1*.

7.14 Asymptotic Formulas for Period Demographic Indicators in the Time Homogeneous Case

When a population process is governed by time homogeneous laws of evolution, the results of Theorem 7.12.1 may be used to deduce asymptotic formulas for most of the period demographic indicators defined in the previous section. A function that enters into these formulas is

$$V_p(t) = \sum_{x=0}^{r} n(t, x) V(x), \tag{7.14.1}$$

the reproductive value of the population based on the female age structure of epoch t. All asymptotic formulas for period demographic indicators displayed in this section can be deduced from three key limit formulas for the means $v(t, y, s)$, $v_B(t, y, s)$, and $v_D(t, y, s)$ defined in (7.5.1), (7.5.2), and (7.5.3).

These three formulas are:

$$\lim_{s \uparrow \infty} \exp[-\alpha s] v(t, y, s) = d^{-1} V_p(t) S_c(y) \exp[-\alpha y]; \tag{7.14.2}$$

$$\lim_{s \uparrow \infty} \exp[-\alpha s] v_B(t, y, s) = d^{-1} V_p(t) q \varphi_c(y) \exp[-\alpha y]; \tag{7.14.3}$$

and

$$\lim_{s \uparrow \infty} \exp[-\alpha s] v_D(t, y, s) = d^{-1} V_p(t) g_c(y) \exp[-\alpha y]. \tag{7.14.4}$$

As an aid to notational compactness, the Laplace transforms,

$$S_c^*(\lambda) = \sum_{y=0}^{r} S_c(y) \exp[-\lambda y], \tag{7.14.5}$$

$$\Phi^*(\lambda) = \sum_{y=15}^{50} \varphi_c(y) \exp[-\lambda y], \tag{7.14.6}$$

and

$$g_c^*(\lambda) = \sum_{y=0}^{r} g_c(y) \exp[-\lambda y], \tag{7.14.7}$$

will also be useful. At this juncture, it will be helpful to recall that these key formulas have been deduced from renewal theory rather than from the spectral structure of the Leslie matrix, as represented by its eigenvalues and eigenvectors.

Considered first are asymptotic formulas for the period distributional indicators defined in *Sect.* 7.13.

Theorem 7.14.1: Suppose a population process is governed by time homogeneous laws of evolution.

(i) Then, the asymptotic formula for the period age distribution is given by

$$\lim_{s\uparrow\infty} a(t, y, s) = \frac{S_c(y) \exp[-\alpha y]}{S_c^*(\alpha)}, \tag{7.14.8}$$

for $1 \leq y \leq r$.

(ii) Similarly, the asymptotic formula for the period age distribution of childbearing is given by

$$\lim_{s\uparrow\infty} a_B(t, y, s) = \frac{\varphi_c(y) \exp[-\alpha y]}{\Phi_c^*(\alpha)}, \tag{7.14.9}$$

for $15 \leq y \leq 50$.

Proof: By noting that $a(t, y, s) = \exp[-\alpha s] v(t, y, s)/\exp[-\alpha s] v(t, s)$ and letting $s\uparrow\infty$, formula (7.14.8) follows from the key limit formula in (7.14.2). A similar argument may be used to deduce (7.14.9) from the key limit formula in (7.14.3). □

Theorem 7.14.1 is mathematically valid for any value of α, the unique real root of the equation $\Phi_c(\alpha) = 1$ [see (7.12.7)]. If $\alpha < 0$, then population size decreases; if $\alpha > 0$, then population size increases. The situation in which $\alpha = 0$ is known in branching processes as the critical case; in demography, this situation is known as the stationary case, indicating that population size would, in theory, remain constant. From the theory of branching processes with time homogeneous laws of evolution, however, it is known that $\alpha \leq 0$ implies that a population will eventually become extinct with probability one. Thus, population size would eventually decrease even if $\alpha = 0$.

These results from the theory of branching processes suggests that when $\alpha \leq 0$, the asymptotic formulas in *Theorem* 7.14.1, as well as those derived below, should be interpreted with caution, because means of stochastic processes and their ratios may not reflect the behavior of actual population numbers. On the other hand, when $\alpha > 0$, so that population size is increasing, one would expect that the ratios of means $a(t, y, s)$ and $a_B(t, y, s)$ would adequately reflect the state of population numbers, particularly in the case of one evolving from a large initial population at some epoch t. The asymptotic formulas in *Theorem* 7.14.1 should thus be more reliable in this case. Problem 11 may be consulted for a period asymptotic distribution of age of death; Problem 12 may also be consulted for an alternative definition of the period age distribution of childbearing.

Formula (7.14.8) is well known in demography as the stable age distribution. Priorities as to who should be credited with first deriving this and other formulas in stable population theory has been the subject of controversy. Samuelson (1976) – see pages 109–129 in Smith and Keyfitz (1977) – has given an account of a controversy that arose between A.J. Lotka and R.R. Kuczynski, two eminent workers in demographic research during the period 1890 to 1950. All of Samuelson's exposition is phrased in terms of a kind of deterministic mathematics, which differs fundamentally from that used in the stochastic formulations considered in this book. For example, rather than deriving the formula for the stable age distribution as in (7.14.8), it is arrived at through a type of formal manipulation. Samuelson's Theorem 1A and 1B, see pages 114 and 115 of Smith and Keyfitz (1977), may be consulted for the spirit in which these manipulations have been carried out. It is of interest to note that, within the branching process-renewal theoretic framework considered in this book, all asymptotic formulas may be rigorously derived by elementary arguments, requiring a minimal number of conditions for their validity.

Unlike the distributional indicators discussed above, the age-specific rate functions displayed in (7.13.10) and (7.13.11) do not depend on s for population processes evolving according to time homogeneous laws of evolution. There will thus be no interesting limit theorems for these indicators. Attention will, therefore, be directed to the scalar indicators defined in the previous section.

Theorem 7.14.2: Suppose a population process is governed by time homogeneous laws of evolution. Then, the asymptotic formulas for the rate of population growth, the crude birth and death rates, and the rate of infant mortality are given by

$$\lim_{s \uparrow \infty} R(t,s) = e^{\alpha} - 1 \simeq \alpha; \qquad (7.14.10)$$

$$\lim_{s \uparrow \infty} R_B(t,s) = \frac{1}{S_c^*(\alpha)}; \qquad (7.14.11)$$

$$\lim_{s \uparrow \infty} R_D(t,s) = \frac{g_c^*(\alpha)}{S_c^*(\alpha)}; \qquad (7.14.12)$$

and

$$\lim_{s \uparrow \infty} D_I(t,s) = g_c(1) \exp[-\alpha]. \qquad (7.14.13)$$

The approximation in (7.14.10) will be good when α is small.

Proof: From (7.14.2), it follows that

$$\lim_{s \uparrow \infty} \exp[-\alpha s] v(t,s) = d^{-1} V_p(t) S_c^*(\alpha). \qquad (7.14.14)$$

Equation (7.14.10) then follows by observing that,

$$R(t, s) = \frac{e^\alpha \exp\left[-\alpha s\right] v(t, x) - \exp\left[-\alpha(s-1)\right] v(t, s-1)}{\exp\left[-\alpha(s-1)\right] v(t, s-1)},$$

and letting $s \uparrow \infty$.

Equation (7.14.3) implies that

$$\lim_{s \uparrow \infty} \exp\left[-\alpha s\right] v_B(t, s) = d^{-1} V_p(t) \Phi_c(\alpha) = d^{-1} V_p(t). \qquad (7.14.15)$$

Given (7.14.14) and (7.14.15), (7.14.11) follows by noting that

$$R_B(t, s) = \frac{\exp\left[-\alpha s\right] v_B(t, s)}{\exp\left[-\alpha s\right] v(t, s)},$$

and letting $s \uparrow \infty$. The proofs of (7.14.12) and (7.14.13) are similar. ∎

Of historical interest is the observation that asymptotic formulas (7.14.11) and (7.14.12) appear in the discussion of Samuelson, see pages 114 and 115 in Smith and Keyfitz (1977). In a subsequent section, the results of some computer experiments designed to study the rate of convergence to the asymptotic formulas derived in this section will be described.

One would expect on intuitive grounds that in populations growing slowly, the crude birth and death rates would be nearly equal. According to (7.14.10), the rate of population growth is approximately α. Let $E[X]$ be the expectation of life at birth in a population process evolving according to time homogeneous laws. Then, by (7.14.11),

$$\lim_{\alpha \downarrow 0} \frac{1}{S_c^*(\alpha)} = \frac{1}{E[X]}, \qquad (7.14.16)$$

and, by (7.14.12),

$$\lim_{\alpha \downarrow 0} \frac{g_c^*(\alpha)}{S_c^*(\alpha)} = \frac{1}{E[X]}. \qquad (7.14.17)$$

Just as one would expect, these rates are indeed nearly equal when $\alpha > 0$ is near zero.

7.15 A Female Dominant Two-Sex Population Process

Despite the fact that human populations are composed of two sexes, attention up until now has been focused on only the female sex. Outlined in this section is a scheme for projecting both sexes in a population process with time inhomogeneous laws of evolution. The process will be called female dominant, because the potential birth process for every female birth cohort may be stopped only by the deaths of females. Not taken into account, for example, in this simplified formulation is the impact of male mortality on the flow of births into a population. Neither are other factors which may affect the flow of births into a population, such as temporary separations of spouses and divorce. The formulation and computer implementation of population processes, taking such factors into account, would require further research.

An extension of the representation of the population process outlined in Sect. 7.8 will be used to formulate a female dominant two-sex population process. Suppose that at epoch t, a population consists of $n_f(t, x)$ and $n_m(t, x)$ females and males of age $x = 0, 1, 2, \ldots, r$. Let $Y_{Bf}(t, y, s)$ be a random function giving the number of female births entering the population at epoch $t + s$ from mothers of age y at epoch $t + s$. The random function $Y_{Bm}(t, y, s)$ for male births is defined similarly. Given the definitions of these random functions, Eqs. (7.8.1), (7.8.2), and (7.8.3) apply separately for the random functions $Y_{Bf}(t, s)$ and $Y_{Bm}(t, s)$, representing the number of female and male births entering the population at epoch $t + s$, $s \geq 1$.

The next step in extending the population process outlined in Sect. 7.8 to the two-sex case is to redefine Eqs. (7.8.4) and (7.8.5). For a cohort of females born at epoch t, let $N(t, s)$ be a random function representing the actual number of births of either sex experienced by any female in the cohort by epoch $t + s$. To take the sex of each child into account define a pair of random variables $\langle X(f), X(m) \rangle$ such that either $\langle X(f), X(m) \rangle = \langle 1, 0 \rangle$ or $\langle X(f), X(m) \rangle = \langle 0, 1 \rangle$, depending on whether the child is female or male. As before, let $q(p)$ be the probability a child is female (male), where $p + q = 1$. Then, if $\langle X_k(f), X_k(m) \rangle$, $k \geq 1$, is a sequence of independent and identically distributed copies of the random variable $\langle X(f), X(m) \rangle$, the random functions

$$N_f(t, s) = \sum_{k=1}^{N(t, s)} X_k(f)$$

and (7.15.1)

$$N_m(t, s) = \sum_{k=1}^{N(t, s)} X_k(m)$$

represent the number of offspring of each sex experienced by epoch $t + s$ by any member of a cohort of females born at epoch t.

With the random functions $N_f(t, s)$ and $N_m(t, s)$ defined as in (7.15.1), the next step is to consider increment functions for each sex defined by

$$\Delta_f(t, x, s) = N_f(t, x + s) - N_f(t, x + s - 1)$$

and (7.15.2)

$$\Delta_m(t, x, s) = N_m(t, x + s) - N_m(t, x + s - 1).$$

Let $\Delta_{fk}(t, x, s)$ and $\Delta_{mk}(t, x, s)$, $k \geq 1$, be independent and identically distributed copies of the random functions in (7.8.4) and (7.8.5). To transpose Eqs. (7.8.4) and (7.8.5) to the two-sex case, all that needs to be done for the case of females is to attach a subscript "f" to all symbols. Since all births are credited to females in a female dominant process, the analogues of Eqs. (7.8.4) and (7.8.5) for males take the forms

$$Y_{Bm}(t, x+s, s) = \sum_{k=1}^{n_f(t, x)} \Delta_{mk}(t-x, x, s);$$ (7.15.3)

if $1 \le s \le 15$; and

$$Y_{Bm}(t, z, s) = \sum_{k=1}^{Y_{Bm}(t, s-z)} \Delta_{mk}(t+s-z, 0, z),$$ (7.15.4)

if $s \ge 16$.

Clearly, all the period probabilities set forth in Sect. 7.2 may be defined separately for each sex. Let $p_f(t, x)$ and $p_m(t, x)$, for example, be the period conditional survival probabilities for females and males at epoch t and ages $x = 0, 1, 2, \ldots, r$; let $q_f(t, x) = 1 - p_f(t, x)$ and $q_m(t, x) = 1 - p_m(t, x)$ be defined as before. Furthermore, let $S_{pf}(t, x)$ and $S_{cf}(t, x)$ be the period and cohort survival functions for females and define the survival functions $S_{pm}(t, x)$ and $S_{cm}(t, x)$ males similarly. Given these definitions, the U and V indicator random variables on the right in (7.8.6), (7.8.7), (7.8.8), and (7.8.9) may be defined for each sex.

As defined, Eqs. (7.8.6) through (7.8.9) refer to females; hence, to redefine them for the two-sex case all that needs to be done is to attach a subscript "f" to each symbol. At the risk of some redundancy but in the interests of clarity, analogues of these equations for males will be set down. For the female dominant population process under consideration, the analogues of Eqs. (7.8.6) and (7.8.7) for males are

$$Y_m(t, y, s) = \sum_{k=1}^{n_f(t, y-s)} U_{mk}(t+s-y, y-s, s),$$ (7.15.5)

if $1 \le s \le r$ and $y \ge s$; and

$$Y_m(t, y, s) = \sum_{k=1}^{Y_{Bf}(t, s-y)} U_{mk}(t+s-y, 0, y),$$ (7.15.6)

if $1 \le y < s \le r$. If $s > r$, Eq. (7.15.6) holds for all ages $1 \le y \le r$. Similarly, the analogues of Eqs. (7.8.8) and (7.8.9) for males are

$$Y_{Dm}(t, y, s) = \sum_{k=1}^{n_f(t, y-s)} V_{mk}(t+s-y, y-s, s),$$ (7.15.7)

if $1 \le s \le r$ and $y \ge s$; and

$$Y_{Dm}(t, y, s) = \sum_{k=1}^{Y_{Bf}(t, s-y)} V_{mk}(t+s-y, 0, y),$$ (7.15.8)

if $1 \le y < s \le r$. If $s > r$, then Eq. (7.15.8) holds for all ages $1 \le y \le r$.

Let $v_{Bf}(t, y, s)$ and $v_{Bm}(t, y, s)$ be the means of the random functions $Y_{Bf}(t, y, s)$ and $Y_{Bm}(t, y, s)$. Denote the means of the random functions $Y_f(t, y, s)$, $Y_m(t, y, s)$, $Y_{Df}(t, y, s)$ and $Y_{Dm}(t, y, s)$, which have been defined either explicitly or implicitly in the above discussion, by $v_f(t, y, s)$, $v_m(t, y, s)$, $v_{Df}(t, y, s)$, and

$v_{Dm}(t, y, s)$, respectively. Finally, let $v_{Bf}(t, s)$ be the mean of the random function $Y_{Bf}(t, s)$. Just as one would expect, there is an analogue of Theorem 7.5.1 for the female dominant two-sex population process under consideration. As an added bonus, the proof of the next theorem implies a more transparent proof of Theorem 7.5.1 based on the birth cohort representation of the population process under consideration.

Theorem 7.15.1: For every $y = 0, 1, 2, \ldots, r$ and $s \geq 0$, the mean functions $v_f(t, y, s)$, $v_m(t, y, s)$, $v_{Bf}(t, y, s)$, $v_{Bm}(t, y, s)$, $v_{Df}(t, y, s)$, and $v_{Dm}(t, y, s)$ are finite and satisfy the following recursive system.

(i) For $s = 0$,

$$v_f(t, y, 0) = n_f(t, x)$$

and $\qquad\qquad\qquad\qquad\qquad\qquad\qquad\qquad\qquad\qquad\qquad$ (7.15.9)

$$v_m(t, y, 0) = n_m(t, x)$$

for $y = 0, 1, 2, \ldots, r$.

(ii) If $y \geq 1$ and $s \geq 1$, then the recursive system for live individuals is

$$v_f(t, y, s) = p_f(t + s, y) v_f(t, y - 1, s - 1)$$

and $\qquad\qquad\qquad\qquad\qquad\qquad\qquad\qquad\qquad\qquad\qquad$ (7.15.10)

$$v_m(t, y, s) = p_m(t + s, y) v_m(t, y - 1, s - 1).$$

(iii) If $y \geq 1$ and $s \geq 1$, then the recursive system for births entering the population from mothers of age y is

$$v_{Bf}(t, y, s) = q \, \varphi_c(t + s - y, y - 1, 1) v_f(t, y - 1, s - 1)$$

and $\qquad\qquad\qquad\qquad\qquad\qquad\qquad\qquad\qquad\qquad\qquad$ (7.15.11)

$$v_{Bm}(t, y, s) = p \, \varphi_c(t + s - y, y - 1, 1) v_f(t, y - 1, s - 1).$$

(iv) If $y \geq 1$ and $s \geq 1$, then the recursive system for deaths to individuals age y is

$$v_{Df}(t, y, s) = q_f(t + s, y) v_f(t, y - 1, s - 1)$$

and $\qquad\qquad\qquad\qquad\qquad\qquad\qquad\qquad\qquad\qquad\qquad$ (7.15.12)

$$v_{Dm}(t, y, s) = q_m(t + s, y) v_m(t, y - 1, s - 1).$$

(v) For every $s \geq 1$,

$$v_f(t, 0, s) = \sum_{y=15}^{50} v_{Bf}(t, y, s)$$

and $\qquad\qquad\qquad\qquad\qquad\qquad\qquad\qquad\qquad\qquad\qquad$ (7.15.13)

$$v_m(t, 0, s) = \sum_{y=15}^{50} v_{Bm}(t, y, s).$$

Proof: Just as in the proof of Theorem 7.5.1, the finiteness of all mean functions at all points $\langle t, y, s \rangle$ follows from the validity of the recursive relationships.

To illustrate the ideas underlying the proof of the theorem, consider the equation for males in (7.15.10). Suppose y and s are such that Eq. (7.15.6) holds. Taking expectations in (7.15.6) and using the condition that $Y_{Bf}(t, s-y)$ is independent of $U_{mk}(t+s-y, 0, y)$, $k \geq 1$, it follows that

$$v_m(t, y, s) = v_{Bf}(t, s-y) S_{cm}(t+s-y, y)$$
$$= p_m(t+s, y) v_m(t, y-1, s-1). \qquad (7.15.14)$$

It can also be shown that Eq. (7.15.14) is valid when s and y are such that Eq. (7.15.5) holds.

To demonstrate the validity of the second equation in (7.5.11), suppose s and y are such that Eq. (7.15.4) holds for $y = z$. By taking expectations, it follows that

$$v_{Bm}(t, y, s) = p \, v_{Bf}(t, s-y) \varphi_c(t+s-y, y)$$
$$= p \varphi_c(t+s-y, y-1, 1) v_f(t, y-1, s-1). \qquad (7.15.15)$$

Equation (7.15.15) may also be shown to be valid when $y = x+s$ and s are such that Eq. (7.15.3) holds.

The techniques for establishing the validity of the other equations are similar to those just outlined. □

From the recursive relations for the mean functions presented in Theorem 7.15.1, it is clear that a female dominant two-sex population process presents no additional serious problems from the point of view of computer implementation. To be sure, the number of arrays that need to be manipulated in a computer are nearly double that of a one-sex model; but, the recursive relations given above make their computation easy. When two sexes are considered simultaneously, all the results in Sects. 7.13 and 7.14 would have to be reformulated, because there are several approaches to extending them to the two-sex case. Alternative formulations for two-sex population processes may be found in Mode (1972, 1974c, d, and 1977). From the discussion in this section, it is clear that the results in Sects. 7.9 and 7.10 for covariances may be extended to the female dominant two-sex population process described in this section, with some minor modifications for males, but no details will be given here.

7.16 An Overview of a Computer Software Design Implementing Population Projection Systems

For the most part, the mathematics underlying the population projection systems developed in this chapter is more complex than that used in the development of the cohort models of maternity histories described in Chaps. 5 and 6. Despite this complexity, their computer implementation can be made very efficient computationally, thanks to the Leslie matrix-type recursive algorithms described in Sects. 7.5 and 7.15 for the mean age structure of a population process. In its present state of development, the software is such that only one sex, usually the female, may be considered in any computer run. But, by making separate runs for each sex, the female dominant two-sex population process described in Sect. 7.15 could be implemented.

The computer software design consists of two sets of FORTRAN code, corresponding to population processes with either time homogeneous or time inhomogeneous laws of evolution. Each set of code has a user-friendly menu analogous to that for MATHIST described in Chap. 5. Called in each of these menus are eighty to ninety subroutines. Some of these subroutines are common to both the time homogeneous and inhomogeneous processes; while others are used for only one process. Three types of computer input – an initial age structure, a fertility component, and a mortality component – are needed to implement both sets of code.

The initial age structure refers to the numbers $n_f(t, x)$ or $n_m(t, x)$, $x = 0, 1, 2, ..., r$, given in (7.15.9), depending on whether the female or male population is under consideration. Quite frequently, these numbers are listed in published documents by some grouping of ages. In order to obtain reliable numerical results, the software was designed to carry out population projections with yearly age groupings on a yearly time scale. Consequently, any grouped initial age structure will have to be interpolated down to a single-year age grouping before it can be used as input to a population projection menu. Contained in the software are utility programs for carrying out such interpolations, using spline procedures. When it is assumed that the initial age structure is determined by some stable age distribution with a formula given by (7.14.8), then the initial age structure of the population may be obtained by multiplying this distribution by some initial population size.

Depending on whether the time homogeneous or time inhomogeneous case is under consideration, the fertility component refers to a single cohort gross maternity function or a set of such functions. Mathematically, the cohort gross maternity function (the cohort fertility density function) is the function $b(t)$ given in (5.3.17); see also the function defined in (4.3.5) for a continuous time formulation. Either an assumed parametric form of $b(t)$ or computer output from MATHIST may be used as input to the menus in the population projection software. For population projections with time homogeneous laws of evolution, only one cohort fertility density function is required as input; for projections with time inhomogeneous laws of evolution, a different cohort fertility density function may be entered into the menu for each birth cohort in a projection. A user may also elect to consider less ambitious schemes depicting the evolution in fertility in some hypothetical population. For example, it may be assumed that one cohort fertility density function governs the fertility of all birth cohorts in a projection up to some fixed epoch. Furthermore, it may be assumed that after this epoch, there is a sudden change in fertility so that all subsequent birth cohorts are governed by another cohort fertility density function.

The point of view used in the software for the time inhomogeneous case is that a set of cohort fertility density functions determine period fertility density functions according to formula (4.9.19). Entering a large number of cohort fertility density functions into a program by typing answers to questions in a menu appearing on a video screen can be a prohibitively laborious task. To overcome such difficulties, a command file may be created which automates the reading of the menu and calls the desired input into the program for a population projection. An analogous command file may be created to auto-mate the reading of the menu for MATHIST, thus making it feasible to compute many cohort fertility density functions. The term, command file, refers to a software structure for the operating system of a PRIME 550 computer, the brand of computer used to carry out all computations reported in this chapter. Unfortunately, command files for the PRIME operating system cannot, in general, be used on other brands of computers. Most operating systems do, however, have capabilities for designing software structures analogous to com-mand files. Although menus governing the operation of software can lead to heavy labor at a video terminal, particularly if many computer runs are to be made, such software structures are still very useful in exploring the numerical implications of the population processes under consideration.

With regards to the mortality component of the population projection system under consideration, only one cohort survival function is needed as input to the menu for a population projection with time homogeneous laws of evolution. For projections with time inhomogeneous laws of evolution, the menu governing the operation of the software is such that a period survival function may be called into the menu for each epoch in a projection. Just as in the case of the fertility component, less complicated schemes depicting the evolution of mortality in a hypothetical population with time inhomogeneous laws of evolution may also be considered.

As set down formally in Sect. 7.2, the point of view taken in the time inhomogeneous case is that period survival probabilities, the p-probabilities, determine cohort survival functions according to formula (7.2.2). Sometimes only period p- or q-probabilities of a life table may be available as computer input. Consequently, the software system has utility programs for converting q-probabilities into p-probabilities and vice-versa. Given period p-probabilities, other programs may be used to compute either period or cohort survival functions as indicated in formulas (7.2.1) and (7.2.2).

After the three types of input have been read into the menu driving the population projection software, there is a few seconds pause before computer output will appear on the video screen. There is a rather large variety of computer output common to either the time homogeneous or time inhomo-geneous cases. Presented in Table 7.16.1 is a list of period age-specific com-puter output common to both the time homogeneous and inhomogeneous cases; Table 7.16.2 contains a list of period scalar demographic indicators for both cases. Finally, Table 7.16.3 contains lists of scalar and age-specific com-puter output connected with asymptotic formulas for population processes with time homogeneous laws of evolution.

With reference to Table 7.16.1, items 1, 2, and 3 refer to values of the mean functions in (7.15.10), (7.15.11), and (7.15.12) for a given sex at epochs $t+s$, $s \geq 1$, for those ages y for which these functions are defined. Item 4 in Table 7.16.1 refers to the period age density function for a given sex appearing on the

Table 7.16.1. Period age-specific computer output common to time homogeneous and inhomogeneous systems

1. Mean population numbers by ages
2. Mean numbers of births by ages of mothers
3. Mean numbers of deaths by ages
4. Period age distributions
5. Period age distribution of deaths
6. Period age-specific fertility rates
7. Period age-specific mortality rates
8. Period survival function

Table 7.16.2. Computer output of period scalar demographic indicators common to time homogeneous and inhomogeneous systems

1. Crude birth rate
2. Crude death rate
3. Infant mortality rate
4. Rate of population growth
5. Total fertility rate
6. Net fertility rate
7. Mean total population
8. Mean total births
9. Mean total deaths

Table 7.16.3. Computer output connected with asymptotic formulas in time homogeneous case

Scalar Output
1. Malthusian parameter
2. Asymptotic population growth rate
3. Asymptotic crude birth rate
4. Asymptotic crude death rate
5. Asymptotic infant mortality rate

Age-specific output
1. Fisherian reproductive values
2. Stable age distribution
3. Stable age distribution of deaths

right in (7.13.3); the period age distribution in item 5 refers to the function $a_D(t, y, s)$ defined in part (b) of Problem 11. The period age-specific fertility rates referred to in item 6 of Table 7.16.1 are values of the function $R_B(t, y, s)$ in (7.13.10). As implemented in the population projection systems under consideration, the cohort net fertility density function in approximated in such a way that

$$R_B(t, y, s) = q b_c^*(t + s - y, y);$$ (7.16.1)

see Problem 5 for details. Observe that Eq. (7.16.1) is merely another way of
expressing the relationship between cohort and fertility density functions. The
period age-specific mortality rates referred to in item 7 are values of the
function in (7.13.11). Lastly, given the q-probabilities in (7.13.11), a period
survival function referred to in item 8 of Table 7.16.1 may be calculated. Items
6, 7, and 8 are of most interest for population projections involving time
inhomogeneous laws of evolution; for, in the time homogeneous case, these
items are merely an echo of constant computer input.

With respect to Table 7.16.2, items 1, 2, 3, and 4 refer to values of the
functions $R_B(t, s)$, $R_D(t, s)$, $D_I(t, s)$, and $R(t, s)$, respectively. Recall that these
functions were defined in Sect. 7.13 and are large population approximations to
the random functions defined in (7.13.12), (7.13.13), (7.13.14), and (7.13.16). The
period total fertility rate in item 5 of Table 7.16.2 is merely the sum of the age-
specific fertility rates mentioned in item 6 of Table 7.16.1. As required by its
definition, the period net fertility rate in item 6 of Table 7.16.2 is the sum of
the products referred to in items 6 and 8 of Table 7.16.1. In the computer
output, items 5 and 6 in Table 7.16.2 refer only to female offspring. When only
the female population is being projected, items 7 and 8 in Table 7.16.2 are
values of the mean functions

$$v_f(t, s) = \sum_{y=1}^{r} v_f(t, y, s)$$

and   (7.16.2)

$$v_{Df}(t, s) = \sum_{y=1}^{r} v_{Df}(t, y, s),$$

for $s \geq 1$, see Eqs. (7.15.10) and (7.15.12). Finally, mean total births in item 8
refer to the function $v_{Bf}(t, s) = v_f(t, 0, s)$ as defined in (7.15.13). Similar statements hold for a projection of the male population.

When a population projection is being made under the assumption that
laws governing the evolution of a population are time homogeneous, the
computer output will contain values of scalars and functions arising in the
asymptotic theory set forth in Sects. 7.12 and 7.14. For example, item 1 under
scalar output in Table 7.16.3 refers to α, the Malthusian parameter, which is
computed by finding the unique real root of the equation $\Phi_c(\alpha) = 1$ – see Eq.
(7.12.7). Item 2 under scalar output is the number $\exp[\alpha] - 1$ on the right in
(7.14.10); while items 3, 4, and 5 are values of the functions (7.14.11), (7.14.12),
and (7.14.13) for a given value of α. Item 1 under age-specific output in Table
7.16.3 are values of the function $V(x)$ defined in (7.12.8) for a given α; item 2
refers to values of the stable age distribution in (7.14.8). Part (c) of Problem 11
contains the definition of the asymptotic function whose values are referred to
in item 3 under the age-specific output in Table 7.16.3.

Having described the types of computer output that may be generated by
the population projection systems under consideration, it is appropriate to
briefly outline ways in which this output may be arranged in forms useful to
perception by human eyes. In both the time homogeneous and inhomogeneous

cases, population projections may be carried out for time spans up to 100 years, which seems adequate for most practical purposes. The software is such that all or any subset of the items of computer output listed in Tables 7.16.1 and 7.16.2 may be saved for every year in a projection. If the storage of large digital arrays in the computer is problematic, only that output corresponding to some fixed multiple of years, such as 2 or 5, may be saved. For projections with time homogeneous laws of evolution, a user may wish to save all the output listed in Table 7.16.3. If tabular hard copy from a printing device is desired, it may be formatted by single year age groups or by age groups of greater length such as five years. Alternatively, if a user wishes to study the computer output from a projection in graphical form, the output may be stored as disk files which in turn may be subsequently called in a menu driving a plotting device. Illustrations of such graphs will be given in subsequent sections.

Of importance in population projections designed to study the demographic impact of a family planning program are cohort fertility density functions, reflecting changing patterns of contraceptive use in a population. Included in the software are implementations of the semi-Markovian model for waiting times to conception described in Sect. 5.9. By using the output of such software as input to MATHIST, a set of cohort fertility density functions, depicting an evolving pattern of contraceptive usage in a hypothetical population, could be computed and used as input to the time inhomogeneous population projection system. Thus, the demographic impact of changing patterns of contraceptive usage could be linked to the evolution of the demographic indicators described in this section.

As yet, software for the algorithms outlined in Sects. 7.9, 7.10, and 7.11 for the covariance functions of a population process has not been sufficiently developed to be included in this overview. Further research on the computer implementation of these algorithms, not described in this chapter, has demonstrated, however, that they may indeed be implemented in a computationally efficient way. Results from computer experiments, using these algorithms, will be published elsewhere.

7.17 Four Computer Runs in the Time Homogeneous Case – A Study of Population Momentum

Studies in population momentum are usually concerned with the following type of computer scenario. Suppose a population has undergone a long period of evolution under constant fertility and mortality regimes. Under such circumstances, the asymptotic stable age distribution on the right in (7.14.8), as well as other asymptotic results in Sect. 7.14, would apply. Next suppose that at a given epoch there is a sudden change in fertility and mortality which continues thereafter so that a population evolves according to these new time homogeneous laws of evolution. At least two types of questions are worthy of consideration in such computer scenarios. One of these questions is: given some initial population number, what will be the total population size at subsequent epochs? Another question is: how long must a population evolve before the asymptotic formulas of Sect. 7.14 apply?

A common practice in demography is that of applying stable population formulas, of the types derived in Sect. 7.14, under the assumption they apply in a stable population. Chapter 7 in Keyfitz (1968) may be consulted for examples of this type of analysis. One of the difficulties underlying such analyses is that any transient behavior of a population is not accounted for. Transient behavior refers to the time taken for a population to reach an equilibrium in which the asymptotic stable formulas apply. An advantage of the population projection system under consideration is that it makes the study of this transient behavior feasible and provides a capability for answering questions of the second type mentioned above. Another purely operational reason for studying questions of population momentum is to obtain validity checks on the software. For, one may use the principle that if period demographic indicators do not converge to their asymptotic stable forms, as required by the mathematical results in Sect. 7.14, then there are errors in the software. Indeed, some minor software errors were detected and then corrected by applying this principle.

Selected for consideration in this section were four computer runs designed to study questions of population momentum. To ease the problem of documenting the fertility components of these runs, five runs from MATHIST, described in Chap. 5, were selected for study. It was assumed that the initial population in all four runs was governed by a fertility regime characterized by the MATHIST run $A_1B_1C_1D_2$, see Tables 5.7.4 and 5.7.5. By consulting Table 5.7.4, it can be seen that this run was determined by early marriage (A_1), short postpartum sterile periods (B_1), no induced abortion (C_1), and high fecundability (D_2). From Table 5.7.5, it may be seen that the cohort total fertility rate associated with this run was 7.14 live births per woman. Since all computer runs reported in this section refer only to the female population, a survival function for females was used as computer input. Chosen as the cohort survival function of the initial population was the ten-parameter survival function described in Sect. 2.10 and based on period survival data for Swedish females at about 1850. At that time, the period expectation of life at birth for Swedish females was about 46 years, indicating a rather high level of mortality.

Four other runs from MATHIST, described in Sect. 5.7, were chosen to characterize sudden changes in the fertility regime of the initial population. Used as fertility regimes in population projection runs 1, 2, 3, and 4 of this section were the MATHIST runs $A_2B_1C_1D_2$, $A_1B_1C_1D_3$, $A_2B_1C_1D_3$, and $A_2B_2C_2D_3$. Table 5.7.4 and Sect. 5.7 may again be consulted for more detailed definitions of the symbols. Listed in Table 7.17.1 are the four computer runs along with brief descriptions of their fertility regimes, which are sufficient for the purposes of this section. Note that short postpartum sterile periods (B_1) and no induced abortions (C_1) are common to the fertility regimes in runs 1, 2, and 3. Also observe that when long postpartum sterile periods (B_2) and induced abortions (C_2) are added as in run 4, the MATHIST run is described as a regime with high fertility control. The low level of mortality common to

Table 7.17.1. Four fertility regimes depicting transitions from early marriage, low fertility control, and high mortality to low mortality[a]

Computer run	MATHIST run	Brief description
1	$A_2 B_1 C_1 D_2$	Late marriage – low fertility control
2	$A_1 B_1 C_1 D_3$	Early marriage – fertility control
3	$A_2 B_1 C_1 D_3$	Late marriage – fertility control
4	$A_2 B_2 C_2 D_3$	Late marriage – high fertility control

[a] Low mortality common to all four runs; expectation of life at birth 72 years.

the four computer runs in Table 7.17.1 was that for the ten-parameter survival function described in Sect. 2.10 and based on period mortality for Swedish females around 1950. At that time the period expectation of life at birth for Swedish females was about 72 years.

Four types of period scalar demographic indicators were chosen for graphic display. The indicators chosen were mean total population, the rate of population growth, the crude birth rate, and the crude death rate. Presented in Table 7.17.2 are the asymptotic values of the rate of population growth, (the Malthusian parameter), and the crude birth and death rates for the stable populations corresponding to computer runs 1, 2, 3, and 4. For ease of reference, the asymptotic values for these indicators in the stable initial population, with high fertility and mortality regimes, are presented in Table 7.17.3. Figures 7.17.1 through 7.17.4 contain the graphs of period mean total population, the rate of population growth, the crude birth rate, and the crude death rate for each of the four computer runs, which lasted for 100 epochs. In all these runs, the size of the initial female population was 4×10^6.

Table 7.17.2. Cohort total fertility rates, the Malthusian parameters, and the crude birth and death rates in the stable populations corresponding to computer runs 1, 2, 3, and 4

Computer run	TFR[a]	α[b]	CBR[c]	CDR[d]
1	4.62	0.0216	0.0279	0.0065
2	4.34	0.0241	0.0298	0.0069
3	3.17	0.0111	0.0203	0.0093
4	2.36	0.0021	0.0148	0.0127

[a] Cohort total fertility rate.
[b] Malthusian parameter.
[c] Crude birth rate in stable population.
[d] Crude death rate in stable population.

Table 7.17.3. Cohort total fertility rates, the Malthusian parameter, and the crude birth and death rates in stable initial population with high mortality[a]

MATHIST run	TFR	α	CBR	CDR
$A_1 B_1 C_1 D_2$	7.14	0.0272	0.0449	0.0180

[a] Expectation of life at birth 46 years.

Of fundamental interest in the study of population momentum is the rate of convergence of period age distributions to their asymptotic stable form. An evolutionary development of period age distributions is difficult to represent graphically, because three dimensional graphs are required. One approach to reducing the problem to two dimensions is to consider a distance function, measuring the distance of a period age distribution from its asymptotic stable form. One such distance function may be defined as follows. Let $a(t, y, s)$ be a period age density at epoch $t+s$; let

$$A(t, y, s) = \sum_{x=1}^{y} a(t, x, s), \qquad (7.17.1)$$

$y \geq 1$, be its distribution function. Similarly, let $A(y)$ be the distribution function of its asymptotic stable form. Then

$$D(t, s) = \max_{y} |A(t, y, s) - A(y)|, \qquad (7.17.2)$$

is a distance function, measuring the distance between a period age distribution and its stable asymptotic form at epoch $t+s$. Plotted in Fig. 7.17.5 are values of the distance function in (7.17.2) for the four computer runs under consideration.

Computer generated graphs provide a means for condensing large amounts of digital information into compact forms, which observers may perceive and interpret differently. Rather than attempting a detailed discussion of each figure, only brief guidelines for interpreting them will be suggested. Before setting down these suggestions, however, it is of interest to ask whether a different initial stable age distribution, based on a lower level of mortality, would have a profound impact on the appearances of the graphs in Figs. 7.17.1 through 7.17.5. In an attempt to obtain a partial answer to this question, an initial stable age distribution, corresponding to the MATHIST run $A_1 B_1 C_1 D_2$ and the period survival function for Swedish females with an expectation of life at birth of 72 years, was used to determine an initial age structure for 4×10^6 females. The four computer runs under discussion were then repeated but the resulting graphs were not essentially different from those presented here. Evidently, initial age structures differing markedly from the two under discussion would be required to substantially change the shapes of the graphs presented in Figs. 7.17.1 through 7.17.5.

7.17.1 Guidelines for Interpreting Graphs of Period Mean Total Population

1. From inspecting Fig. 7.17.1, it can be seen that mean total population size in runs 1, 2, 3, and 4 had grown from 4 million initially to approximately 40, 50, 14, and 7 million, respectively, by the end of 100 epochs. By arranging these numbers from the largest to the smallest, the ordering – 2, 1, 3, 4 – for the four runs results. Of fundamental interest is the observation from Table 7.17.2 that values of α, the Malthusian parameter, and CBR, the crude birth rate in a stable population, have the same ordering. It is also of interest to observe from Table 7.17.2 that the ordering of the values of TFR, the cohort total fertility rate, for runs 1, 2, 3, and 4 was not the same as that for α and CBR.

2. From Table 7.17.2, it can be seen that the Malthusian parameters for runs 1 and 2 are 0.0216 and 0.0241. By inspecting the graphs in Fig. 7.17.1 for these two runs, it can be seen that a rather small difference in Malthusian parameters will produce significant differences in population sizes attained after 100 epochs.

3. According to the specifications given in Table 7.17.1, the fertility regimes for runs 2 and 3 differ only by the distribution of age of first marriage in a cohort of females. By comparing the graphs for mean total population in Fig. 7.17.1 for these runs, it can be seen that delaying marriage can have a profound impact on slowing population growth.

4. As expected, the fertility regime in run 4 with late marriage and high fertility control produced the slowest population growth; compare the graph for run 4 with those for the other runs in Fig. 7.17.1.

7.17.2 Guidelines for Interpreting Graphs of Period Rates of Population Growth, Crude Birth Rates, and Crude Death Rates

1. As can be seen from Figs. 7.17.2 and 7.17.3, the graphs of period rates of population growth and crude birth rates follow a similar pattern. After a sharp decline from their values in the initial stable population, these period rates rise for spans of time ranging from about 20 to 30 epochs. Thereafter, they decline in an oscillatory fashion to their asymptotic stable values.

2. The most rapid convergence to the asymptotic values in Tables 7.17.2 and 7.17.3 was for run 2, the run with the largest Malthusian parameter. Unlike runs 1, 3, and 4, the graphs for run 2 exhibit little oscillatory behavior in their approaches to the asymptotic stable values. The reasons for the oscillatory behavior of the curves for runs 1, 3, and 4 in Figs. 7.17.2 and 7.17.3 are not easy to discern; but, they may be related to the nature of the complex roots of the equation determining the Malthusian parameter.

3. The shapes of the curves for the period crude death rates in Fig. 7.17.4 are intrinsically different from those for the rate of population growth and crude birth rates in Figs. 7.17.2 and 7.17.3. After a sharp decline from the values for the initial stable population, the curves for the period crude death rates rise, for the most part, to their asymptotic stable values. The curves for runs 1 and 2 in Fig. 7.17.4 exhibit little oscillatory behavior; while those for runs 3 and 4 are more oscillatory. As would be expected on intuitive grounds,

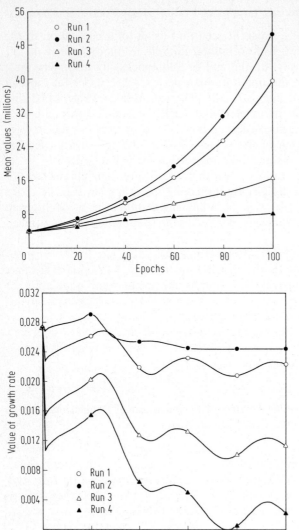

Fig. 7.17.1. Period mean total population

Fig. 7.17.2. Period population growth rates

the graphs in Fig. 7.17.4 and the values of CDR in Table 7.17.2 tend to confirm the notion that in populations with similar levels of mortality, the more rapidly growing ones will have the smaller crude death rates.

7.17.3 Guidelines for Interpreting Graphs of Distances of Period Age Distributions from Their Asymptotic Stable Forms

1. A useful approach to the study of population momentum is to examine changes in the period age distribution in terms of its distance from its asymptotic stable form. As can be seen from an inspection of the graphs of this

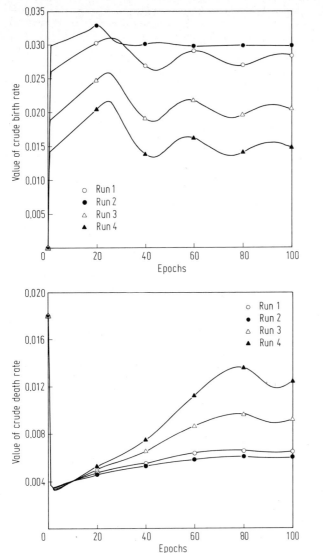

Fig. 7.17.3. Period crude birth rates

Fig. 7.17.4. Period crude death rates

distance by epochs in runs 1, 2, 3, and 4 in Fig. 7.17.5, when each cohort has an expectation of life at birth of 72 years, the rate convergence of period age distributions to their asymptotic stable forms is slow in terms of human lifespans.

2. A distance of zero indicates two distributions coincide. By using this criterion, it can be seen from Fig. 7.17.5 that the period age distribution for run 1 had essentially converged to its stable form after 100 epochs; the other runs has a nonzero distance function after 100 epochs, indicating that convergence had not as yet occurred.

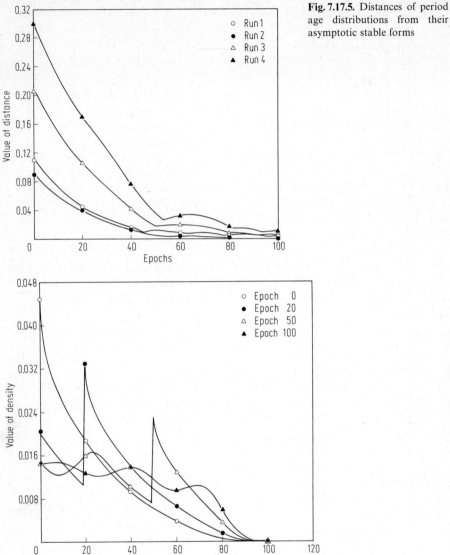

Fig. 7.17.5. Distances of period age distributions from their asymptotic stable forms

Fig. 7.17.6. Period age densities at selected epochs for Run 4

3. At almost every epoch in Fig. 7.17.5, the ordering of the distances of the period age distributions from their asymptotic stable forms was the reverse of that for the Malthusian parameters in the four runs. Thus, the more rapid the rate of population growth, the more rapid was the convergence of the age distribution to its asymptotic stable form.

4. Distances of period age distributions from their stable asymptotic form supply virtually no information on the shape of the corresponding density

functions. In order to get some insights into the shapes of these functions, period age densities for run 4 are plotted in Fig. 7.17.6 at selected epochs. From an inspection of Fig. 7.17.6, it can be seen that sudden changes in a fertility regime produce a large moving wave in the period age distribution; see the densities at epochs 20 and 50. Following the large wave are smaller waves, which are not completely damped out even at epoch 100.

7.17.4 Implications for Population Policy

1. The advocacy and implementation of a population policy for delaying marriage or childbearing to the late twenties could have a profound long term effect in slowing population growth, even with only moderate fertility control.

2. In rapidly growing populations with high levels of fertility, the success or failure of implementing a policy for decreasing fertility should not be judged solely in terms of the short term behavior of period demographic indicators, such as the rate of population growth or the crude birth rate. As population projection runs 1, 2, 3, and 4 have demonstrated, when sudden fertility changes are introduced into a rapidly growing population, these period demographic indicators may actually increase before they decrease. Moreover, the period of increase and decrease can be long in terms of human life spans.

7.18 A Study of Changing Mortality and Constant Fertility in the Time Inhomogeneous Case

Inspections of historical population data frequently suggest that an observer is viewing realization of some stochastic process with time inhomogeneous laws of evolution. It is for this reason that population processes with time in-homogeneous laws of evolution seem to be more realistic representations of actual populations than processes with time homogeneous laws. Discussed in this section is a computer scenario in which it was assumed that fertility remained constant while a population experienced a large decline in mortality. Such a scenario seems to correspond roughly to the large increases in population that have occurred in the less developed countries of the world during the past several decades. Among the authors who have discussed this phenomenon at some length is Coale (1983). The main policy implication of the computer run discussed in this section is that to avoid explosive population growth when mortality declines, policies aimed at diminishing fertility need to be implemented.

An example of a population that has experienced a significant decline in mortality since 1900 was that of nonwhite U.S. females, a decline that has been documented by the National Center for Health Statistics.[1] According to Table 5.5 of this document, expectation of life at birth for nonwhite females in 1900

---

[1] Vital Statistics of the United States, 1978, Vol. II – Sect. 5, Life Tables. DHHS Publication No. (PHS) 81–1104.

was 33.5 years; by 1978, this expectation had risen to 73.6 for an average gain of about 0.51 years per year. The time series of expectations of life at birth for these females did not increase monotonely; for, during some epochs there were rather wide fluctuations in the level of mortality. For example, in 1918, the year of a world wide influenza epidemic, the expectation of life at birth for these females was 32.5; in 1917 and 1919, this expectation was 40.8 and 44.4, respectively.

Thanks to an evolutionary computer model of mortality developed by M.E. Jacobson, a graduate student in Mathematical Sciences at Drexel University, it was possible to use the time series of expectations of life at birth for U.S. nonwhite females to generate computer input to the mortality component of the time inhomogeneous population projection system. A detailed description of this evolutionary model of human mortality will be published elsewhere. For purposes of this section, it is sufficient to point out that the model was based on the ten parameter system described in Sect. 2.10 and the mortality experience of Swedish females during the period spanning the years 1780 to 1965. By passing a time series of expectations of life at birth into this model, it was possible to compute, with the help of a command file, 100 model period survival functions. The first 79 of these functions were based on the actual experience for U.S. nonwhite females from 1900 to 1978; the remaining model period survival functions were copies of the 79-th one. To dramatize the impact of unchanging fertility when mortality declines on the speed of population growth, the fertility regime, characterized by the MATHIST run $A_1B_1C_1D_3$, see Table 7.17.1, was used throughout the population projection run. Recall, that among the four fertility regimes considered in Sect. 7.17, that for $A_1B_1C_1D_3$ lead to the most rapid rates of population increase.

Any population projection run may not be particularly informative by itself unless it can be contrasted with at least one baseline run. Used to provide a contrast to the time inhomogeneous run was a baseline run for a hypothetical population in which both fertility and mortality were held constant. The level of constant fertility was again that for the same MATHIST run, $A_1B_1C_1D_3$, described above; the level of cohort mortality was determined by an expectation of life at birth of 33.5 years, the period expectation of life at birth experienced by U.S. nonwhite females in 1900. In both population projection runs, the initial age structure of the population was determined by the age distribution with a stable population corresponding to the fertility regime of the MATHIST run, $A_1B_1C_1D_3$, and a cohort survival function based on an expectation of life at birth of 33.5 years. An initial population size of 4×10^6 females was used in both runs. Any differences in the two projections runs are, therefore, attributable to a decline in mortality.

Presented in Figs. 7.18.1 and 7.18.2 are graphs of the mean total population and mean total births and deaths for 100 epochs in the two population projection runs. Figures 7.18.3 and 7.18.4 contain graphs of the period crude birth and death rates and the period rates of population growth. Finally, Fig. 7.18.5 is devoted to graphs of selected period age densities for the time inhomogeneous run with declining mortality. Just as in Sect. 7.17, the discussion of these figures will be limited to some suggested guidelines.

7.18.1 Guidelines for Interpreting Mean Total Population and Mean Total Births and Deaths

1. The Malthusian parameter for the baseline run turned out to be negative. Consequently, as can be seen from Fig. 7.18.1, mean total population size decreased slowly in the baseline run from 4×10^6 initially to somewhat over 3×10^6 by the end of 100 epochs.

  2. Mean total population size in the inhomogeneous run also decreased for about 30 epochs but thereafter increased dramatically to about 14×10^6 by the end of 100 epochs. The inhomogeneous run illustrates the type of explosive

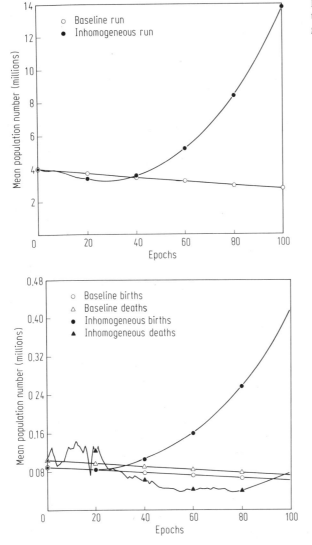

Fig. 7.18.1. Mean total population – baseline versus inhomogeneous run

Fig. 7.18.2. Mean total births and deaths – baseline versus inhomogeneous runs

population growth that will occur if declines in mortality are not accompanied by declines in fertility.

3. Like total population size, the mean number of births and deaths decrease in the baseline run, as can be seen by inspecting Fig. 7.18.2. Observe that throughout the baseline run, the mean number of deaths exceeds the mean number of births at each epoch. Thus, population size must decline.

4. The curve in Fig. 7.18.2, depicting the mean number of deaths in the inhomogeneous run, is irregular, due to the irregularities in the time series of period expectations of life at birth for U.S. nonwhite females. After about 30 epochs, the curve for mean total births for the inhomogeneous run in Fig.

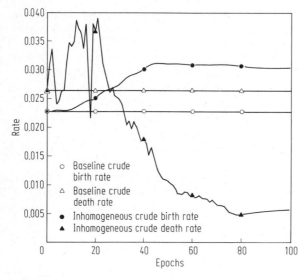

Fig. 7.18.3. Period crude birth and death rate – baseline versus inhomogeneous runs

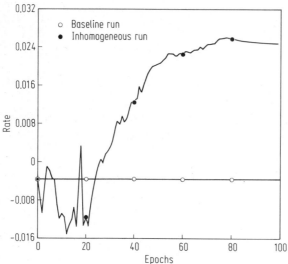

Fig. 7.18.4. Period rates of population growth – baseline versus inhomogeneous run

7.18.2 begins to greatly exceed the curve for mean total deaths, which is also indicative of a rapidly growing population.

7.18.2 Guidelines for Interpreting Period Crude Birth and Death Rates and Rates of Population Growth

1. From Figs. 7.18.3 and 7.18.4, it can be seen that the curves for the period crude birth and death rates, as well as rates of population growth, remained constant in the baseline run. This constancy is consistent with a population evolving according to time homogeneous laws of evolution. Just as would be expected in a declining population, the line for the crude death rate exceeds that for the crude birth rate in Fig. 7.18.3.

2. Unlike the curve for the period crude death rate for the inhomogeneous run in Fig. 7.18.3, that for the period crude birth rate exhibits few irregularities, due primarily to the assumption of a constant fertility regime. As mortality declines, the curves for the period crude birth and death rates diverge with crude birth rates being dominant, which is another indication of an increasing population.

3. Due to comparatively wide fluctuations in mortality, the curve for the period rates of population growth displays marked fluctuations about zero during the first 25 epochs of the inhomogeneous run. Thereafter, as mortality declines, the period rates of population growth increase sharply, which is again indicative of a rapidly growing population.

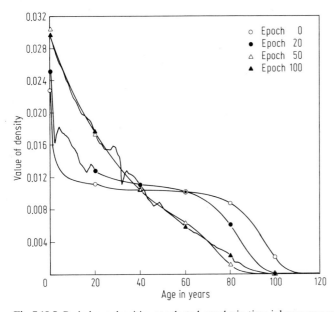

Fig. 7.18.5. Period age densities at selected epochs in time inhomogeneous run

7.18.3 Guidelines for Interpreting Period Age Densities in the Time Inhomogeneous Run

1. The period age density at epoch 0 in Fig. 7.18.5 is that determined by the fertility regime for the MATHIST run, $A_1 B_1 C_1 D_3$, and the evolutionary model of Swedish female mortality with an expectation of life at birth of 33.5 years.

2. The trend towards skewness to the left as the epochs increase in Fig. 7.18.5 is consistent with the period age densities of a rapidly growing population. That is, a continued rapid influx of births into a population tends to skew period age densities to the left.

7.19 Further Reading

As has been shown in the preceding sections, when the one-sex population processes considered in this chapter are viewed from the perspective of lines of descent, a generalized age-dependent branching process in discrete time arises. There is an extensive literature on branching processes. In addition to the books, Harris (1963), Mode (1971), and Jagers (1975), already mentioned, other books on the subject include: Athreya and Ney (1972), Sevast'yanov (1974), Joffe and Ney (1978), and Asmussen and Hering (1983). Apart from the literature on Galton-Watson processes, which are a class of branching process in discrete time, most of the literature is concerned with formulations in continuous time. Furthermore, it is usually assumed that the laws governing the evolution of the process are time homogeneous, leading to an emphasis on limit theorems. Throughout most of the literature on branching processes, little attention has been given to problems of computer implementation.

The material presented in this chapter differs from most literature on branching processes in at least three ways. Firstly, as an aid to developing models suitable for computer implementation, attention has been confined to formulations in discrete time. Secondly, since the emphasis has been on formulations that may be implemented on a computer, it has been possible to deal effectively with time inhomogeneous laws of evolution. Thirdly, in an attempt to keep the mathematics at a level that may be communicated to a wide audience, little attention has been given to the subject of analyzing the random functions themselves, a subject that has been given much attention in the literature on branching processes but requires more advanced mathematical analysis.

Branching processes first appeared in the literature on the mathematics of population in connection with the extinction of family surnames. It is not surprising, however, that many other applications of this theory have been found. An interesting case in point is the study of kinship, Pullum (1982). Other recent papers dealing with similar problems are those of Jagers (1982) and Joffe and Waugh (1982). Another subject that has been given considerable attention in the literature on branching processes is that of immigration. A research problem of some interest would be that of extending the theory of the present chapter to include both immigration and emigration. Other research topics worthy of investigation would include the development of methodologies taking into account migration and the spatial distribution of populations.

A subject that has been given essentially no attention in this book is that of gathering data on members of a population. In principle, the counting or enumeration of the members of a population would appear to be straightforward; for, all that needs to be done is to apply the elementary process of counting. When populations are large and their members move about, however, a little reflection leads to the conclusion that their enumeration is far from straightforward. An interesting article in which the intricacies in taking and applying the results of the 1980 U.S. Census are discussed is that of Barabba et al. (1983).

Essentially none of the literature on population momentum was touched upon in Sect. 7.17. Included in the papers belonging to this literature are those of Keyfitz (1971), Mitra (1976), and Krishnamoorthy et al. (1981).

Problems and Miscellaneous Complements

1. Suggested in this complement are procedures for the computer implementation of formulas (7.2.1), (7.2.2), and (7.2.3).

(a) Basic to the computer implementation of formulas (7.2.1), (7.2.2), and (7.2.3) is the calculation of period survival probabilities appearing on the right in (7.2.1) for some finite set of epochs $t, t+1, \dots$. One approach to calculating these probabilities is to utilize the Coale-Demeny system of model life tables described in Sect. 2.8. By specifying period expectations of life at birth for some epochs $t, t+1, \dots$, the period survival probabilities $p(t+s, x)$ could, in principle, be calculated at epoch $t+s, s=0, 1, 2, \dots$. A disadvantage of the Coale-Demeny system of model life tables is that the expectation of live at birth is confined to the ages 20 to 80. Period expectations of life at birth for females is now approaching the upper limit of 80 in some populations and may soon exceed it.

2. Let $X(t)$ be a random variable whose survival function is defined in (7.2.2). The distribution function of $X(t)$ is defined by

$$P[X(t) \leq x] = G_c(t, x) = 1 - S_c(t, x)$$

for $x \geq 0$. Because $X(t) \geq 0$ with probability one,

$$P[X(t) = 0] = 1 - S_c(t, 0) = 0,$$

for all t.

(a) Show that the p.d.f. of $X(t)$ may be written in the form

$$P[X(t) = x] = S_c(t, x-1) - S_c(t, x)$$

for $x \geq 1$.

(b) Use the result in part (a) to deduce the formula in (7.2.3).

3. This problem is devoted to a more precise discussion of the random function defined in (7.3.4).

(a) For any member of a cohort of females born at epoch t, let $N(t, s)$ be a random function representing the actual number of live births experienced by epoch $t+s$. Let X_k, $k \geq 1$, be a sequence of independent and identically distributed random variables such that, for every $k \geq 1$, $X_k = 1$, if the k-th live birth is female; and let $X_k = 0$ otherwise. For every $k \geq 1$, $E[X_k] = q$, the probability a newborn is female. Then, the random function

$$N_f(t, s) = \sum_{k=1}^{N(t, s)} X_k$$

represents the actual number of female births experienced up to epoch $t+s$, $s \geq 1$, for any member of a cohort born at epoch t. It will be assumed that the random function $N(t, s)$ and the sequence $\langle X_k \rangle$, $k \geq 1$, are independent. Show that

$$E[N_f(t, s)] = q E[N(t, s)].$$

(b) The precise definition of the random function in (7.3.4) is

$$\Delta(x, i) = \Lambda(t - x, x, i) = N_f(t - x, x + i) - N_f(t - x, x + i - 1).$$

(c) Needed in the derivation of the covariance function for the population birth process is the product moment function $E[N_f(t, s) N_f(t, s+\tau)]$, defined for all t, $s \geq 1$, and $\tau \geq 0$. Because the function $N_f(t, s)$ is increasing in s with probability one, $N_f(t, s+\tau)$ may be written in the form

$$N_f(t, s+\tau) = N_f(t, s) + N_f(t, s+\tau) - N_f(t, s),$$

where the increment on the right is always nonnegative and independent of $N_f(t, s)$, given $N(t, s)$ and $N(t, s+\tau)$. With the help of this observation, show that

$$E[N_f(t, s) N_f(t, s+\tau)] = q(1 - q) E[N(t, s)]$$
$$+ q^2 E[N(t, s) N(t, s+\tau)].$$

4. Because $\varphi_c(t, s) = 0$ for every t when s is outside the interval $15 \leq s \leq 50$, the renewal density $m_c(t, s)$ used in Theorems 7.4.1 and 7.4.2 has special properties. These properties are useful when writing down explicit formulas for solutioms to renewal type equations.

(a) Show that $m_c(t, s) = 0$ for all t when $1 \leq s \leq 14$.

(b) Show that $m_c(t, s) = q \varphi_c(t, s)$ for all t when $15 \leq s \leq 29$.

5. There is a simple approximation to $L_{0j}(t, s)$ as given in (7.5.33) that does not require the calculation of the cohort survival function.

(a) Let

$$b_c^*(t, x) = \int_{x-1}^{x} b_c(t, y) dy.$$

Use formula (7.2.8) to show that $qp(t+j,j)b^*(t+s-j,j)$ is an upper bound of $L_{0j}(t,s)$. This upper bound is a good approximation when $S_c(t+s-j,j-1)$ and $S_c(t+s-j,j)$ do not differ greatly.

6. As one would expect, there is a close connection between the random functions $\Delta(t,x,s)$, $U(t,x,s)$, and $V(t,x,s)$ defined in Sect. 7.8 and those in Def. 7.3.2.

(a) Show that

$$Z_B^{(0)}(t,x,y,s) = \Delta(t-x,x,s)\delta_0(x+s-y),$$

$$Z^{(0)}(t,x,y,s) = U(t-x,x,s)\delta_0(x+s-y),$$

and

$$Z_D^{(0)}(t,x,y,s) = V(t-x,x,s)\delta_0(x+s-y)$$

at all points $\langle t,x,y,s\rangle$.

(b) Then, conclude that

$$E[\Delta(t-x,x,s)] = q\varphi_c(t-x,x,s),$$

$$E[U(t-x,x,s)] = S_c(t-x,x,s),$$

and

$$E[V(t-x,x,s)] = g_c(t-x,x,s).$$

7. As can be seen from the representation of the population structure based on birth cohorts presented in Sect. 7.8, properties of random sums of random variables will be needed in the derivation of variance and covariance functions for the population process.

(a) Let $N \geq 0$ be an integer valued random variable with a finite mean and variance; let X be a random variable with a finite mean and variance and suppose the sequence $\langle X_k \rangle$, $k \geq 1$, is independent and indentically distributed copies of X that is also independent of N. Define a random variable Y by $Y = 0$, if $N = 0$, and by

$$Y = \sum_{k=1}^{N} X_k,$$

if $N \geq 1$. Show that

$$E[Y] = E[N]E[X]$$

and

$$\text{var}[Y] = E[N]\,\text{var}[X] + (E[X])^2\,\text{var}[N].$$

(b) Let N_1 and N_2 be random variables taking values in the nonnegative integers and suppose that they have finite means and variances. Let $\langle X_1, X_2 \rangle$ be a pair of random variables with finite means and variances and let $\langle X_{1k}, X_{2k} \rangle$, $k \geq 1$, be a sequence of independent and identically distributed copies of $\langle X_1, X_2 \rangle$ which is also independent of N_1 and N_2. Define a pair of random sums Y_1 and Y_2 be

$$Y_1 = \sum_{k=1}^{N_1} X_{1k} \quad \text{and} \quad Y_2 = \sum_{k=1}^{N_2} X_{2k},$$

for $N_1 \geq 1$ and $N_2 \geq 1$.

(c) Verify that if the random variables X_1 and X_2 in part (b) are independent, then

$$\text{cov}[Y_1, Y_2] = \text{cov}[N_1, N_2] E[X_1] E[X_2].$$

(d) Suppose there is a random variable N such that the random variables in part (b) satisfy the equations $N_1 = N_2 = N$ with probability one. Also suppose that the random variables X_1 and X_2 in part (b) are not independent. Show that under these conditions

$$\text{cov}[Y_1, Y_2] = E[N] \text{cov}[X_1, X_2] + \text{var}[N] E[X_1] E[X_2].$$

(e) Suppose there is a random variable X such that $X_1 = X_2 = X$ with probability one and the conditions in part (d) also hold. Verify that under these conditions, the variance formula in part (a) is a special case of the covariance in part (d). Note also that when X_1 and X_2 are independent, then the formula in (d) reduces to that in (c).

8. During the proof of Theorem 7.9.3, a formula for the covariance on the right in (7.9.16) was used.

(a) To derive the formula on the right in (7.9.11), show that

$$E_x[\Delta(t-x, x, s) U(t-x, x, s)]$$
$$= S_c(t-x, x, s) E_x[\Delta(t-x, x, s) | U(t-x, x, s) = 1],$$

where E_x stands for conditional expectation given that $X(t-x) > x$.

(b) From the equations connecting the potential and actual birth processes, see Sect. 4.3, show that

$$E_x[\Delta(t-x, x, s) | U(t-x, x, s) = 1] = q b_c^*(t-x, x+s),$$

where $b_c^*(t-x, x+s)$ is defined as in Problem 5(a). Formula (7.9.11) now follows from the results in Problem 6(b) and the formula for a covariance.

9. Outlined in this problem is a method for proving assertion (ii) in Lemma 7.12.1.

(a) When $C(\lambda)$, the Laplace transform defined in (7.12.2), satisfies the condition $C(0) < 1$ and the function $b(s)$ is bounded, then the solution to Eq. (7.12.1) may be represented as the series

$$a(s) = b(s) + \sum_{n=1}^{\infty} \sum_{v=0}^{s} c^{(n)}(v) b(s-v),$$

which converges for all $s \geq 0$. In the above, $c^{(n)}(v)$, $v \geq 0$, is the n-fold convolution of $c(v)$ with itself.

(b) From the theory of Laplace transforms, it is known that

$$C^{(n)}(\lambda) = \sum_{v=0}^{\infty} c^{(n)}(v) \exp[-v\lambda] = [C(\lambda)]^n,$$

for all $\lambda \geq 0$.

(c) Let B be a number such that $|b(s)| \leq B$ for all $s \geq 0$. Use the result in part (b) to show that the series in part (a) is bounded uniformly in $s \geq 0$ by the convergent geometric series

$$B \sum_{n=1}^{\infty} [C(0)]^n.$$

(d) Show that if

$$\lim_{s \uparrow \infty} b(s) = b,$$

then,

$$\lim_{s \uparrow \infty} \sum_{v=0}^{s} c^{(n)}(v) b(s-v) = [C(0)]^n b$$

for every $n \geq 1$.

(e) Use the results in parts (c) and (d) to conclude that Eq. (7.12.6) is valid.

10. In the proof of Theorem 7.12.2, the condition that all the functions in (7.11.11) converge to zero as $s \uparrow \infty$ was used.

(a) According to part (a) of Problem 6,

$$Z^{(0)}(t, x, y, s) = U(t-x, x, s)\delta_0(x+s-y).$$

Hence, for s large this random function is identically zero with probability one. All expectations are thus zero for s sufficiently large.

11. Another distributional indicator of interest is the period age distribution of death. It has been used in the study of mortality in historical demography in cases where a sample of ages of death is available from the historical records.

(a) In a population process evolving according to time inhomogeneous laws of evolution, the period age distribution of age of death at epoch $t+s$ is defined by the random function

$$\hat{a}_D(t, y, s) = \frac{Y_D(t, y, s)}{Y_D(t, s)}.$$

The numerator in $\hat{a}_D(t, y, s)$ is the random function in (7.7.3); the denominator is

$$Y_D(t, s) = \sum_{y=1}^{r} Y_D(t, y, s).$$

(b) Let $v_D(t, y, s)$ and $v_D(t, s)$ be the means of the random functions in $a_D(t, y, s)$; and, let $\hat{a}_D(t, y, s) = v(t, y, s)/v_D(t, s)$. Just as in Theorem 7.13.1, show that when $n(t, x)$ is large for all $x = 0, 1, 2, \ldots, r$, there exists a random function

$o(t, y, s)$ such that

$$\hat{a}_D(t, y, s) = a_D(t, y, s) + o(t, y, s).$$

(c) In a population process with time homogeneous laws of evolution, use arguments similar to those in Theorem 7.14.1 to show that

$$\lim_{s \uparrow \infty} a_D(t, y, s) = \frac{g_c(y) \exp[-\alpha y]}{g_c^*(\alpha)}$$

for $1 \leq y \leq r$. The function on the right is known as the age distribution of death in a stable population. Extensive tabulations of this distribution have been made by Coale and Demeny (1966).

12. One approach to interpreting age-specific fertility (birth) rates is that they provide a method for estimating a period fertility density function; see, for example, Eq. (4.9.19). Given an estimate of a period fertility density function, an estimate of a period total fertility rate may be obtained.

(a) In reference to formula (7.13.10), use (7.2.9) to show that

$$R_B(t, y, s) = \frac{q \varphi_c(t + s - y, y)}{S_c(t + s - y, y)}$$

for $15 \leq y \leq 50$.

(b) Suppose a period total fertility rate is estimated by the random function

$$\hat{T}_p(t, s) = \sum_{y=15}^{50} \hat{R}_B(t, y, s)$$

in a population evolving from a large initial population at epoch t, according to time inhomogeneous laws of evolution. Use results similar to those in Theorem 7.13.1 and the formula in part (a) to investigate how well this random function estimates the theoretical period total fertility rate. Give special attention to the case where the cohort survival function does not change much throughout the childbearing ages.

(c) Sometimes a period age distribution of childbearing is defined as the random function

$$\hat{a}_B^*(t, y, s) = \frac{\hat{R}_B(t, y, s)}{\hat{T}_p(t, s)}.$$

The notion underlying this definition is that this function is supposed to estimate the normalized period fertility density function, i.e., the period fertility density function divided by the corresponding total fertility rate.

(d) For population processes evolving according to time homogeneous laws of evolution, compare the period distribution in part (c) with the limit formula in (7.14.9).

References

Asmussen, S. and Hering, H. (1983): *Branching Processes*. Birkhäuser, Boston, Basel, Stuttgart

Athreya, K.B. and Ney, P. (1972): *Branching Processes*. Springer-Verlag, Berlin, Heidelberg, New York

Baraabba, V.P., Mason, R.O. and Mitroff, I.I. (1983): Federal Statistics in a Complex Environment. *The American Statistician* 37:203–212

Chandrasekaran, C. and Hermalin, A.I. (Editors) (1975): *Measuring the Effect of Family Planning Programs on Fertility*. Ordina Editions, 4830 Dolhain, Belgium

Coale, A.J. and Demeny, P. (1966): *Regional Model Life Tables and Stable Populations*. Princeton University Press, Princeton, New Jersey

Coale, A.J. (1972): *The Growth and Structure of Human Populations*. Princeton University Press, Princeton, New Jersey

Coale, A.J. (1983): Recent Trends in Fertility in Less Developed Countries. *Science* 221:828–832

Cohen, J.E. (1976): Ergodicity of Age-Structure in Populations with Markovian Vital Rates I: Countable States. *J. Amer. Statist. Assoc.* 71:335–339

Cohen, J.E. (1977a): Ergodicity of Age-Structure in Populations with Markovian Vital Rates II: General States. *Advances in Appl. Probability* 9:18–37

Cohen, J.E. (1977b): Ergodicity of Age-Structure in Populations with Markovian Vital Rates III: An Illustration. *Advances in Appl. Probability* 9:462–475

Crump, K.S. and Mode, C.J. (1968): A General Age-Dependent Branching Process I. *Jour. Math. Anal. Appl.* 24:494–508

Crump. K.S. and Mode C.J. (1969): A General Age-Dependent Branching Process II. *J. Math. Anal. Appl.* 25:8–17

Feller, W. (1968): *An Introduction to Probability Theory and Its Applications*. Vol. I, 3rd ed., John Wiley and Sons, New York

Fisher, R.A. (1958): *The Genetical Theory of Natural Selection*. Dover, New York

Harris, T.E. (1963): *The Theory of Branching Processes*. Springer-Verlag, Berlin Göttingen, Heidelberg

Jagers, P. (1969): A General Stochastic Model for Population Development. *Skandinavisk Aktuarietidskift* 84–103

Jagers, P. (1975): *Branching Processes with Biological Applications*. John Wiley and Sons, London, New York, Sydney, Toronto

Jagers, P. (1982): How Probable Is It To Be Birst Born? – And Other Branching Process Applications to Kinship Problems. *Mathematical Biosciences* 59:1–15

Joffe, A. and Ney, P. (Editors) (1978): *Branching Processes*. Marcel Dekker, New York

Joffe, A. and Waugh, W.A.O'N. (1982): Exact Distributions of Kin Numbers in a Galton-Watson Process. *Jour. Appl. Prob.* 19:767–775

Keyfitz, N. (1968): *Introduction to the Mathematics of Population*. Addison-Wesley, Reading, Mass.

Keyfitz, N. (1971): On the Momentum of Population Growth. *Demography* 8:71–80

Keyfitz, N. (1981): The Limits of Population Forecasting. *Population Development and Review* 7:579–593

Keyfitz, N. (1982): Can Knowledge Improve Forecasts? *Population Development and Review* 8:729–751

Krishnamoorthy, S., Potter, R.G. and Pickard, D.K. (1981): Population Momentum – Its Relation to the Moments of Replacement Level Fertility. *Mathematical Biosciences* 53:41–51

Leslie, P.H. (1945): On the Use of Matrices in Certain Population Mathematics. *Biometrika* 33:183–212

Leslie, P.H. (1948): Some Further Notes on the Use of Matrices in Population Mathematics. *Biometrika* 35:213–245

Lotka, A.J. (1956): *Elements of Mathematical Biology*. Dover, New York, N.Y.

Mitra, S. (1976): Influence of Instantaneous Fertility Decline to Replacement Level on Population Growth. *Demography* 13:513–520

Mode, C.J. (1971): *Multitype Branching Processes – Theory and Applications*. American Elsevier, New York

Mode, C.J. (1972): A Bisexual Multi-Type Branching Process and Applications in Population Genetics. *The Bulletin of Mathematical Biophysics* 34:13–31

Mode, C.J. (1974a): Discrete Time Age-Dependent Branching Processes in Relation to Stable Population Theory in Demography. *Mathematical Biosciences* 19:73–100

Mode, C.J. (1974b): Applications of Terminating Nonhomogeneous Renewal Processes in Family Planning Evaluation. *Mathematical Biosciences* 22:293–311

Mode, C.J. (1974c): A Discrete Time Stochastic Growth Process for Human Populations Accommodating Marriages. *Mathematical Biosciences* 9:201–219

Mode, C.J. (1974d): On the Evaluation of a Determinant Arising in a Bisexual Stochastic Growth Process. *Journal of Mathematical Analysis and Its Applications* 46:190–211

Mode, C.J. (1976): Age-Dependent Branching Processes and Sampling Frameworks for Mortality and Marital Variables in Non-Stable Populations. *Mathematical Biosciences* 30:47–67

Mode, C.J. (1977): A Population Projection Component of a Computer System-Population Policies and Birth Targets. *Mathematical Biosciences* 36:87–107

Mode, C.J. and Busby, R.C. (1981): Algorithms for a Projection Model of Demographic Indicators Based on Generalized Branching Processes. *Mathematical Biosciences* 57:83–107

Mode, C.J. and Littman, G.S. (1974): Applications of Computerized Stochastic Models of Human Reproduction and Population Growth in Family Planning Evaluation. *Mathematical Biosciences* 20:267–292

Pickens, G.T., Busby, R.C. and Mode, C.J. (1983): Computerization of a Population Projection Model Based on Generalized Branching Processes – A Connection with the Leslie Matrix. *Mathematical Biosciences* 64:91–97

Pollard, J.H. (1966): On the Use of the Direct Matrix Product in Analysing Certain Stochastic Models. *Biometrika* 53:397–415

Pollard, J.H. (1973): *Mathematical Models for the Growth of Human Populations.* Cambridge University Press, Cambridge

Pullum, T.W. (1982): The Eventual Frequencies of Kin in a Stable Population. *Demography* 19:549–565

Ross, J.A. (1975): Acceptor Targets. In *Measuring the Effect of Family Planning Programs on Fertility*, ed. by C. Chandrasekaran and A.I. Hermilin. Ordina Editions, 4830 Dolhain, Belgium

Sevast'yanov, B.A. (1974): *Verzweigungs-Prozesse.* Akademie-Verlag, Berlin.

Smith, D. and Keyfitz, N. (1977): *Mathematical Demography – Selected Papers.* Springer-Verlag, Berlin, Heidelberg, New York

Sykes, Z.M. (1969): On Discrete Stable Population Theory. *Biometrics* 25:285–293

Author Index

Numbers in *italic* indicate pages on which full references occur. The others indicate pages of text where works of authors are cited.

Subject Index